T0143147

THE
COEVOLUTIONARY
PROCESS

THE
COEVOLUTIONARY
PROCESS

John N. Thompson

THE UNIVERSITY OF CHICAGO PRESS

CHICAGO AND LONDON

John N. Thompson is professor in the Departments of Botany and Zoology at Washington State University. He is the author of *Interaction and Coevolution* (1982).

The University of Chicago Press, Chicago 60637
The University of Chicago Press, Ltd., London
© 1994 by The University of Chicago
All rights reserved. Published 1994
Printed in the United States of America

03 02 01 00 99 98 97 96 95 94   1 2 3 4 5

ISBN: 0-226-79759-7 (cloth)
       0-226-79760-0 (paper)

Library of Congress Cataloging-in-Publication Data

Thompson, John N.
      The corevolutionary process / John N. Thompson.
         p.   cm.
      Includes bibliographical references (p.   ) and index.
         1. Coevolution.   2. Insect-plant relationships.   I. Title.
      QH372.T48   1994
      575—dc20   94-14140
                                                   CIP

CONTENTS

PREFACE

The coevolutionary process both requires and produces some degree of specialization within biological communities. These twin themes of specialization and coevolution are at the center of many of the major debates in evolutionary biology: the organization of biological communities, the evolution of life histories, the maintenance of sexual reproduction, the mechanisms of speciation, and the patterns in phylogeny. From the *Origin of Species* onward, evolutionary biologists have probed the ways in which species specialize in their interactions with one another and how these specialized relationships sometimes result in reciprocal evolutionary change. These studies of evolving interactions have produced one of the longest lineages of discussion, observation, and experiment in evolutionary biology. To be sure, organisms specialize not only to one another but also to different physical environments. And that, too, has had a long tradition of study in evolutionary biology. It is, however, the specialized links and networks of interaction among species that have produced much of the diversity of life and the organization of communities. It is the evolution of these linked lives that has produced Darwin's entangled bank.

The relationship between specialization and coevolution is asymmetric. Co- evolution demands, and produces, some degree of specialization in interactions between species, but not all highly specialized interactions are coevolved. It is precisely this asymmetry that generates much of the varied richness we find in the relationships between species and the organization of communities. That richness is amplified as populations become geographically differentiated in their degrees of specialization and coevolve with other species along different evolutionary trajectories. The challenge is to understand just how these asymmetries in specialization and the geographic differences in interactions together shape the coevolutionary process and the evolution of interactions in general.

When a little over a decade ago I finished an earlier book on evolving interactions, *Interaction and Coevolution,* most of us thought about coevolution as either 'specific' or 'diffuse'. Most research was devoted to the problem of how particular forms of interaction evolve (e.g., competition, mutualism) and whether particular interactions showed evidence of coevolution. In that book I asked three questions: Are there patterns in how interactions differ in the selection pressures that they place on species? Are there patterns in how interactions change over evolutionary time? And, are some interactions and ecological conditions more likely to lead to coevolution than others? Based upon what we knew then, I evaluated how parasitism, grazing, predation, competition, and the various forms of mutualism differed in the selection pressures they exerted on populations and the likelihood of long-term coevolution. And I asked how interactions grew and diversified, sometimes causing speciation and sometimes collecting additional species into the interaction. Two insightful edited volumes on coevolution appeared about the same time as *Interaction and Coevolution* (Futuyma and Slatkin 1983; Nitecki 1983), and together these three volumes highlighted the diversity of approaches that was emerging in the study of evolving interactions.

Since then, questions on patterns and processes in evolving interactions have become broader as the methods for studying them have become more refined. The study of the distribution of outcomes in interactions (rather than the mean outcome), the marriage of phylogenetic with ecological approaches, the similar marriage of ecological with genetic approaches, and the realization that the word 'coevolution' is an umbrella for a variety of processes and outcomes of reciprocal evolutionary change have all resulted in a much better understanding of both specialization and coevolution. Moreover, more studies have been designed to address specific hypotheses on how interactions evolve. Since the middle 1980s more than 1,200 papers have been published that use the word 'coevolution', and at least a dozen books use the word in the title. During the same time, thousands of other papers have made some mention of the evolution of interactions as either a major theme or one of a series of themes. The study of evolving interactions has become a growth industry.

To a large extent *The Coevolutionary Process* came to be written because over the past twenty years I have continually been confronted with the related problems of specialization, geographic structure, and coevolution in every question I have asked on how species, and the interactions between them, evolve. The problems of specialization and geographic structure have always been there whether I have been trying to grapple with broad patterns in coevolution, with the origins, evolutionary ecology, and phylogeny of mutualism, or with the evolutionary dynamics of host associations. The result has been a geographic view of coevolution that I develop here in detail.

A second impetus for the book has been the backlash that has followed in recent years from the coevolution-is-everywhere attitude of the past few decades in evolutionary ecology. Frustrated with the lack of extreme specialization (i.e., one-to-one) in many relationships between species, some evolutionary biologists have increasingly taken the tack of ignoring how coevolution may have shaped interactions. Others, using a tack just as fatal to the concept, have taken to handling the complex relationships between specialization and coevolution as gingerly as possible through the catchall designation 'diffuse coevolution'. Neither of these reactions reflects the progress that is being made in coevolutionary studies. Consequently, I wanted to use the arguments in this book to show that we are progressing quickly in our ideas and results on coevolution and that a geographic view of coevolution can alleviate some of the frustration that has resulted from viewing reciprocal change as either one-to-one or diffuse.

The questions I explore here are part of the continuing search for patterns and processes in evolving interactions. If we can understand how interactions evolve, and sometimes coevolve, in different ways under different sets of ecological and genetic conditions, we will have a much more comprehensive understanding of the overall process of evolution. On a shorter time scale, a major imperative for conservation biology in the future will be not only to conserve or restore populations of declining species but also to conserve the interactions that have shaped these species—or at least to understand why the interactions are changing. To do that requires a more comprehensive theoretical framework for the evolution of interactions than we have now. Similarly, the application of ecological principles to such fields as biological control of pest species relies heavily upon how specialization and coevolution may be expressed and evolve in new environments. The continuing evolution of the interaction between rabbits and myxoma virus in Australia is just one example of why we need to understand better the relationship between specialization and coevolution under different ecological conditions.

This book, then, is an extended argument on the interrelationships between specialization, the geographic structure of interactions, and coevolution. How and when does extreme specialization evolve in interactions between species? How do different degrees of specialization create different forms of coevolution? And how does geographic structure in specialization and in the outcomes of interactions govern the dynamics of the coevolutionary process? My overall concern is with the ways in which ecological conditions, life histories, genetics, and the geographic structure of populations influence the evolution of specialization and, in turn, shape the process of coevolution. I wish specifically to develop a view that I will call *the geographic mosaic theory of coevolution.* I use the word 'theory' here not to suggest that this view is already well tested and well established but rather to suggest that the arguments and examples developed in these chapters indicate that there is some empirical support for the different components of this general view of coevolution and that specific hypotheses can be developed from it.

This geographic view of coevolution is developed as a series of major points built chapter by chapter. My intention is to try to bridge the gap between the evolutionary ecological study of coevolution within local populations, where we can observe reciprocal evolution in some populations but not in others, and the broader patterns that we seek when we compare the phylogenies of interacting species.

I began the writing of this book early in 1991, but the first full draft was completed while I was on sabbatical leave for a year as a Fulbright Senior Scholar at the Division of Plant Industry, CSIRO, in Canberra. I am grateful to the Australian–American Educational Foundation, the Division of Plant Industry, and Washington State University for providing me with that opportunity for uninterrupted work, and to the Ecological Studies and Systematics programs of the National Science Foundation and the Competitive Grants program of USDA, which have provided grants for my empirical studies of evolving interactions over the years. Throughout my year in Australia Jeremy Burdon, Tony (A. H. D.) Brown, and I discussed ideas on coevolution and the genetics of interactions at length, and I gained tremendously from their views and knowledge. After I returned to Pullman, the students and colleagues in my laboratory—David Althoff, Hal Hansel, Amitabh Joshi, Laura Patten, Wayne Wehling—read through the completed chapters and challenged me with wonderfully candid comments and criticisms. Hal Hansel worked painstakingly on the figures, as did Wayne Wehling on the photographic work. Several others kindly provided original work: Anders Nilsson provided the photographs of a long-spurred Madagascan orchid and long-tongued hawkmoth, and Lisa Roberts allowed her drawing of *Greya politella* ovipositing into a flower of *Lithophragma parviflorum* to be reproduced on the cover.

Jeremy Burdon, Olle Pellmyr, Peter Price, and Mary Willson each read the antepenultimate version. I am deeply grateful for the insights on the evolution of interactions that each of them has provided during the writing of this manuscript and in other discussions on evolving interactions over the years. I was also very fortunate in receiving thoughtful and careful reviews from Sara Via and from three anonymous readers for the University of Chicago Press, all of whom read the penultimate version. Their insights helped me to sharpen my arguments in the final version. While preparing the book, I also profited from conversations with many other colleagues. Some of these conversations were, at the time, related only tangentially to the book, but they exerted their influences later. I would particularly like to thank the following for discussions or preprints at critical moments in my writing: Craig Benkman, James Bull, Jeff Dangl, Jared Diamond, Hugh Dingle, Douglas Futuyma, Richard Lenski, Michael Singer, and Richard Stouthamer. My wife, Jill,

has provided a synergistic enthusiasm for my interests in things coevolutionary that merits far more than the usual acknowledgment of spousal support.
Finally, Susan Abrams of the University of Chicago Press has extended the
kind of unfailing help and encouragement that has made the whole process
of turning an outline into a book an enjoyable one.

OVERVIEW

The argument of this book is built in four parts, leading to the geographic mosaic view of coevolution developed in Part IV.

Part I (chapters 1 through 3) places the study of the relationship between specialization and coevolution into a broader ecological and historical perspective. It begins with a discussion of how our views on specialization in interactions shape almost every aspect of our perception of the ecology and evolution of biological communities. This is followed by a historical perspective on how we have reached our current ideas on the evolution of specialization and the relationship between specialization and coevolution. No overall historical treatment of the development of views on specialization and coevolution has been attempted before. There have been historical studies of particular kinds of interaction (e.g., pollination) and of the ways in which our views of coevolution have changed over the past quarter century, but our current overall perceptions of both specialization and coevolution have been shaped in part by arguments that trace all the way back to Darwin and other nineteenth- and early-twentieth-century naturalists. Those arguments have subsequently ramified among the subdisciplines of biology and then anastomosed with newer ones, creating the diversity of ideas we have today. My effort here is admittedly only the beginning of a historical analysis that will ultimately demand a much deeper and longer appraisal. I outline some of the major threads of thought that seem to have led to our present views. Each of those threads is connected to other, major and minor, threads, and as with all historical analyses, the resultant interpretation of the pattern depends upon what one sees as the major, rather than the minor, threads.

Part II (chapters 4 through 6) uses the phylogeny, genetics, and ontogeny of specialization to examine why a range of specialization develops within almost all phylogenetic lineages, why geographic differences in specialization develop among populations within species, and how organisms can evolve to specialize simultaneously on more than one species. These different aspects of the evolution of specialization are the raw material for understanding the diversity of forms of coevolution found in natural communities, and they are also the raw material for a geographic view of the coevolutionary process.

Part III (chapters 7 through 11) takes the overall patterns of specialization discussed in Part II and evaluates how natural selection molds them in different ways in different forms of interaction. These chapters compare specialization in parasites, grazers, predators, symbiotic mutualists, and free-living mutualists; they discuss the problems of the evolution of defense against multiple enemies and the evolution of mutualism with multiple mutualists; and they evaluate how interactions vary in outcome among environments. Together these chapters emphasize that there is a strong geographic structure to the outcomes, adaptations, and patterns of specialization found in many interactions, which forms the basis for the coevolutionary process.

Part IV (chapters 12 through 16) uses the patterns of variation and specialization discussed in the earlier chapters to develop the geographic mosaic theory of coevolution. I argue that the evolution of interactions within local populations is often only the raw material for the broader pattern of coevolution that shapes the relationship between two or more species. Some local populations of a pair or group of species may specialize to one another and coevolve, other populations of the same species may differ in their patterns of specialization and coevolution, and still others may fall completely outside the geographic range of the interactions. The differences in specialization and coevolution among these populations, together with gene flow and sometimes extinction of populations, shape the overall trajectory of coevolution for a pair or group of species.

Within this view, much of the dynamics of coevolution occurs at a geographic scale above the level of local populations and below the level of the fixed characters that we analyze in comparing the phylogenies of interacting species. By ignoring the geographic scale of coevolution, we risk underestimating the importance of the coevolutionary process. I do not argue that all coevolution occurs in this way, only that it is an important level of analysis that bridges studies of local populations and of whole species and incorporates many of the aspects of the genetics and ecology of species that we have discovered in recent decades.

After discussing coevolution between pairs of species from this perspective, I indicate ways in which a geographic view of interactions allows us to replace the catchall phrase 'diffuse coevolution' with more-specific, testable hypotheses on reciprocal change among groups of species. A major purpose of the book, in fact, is to bring together the major components of an overall geographic view of coevolution (and evolving interactions in general) so that we ask more specifically how and why these components interrelate in different ways in different interactions. The book closes with two short chapters:

a Synthesis of the arguments and an Epilogue that is a personal view of the links between specialization, coevolution, and conservation.

In developing the arguments, I marshal support (and note exceptions) from a wide range of taxa, including animals, plants, fungi, and microbial species. But I use two interactions as touchstones, returning to them in several chapters and discussing them in detail. One is the evolution of antagonistic interactions between swallowtail butterflies and plants; the other is the evolution of mutualistic interactions between prodoxid moths (the lineage including yucca moths) and flowers. These two interactions have been the object of much of my experimental work on the evolution of specialization and coevolution in recent years, and together they illustrate how different ecological, genetic, and phylogenetic conditions influence specialization and the coevolutionary process.

PART I
THE ENTANGLED BANK

CHAPTER ONE

Specialization within Darwin's Entangled Bank

The most famous paragraph in Darwin's *Origin of Species,* the one that draws the book to a close, begins with a sentence that revels in both the diversity of species and their interactions:

> It is interesting to contemplate an entangled bank, clothed with many plants of many kinds, with birds singing on the bushes, with various insects flitting about, and with worms crawling through the damp earth, and to reflect that these elaborately constructed forms, so different from each other, and dependent on each other in so complex a manner, have all been produced by laws acting around us. (Darwin 1859, 489)

Time and again in the *Origin,* Darwin emphasized the importance of interactions within and between species—what he usually called mutual relations—in shaping the evolution of life. He realized that one of the major challenges would be to understand how species evolve the ability to exploit other species and flourish in the midst of enemies. He wrote, "It is good thus to try in our imagination to give any form some advantage over another. Probably in no single instance should we know what to do, so as to succeed. It will convince us of our ignorance on the mutual relations of all organic beings" (Darwin 1859, 78).

Darwin understood that the evolution of biological diversity has two equally important aspects: the diversity of species and the diversity of interactions between species. The species and their interactions together form the biological communities that he called the entangled bank. The interactions are not something added on to species and evolution; they are part of the very fabric of what makes a species and a biological community. Up to 98% of flowering plants in tropical lowland rain forests rely upon animals for pollination or dispersal of seeds (Bawa 1990). In some of the tropical rain forests of Australia, Africa, and the Neotropics 70–90% of woody plants have fleshy fruits adapted for dispersal by birds or mammals, and in some local tropical communities 100% of the tree species have fleshy fruits (Howe and Smallwood 1982; Willson, Irvine, and Walsh 1989). All animals require living plants or other animals as hosts or prey. Even those that feed as detriti-vores actually feed primarily on the microorganisms coating the detritus (Be-rrie 1976). Fungi, bacteria, and algae all include taxa that are obligate associates of other living organisms. About 21% of described fungal species form lichen associations, another 8% are mycorrhizal, and many more are pathogens of plants or animals (Hawksworth 1988).

This complex, evolved network of interactions among species is the reason why the origins and phylogeny of species are at best only half the problem in understanding the diversity of life. The diversity of species cannot make sense unless we also understand the diversity of interactions among them. These two halves of evolutionary biology are linked through the processes of specialization and coevolution. Specialization in interactions with other species is the root cause of why the world has millions of species rather than thousands. All forms of life specialize to particular physical environments, but it is specialization to other species that has made the entangled bank so diverse and complex.

This book is about the links between specialization, the geographic struc-ture of interactions, and the coevolutionary process. Specialization and coe-volution have long been linked in discussions of evolving interactions be-tween species, but there is no simple relationship between them. Throughout these chapters I use 'specialization' to mean a limitation in the number of other species with which a particular species interacts. An extreme specialist is a species that relies upon only one or a few closely related other species for survival or reproduction during a major part of its life cycle—that is, species such as yuccas, rust fungi, fig wasps, snail kites, and pandas. The simplest forms of coevolution are between pairs of extreme specialists, but there are also other, more complex, relationships between specialization and coevolution. I use 'coevolution' throughout the book to mean reciprocal evo-lutionary change in interacting species, and the last five chapters consider in detail the many forms it can take in shaping patterns of adaptation and speci-ation. Finally, by 'geographic structure' I mean the combination of the meta-population structure found in species and the variation that occurs among populations in specialization, adaptation, and the outcomes of interactions.

In this first chapter, I consider how our ideas about specialization influence our views about the structure and organization of biological communities—from our estimates of the number of living species to our expectations about the antiquity of interactions, the origins of new interactions, and the fate of interactions over evolutionary time.

Specialization and Biodiversity

Our views about the structure of biological communities and the diversity of life are shaped by our assumptions about specialization in interactions. Differing assumptions about specialization, for instance, are at the root of current disagreements over the number of living species. Somewhere between 1.4 and 1.8 million species have been described (Stork 1993), but estimates of the actual number of species worldwide have soared upward over the past decade. Current estimates range mostly between 3 and 40 million (Hodkinson and Casson 1991; Stork 1988; May 1990, 1991; Pimentel et al. 1992), with some suggestions that the number may even be as high as 100 million (Ehrlich and Wilson 1991). Janzen (1992) estimates that Costa Rica alone has 500,000 species, of which 365,000 are arthropods. Most of that diversity is due to insects, which are the most species-rich detritivores, herbivores, carnivores, prey, competitors, and mutualists in terrestrial and some freshwater communities (Thompson 1984). The higher estimates of worldwide diversity suggested during the past decade result from studies indicating that a significant proportion of the world's species are parasites, commensals, or mutualists specialized to live on one or, at most, a few hosts.

The great disparity among the estimates comes mostly from differences in guesses about just how large a proportion of the world's species is made up of extreme specialists. Estimates have increased since the early 1980s mostly as a result of studies indicating that the canopies of tropical trees may harbor millions of insects specific to particular tree species. Using insecticidal fog to collect canopy-dwelling arthropods from nineteen *Leuhea seemanii* trees in Panama over three years, Erwin and Scott (1980) collected 7712 beetles, excluding weevils. These beetles sorted into at least 945 species spanning 56 families. Over half these beetles were herbivores, with the remaining classified as predators, fungivores, and scavengers. From other studies Erwin (1982) estimated that these trees also probably harbored at least 200 weevils, bringing the total for all beetle species to about 1200. He then guessed that 20% of the herbivores and 5–10% of the other species had life histories tied in one way or another specifically to *L. seemanii*, arriving at an estimate of 162 specialists on this one plant species. Erwin extrapolated these results to an estimate of worldwide diversity by making four assumptions: each species of tropical tree has 162 host-specific beetle species, beetles include 40% of arthropod species, at least one other arthropod species is on or below a tree for every two species found in the canopy, and there are 50,000 tropical tree species (Erwin 1982). Combining these estimates, he arrived at an overall estimate of 30 million tropical arthropod species.

Each of the intermediate guesses influences the estimate, and biologically plausible alternative guesses for these numbers can produce estimates for arthropod diversity ranging from 2 to 80 million species (Stork 1988, 1993; May 1990; Hodkinson and Casson 1991). But the percentage of specialist species has become the lightning rod for the debate. Alternative percentages have been suggested, but these too are guesses based upon very limited or localized evidence. Stork (1988) used a range of estimates for the percentage of specialist species—some as low as 5%—in his reanalysis of Erwin's calculations. Basset (1992), too, used low estimates. Based upon a study of the insects associated with one tree species in northern Australia, in which only 11% of the herbivores and 3–4.5% of all arthropod species appeared to be specialists, Basset argued that polyphagy in tropical rain forests may be common. May (1991) took an even more extreme position, arguing that the percentage of host-specific arthropods on tropical tree species was probably closer to 2%. The basis for May's argument was a conjecture that individual tropical tree species are too sparse to favor extreme specialization in phytophagous insects. Dixon et al. (1987) have also used this kind of argument to explain the lower number of aphids in tropical, as compared with temperate, regions. Similar arguments have been used to explain the lower diversity of some kinds of parasitoids in the Tropics. The more host-specific larval parasitoid taxa are relatively more common in temperate regions and the less host-specific pupal parasitoid taxa are relatively more common in the Tropics (Gauld 1986a,b; Hawkins 1990). For all these taxa, a lower percentage of extreme specialists may occur in tropical forests from either of two causes: selection for decreased specialization within the Tropics or failure of taxa of specialists to become established.

The only way to resolve the question of whether the estimates of the proportion of extreme specialists in the Tropics are too high or too low is through long-term, detailed studies of patterns of specialization within local communities. So far there has been only one: Janzen's (1988) long-term, intensive and extensive study of specialization in moth and butterfly caterpillars in Santa Rosa National Park in northwestern Costa Rica. The 10,800-hectare park includes about 725 species of vascular plants. During the ten-year period from 1977 to 1987, Janzen and his assistants reared more than 10,000 wild-caught caterpillars and collected and pinned more than 50,000 moths. By 1987, the number of known species in the park was 3142, and the rate of capture of new species at light traps had dropped to about 1 per month. More than 95% of the species eat green leaves. From the rearings so far completed, Janzen estimated that at least half of the Santa Rosa caterpillars probably feed on only one plant species and most of the remainder feed on only a few chemically or taxonomically related hosts. More than any other single study, this remarkably detailed analysis of local patterns in specialization emphasizes that extreme host specificity is common in phytophagous insects.

Janzen's estimate of at least 50% for the frequency of extreme specialists is a result of a long-term study of insects in a tropical dry forest. Detailed studies of other particular taxa are providing support for higher, rather than lower, estimates of the proportion of extreme specialists. Among fifteen studies of local host specialization within butterfly or beetle taxa in the Neotropics, the median number of host-plant species used by an insect species ranged from one to four (Marquis 1991). In ten of those studies, the median was one or two. These local studies, however, will need to be tied to broader geographic patterns in specialization in order to arrive at an overall estimate of species diversity. Species commonly specialize locally on one or a few local hosts, but populations of the same species may differ geographically in the hosts they use.

Nevertheless, future studies of taxa are likely to reveal more, rather than less, specialization. Phylogenetic analyses have indicated that adoption of the phytophagous habit in insects—the group responsible for much of the discrepancy in estimates of diversity—has generally resulted in higher rates of diversification and speciation than those of closely related nonphytophagous taxa. Mitter, Farrell, and Wiegmann (1988) estimated that phytophagy has arisen at least fifty times among extant taxa. In a careful analysis of thirteen sister groups of insects, they found in eleven of those groups that the phytophagous lineage (i.e., those feeding on vascular plants) had at least twice the number of species as the nonphytophagous lineage. The evolution of phytophagy, then, has repeatedly opened new adaptive zones into which insect species have radiated and specialized.

Fungi, nematodes, and mites are the other major taxonomic groups besides insects that are both rich in species and often extreme in their specialization to particular hosts. There are more than 13,000 species of nematodes, at least 20,000 species of mites, and three to seventy-five times that number of fungi (Maggenti 1983; Woolley 1988; Hawksworth 1991). Like the estimates for insects, estimates of the number of fungi have recently increased, based upon assumptions about extreme specialization. There are approximately 69,000 known fungal species, but Hawksworth (1991) has estimated that there may be up to 1.5 million. Hawksworth's estimate is based upon the observation that the ratio of vascular plants to fungi in some well-studied communities is about 1 to 6. Assuming that there are about 270,000 vascular plants, and allowing for duplicate names (e.g., fungal species for which different phases of the life cycle have been given different names), Hawksworth suggested a total of 1,504,800 fungi. As he does for insects, May (1991) thinks this estimate is much too high because it assumes that the degree of host specificity in tropical fungi is as high as in temperate fungi. Moreover, May argues that if only 4% of fungi have been described, then 96% of the fungi found in any newly studied area should turn out to be new species. Yet, recent surveys in Africa, South America, and the tropical Pacific have found only 25–30% of the species to be undescribed.

Surveys, however, are probably more prone to lumping than to splitting. More informative may be the percentage of new species in recent monographic revisions written by specialists on particular groups. Hawksworth cites six fungal taxa revised since 1980 in which the number of newly described species was 29–86%. If Hawksworth is even close to being right, then fungi, because of their extreme specialization to particular hosts, are, after insects, the most species-rich taxonomic group in the world. In reviewing estimates of the percentages of known species for various taxa, Hawksworth (1991) suggested that some other taxonomic groups may also undergo upward projection in estimates of their numbers. Although somewhere between 67 and 81% of algae, bryophytes, and vascular plants may be described, perhaps only 4–10% of bacteria and viruses, which include many extreme specialists, may be known.

It is easy to make all this fuss over the worldwide number of species seem like little more than a twentieth-century version of debates over angels and heads of pins. But without a reasonable understanding of the taxonomic distribution of species diversity and the degree of specialization among species, our musings over the organization of communities, the evolution of interspecific interactions, the outcomes of coevolution, and priorities for conservation all have a hollow ring. The debate over the worldwide number of species shows clearly that our views on specialization influence profoundly the way in which we perceive the organization of biological communities.

Antiquity of Specialized Interactions

Our views on specialization also influence our thinking on whether interactions have the opportunity to become fine-tuned over long periods of evolutionary time. One of the most striking aspects of evolving interactions between species is that new associations are continually forming within communities while others seem to remain intact for millennia. The ease with which we can find newly formed interactions has been used by some evolutionary biologists to argue that long-term specialization and coevolution are uncommon. Nevertheless, an increasing number of studies are showing that some interactions are quite old (Boucot 1990), and the antiquity of some of these associations stands as evidence that some specialized relationships have had more than ample opportunity to become fine-tuned over long periods of evolutionary time.

Some lichen associations, for instance, may extend back 100–200 million years (Hawksworth 1988). Associations between some arthropods and their plant or animal hosts are younger but still may be older than 40 million years. Among the oldest recorded potential mutualisms are those between angiosperms and mites, extending from the Eocene (37–49 million years ago) of southern Australia to today. Over eighty plant families produce, along the leaf veins on the underside of their leaves, small structures (domatia) that harbor mites (O'Dowd and Willson 1989). These mites feed on fungi, lichens, or other arthropods and are rarely phytophagous, and O'Dowd and Willson (1989) have argued that these mites are probably mutualistic with their host plants. After finding these domatia on so many kinds of plants in Australia, O'Dowd et al. (1991) searched for similar structures on fossil leaves. Using "mummified" leaves extracted from Eocene clay lenses in Victoria and South Australia, they discovered domatia on fossil leaves of Elaeocarpaceae and Lauraceae similar to those found in extant taxa within those families. Associated with those domatia were oribatid mites similar to those found on the extant taxa. Many of the predatory mites found within domatia are probably not restricted to any one plant species, but some could be: predatory mites include some taxa that are highly specific to particular groups of phytophagous arthropods (e.g., Zhang, Sanderson, and Nyrop 1992), and some of these phytophagous species may themselves be specific to particular plant species. If some of the plants and predatory mites represent a continuous sequence of ancestral-descendent populations, then they are among the oldest known mutualistic associations of any kind between free-living species.

Some associations involving insects may be even older, and at least one association involving aphids appears to be at least 50 million years old. The aphid subtribe Melaphidina includes four Asian genera and one American monotypic genus. All species form galls on sumac (*Rhus*). The Asian and American lineages of sumac have probably been separated for about 48 million years, when climatic changes pushed the distribution of sumac south of the Bering land bridge. The use of sumac would not alone be fully convincing as evidence for the age of this association. It could have been independently colonized by the Asian and American lineages of these aphids. Moran (1989), however, used the complex life cycle of these species to infer that the aphids evolved their current host associations before the Asian and American taxa became geographically separated. One Chinese species is known to alternate between sumac and mosses in the genus *Mnium*. Moran took individuals of the American species and, during three years, initiated long-lived colonies on mosses. It seems unlikely, and unparsimonious, that the unusual ability to form galls on sumac and also survive on mosses would have evolved independently in the Asian and American lineages.

Some other associations between insects and plants seem to trace back at least to the Miocene. The aphid *Longistigma caryae* feeds today on plants in the Fagaceae (e.g., hickories, walnuts) in eastern North America. A fossil aphid almost identical in morphology has been found associated with leaves of this plant family in Miocene deposits dating back 8 million years (Heie and Friedrich 1990). In other associations, the characteristic damage caused to leaves by some extant insects is very similar to that found on fossil leaves. Some fossil oaks from the Miocene of western North America have lepidopteran leaf mines that are almost identical in shape to those found on extant oaks, suggesting that these associations have remained intact for millions of years (Opler 1973).

In a few cases, some unique environments provide indirect evidence for both long-term specialization and coevolution. Lake Tanganyika in Africa's rift system harbors a large number of endemic species that may have been coevolving for as long as 7 million years, which is the estimated maximum age of the lake. The lake's endemic gastropods and the endemic potamonautid crab *Platytelphusa armata* that preys on them are much more robust in their morphology than other freshwater species (West, Cohen, and Baron 1991). Compared with other African and Eurasian gastropods, the Tanganyikan endemics have significantly thicker apertural lips, more sculpturing on their shells, and stronger shells that fracture at considerably higher loads (fig. 1.1). The claws of the endemic predatory crab, which are enlarged in both males and females, are similarly large and robust relative to other freshwater species (fig. 1.2). In addition, the chelae are molariform rather than serrate. The more-robust features of both the gastropods and the crab are due neither to a closer marine origin for these species relative to other freshwater species nor to any peculiarity of the water chemistry of the lake (West, Cohen, and Baron 1991). These robust features are similar to those found in species in marine environments because, in both environments, long-term associations appear to have favored highly armored prey and predators capable of countering those defenses (Vermeij 1987).

The features found in Lake Tanganyikan species seem best explained as a result of long-term coevolution. In both intraspecific and interspecific experiments, West, Cohen, and Baron (1991) found that the more-robust features of the gastropods all contributed to better defense against predation, and that the more-robust claws in the crab increased the frequency of successful predation. The Tanganyikan gastropods have shells that are at least an order of magnitude more resistant to a directed stress, such as that imposed by a crab claw, than those of other freshwater gastropods, and these features have evolved independently in two families of Tanganyikan gastropods. Nevertheless, as West, Cohen, and Baron (1991) noted, their interpretation of long-term coevolution, although compelling, is still preliminary. The evolution of these characters needs to be evaluated using the lake's fossil record, so that

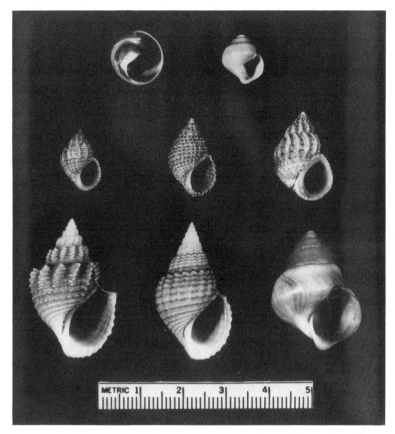

Fig. 1.1. The robust shells of endemic gastropods from Lake Tanganyika. Reprinted, by permission, from West, Cohen, and Baron 1991.

a phylogenetic analysis of the species can indicate which characters are the more recently derived within these lineages.

Fossil evidence providing these kinds of evidence for long-standing inter-actions between particular species, or at least genera, is still uncommon. The problem, of course, is that it is hard enough finding fossils of living taxa. The chance of finding convincing biogeographic and fossil evidence for two taxa and the interaction between them is many times more remote. Fossil sharks and cephalopods (nautiloids and ammonoids) indicate just how hard it is to find interpretable evidence of long-standing interactions. Cephalopods have the kinds of hard body parts that are commonly preserved in the fossil record. Hence, evidence of predation on these taxa should be much easier to find

Fig. 1.2. The unusually robust, molariform chelae of the endemic Lake Tanganyikan crab *Platy-telphusa armata* used to crush gastropod shells. Reprinted, by permission, from West, Cohen, and Baron 1991.

than for many others. Hansen and Mapes (1990) combed through about 5000 fossil cephalopods from the Pennsylvanian (320–260 million years ago) and found convincing evidence of shark bites on 25 specimens; that is, 0.5%. That low percentage may indicate that shark predation on cephalopods was rare, but they argue instead that most cephalopod shells would have been smashed by the sharks during the attack and subsequent feeding. If they are right, then these cephalopod fragments sporting shark-tooth holes are the fortuitously unsmashed bits from an interaction that may have been much more common than indicated by the fossil record.

The interpretation of this or any ancient and persistent association found in the fossil record is not straightforward. Fossils showing similarity in form to currently interacting species do not imply that the interaction has remained unchanged for millions of years. The fossil record for the most part records only morphology, with incomplete glimpses into behavior, such as the movements of insect larvae recorded in leaf mines. The many other genetic changes in behavior and physiology are left unrecorded. Hence, stability in morphology is more difficult to evaluate than change when considering specialization and coevolution. Nevertheless, what the fossils can tell us is that some associations remain intact for millions of years, thereby allowing at least the potential for long-term coevolution.

'Long term', however, means different things to different evolutionary biologists. In analyzing the age of interactions between highly specialized herbivorous molluscs and large marine algae in the cool-temperate North Pacific and North Atlantic oceans, Vermeij (1992) found that at least six of the twelve known specialized associations were established no earlier than the Pliocene, resulting from the biotic interchange between the North Pacific and North Atlantic following the opening of the Bering Strait. He used these results to argue against "the widely held view that specialized ('coevolved') interactions tend to be ancient." Whether these interactions are "ancient" or not depends upon one's perspective on evolution and the processes that shape communities. (Whether the interactions are coevolved is a separate question, which need not be directly linked, as Vermeij has done, to the question of specialization.) There have been millions of generations of these taxa since the Pliocene—and whether 50% (six of twelve associations) is high or low depends upon the point one wants to make. Vermeij's objective was to show that specialization is not a result of a drawn-out evolutionary process that requires many millions of years, and his data do in fact support that view. The results, however, also show that highly specialized associations between molluscs and large marine algae in the northern oceans have evolved repeatedly, and at least some of them appear to be millions of years old.

New Interactions

As Vermeij's (1992) results indicate, the antiquity of interactions and specialization (and, for that matter, coevolution) are not linked in any predictable way. Specialization does not necessarily evolve slowly over time through reciprocal evolutionary change, and it is not necessarily a result of gradual fine-tuning of life histories, morphology, and physiology. We now know that specialization can sometimes evolve quickly in new interactions. The rapid colonization of introduced crop plants and weeds by phytophagous insects indicates just how fast new interactions can form and sometimes evolve in preferences for a new host (Strong, Lawton, and Southwood 1984). We now have convincing evidence that some plant-feeding insects have colonized new hosts in the past several centuries and become partially or completely specialized for feeding on those hosts. One or more populations of the pierid butterfly (*Colias philodice*) (Tabashnik 1983), the checkerspot butterfly (*Euphydryas editha*) (Thomas et al. 1987), and the western anise swallowtail (*Papilio zelicaon*) (Thompson 1993) have all evolved increased preference for plants introduced into North America over the past one hundred to two hundred years. Populations of the apple maggot (*Rhagoletis pomonella*) have become specialized for either native hawthorn or introduced apple (Bush 1969; Feder, Chilcote, and Bush 1988). The soapberry bug (*Jadera haematoloma*) has diversified in North America over the past fifty years into populations that specialize on different introduced sapindaceous plants. These bugs use their tubular mouthparts to pierce the fruits of their hosts and reach the seeds on which they feed. The sapindaceous hosts differ from one another in the sizes of their fruits, and the soapberry bug populations have evolved beak lengths that correspond to the differences in fruit size (Carroll and Boyd 1992). The great age of some interactions and the ability of species to form new interactions as ecological conditions change together guarantee a wide range within any lineage in the antiquity of interactions with other taxa.

Each new interaction offers its own opportunities for specialization and imposes its own limitations. As populations form new interactions, they evolve and specialize in what are sometimes surprising new directions, diverging far from the relationships that their ancestors maintained with other species. Birds of paradise (Paradisaeidae) range in diet from species that are almost entirely frugivorous to others that are almost entirely insectivorous (Beehler 1983a, 1987; Pratt and Stiles 1985; Diamond 1986a). Geometrid moths, whose twiglike caterpillars normally chew on leaves, include in Hawaii some remarkable species in the genus *Eupithecia* whose larvae are ambush predators of other insects (Montgomery 1983). Nematodes have diverged to include predators, plant parasites, and animal parasites, and ciliate protozoa include both free-living species and others that are symbiotic within termites, wood roaches, and mammals (Beaver and Jung 1985).

Many of these novel interactions develop from preexisting relationships that have taken a new evolutionary turn. For instance, the evolution of carnivory in some lycaenid butterfly larvae has been one of the consequences of the evolution of tending of lycaenid butterfly larvae by ants. Lycaenids (blues, coppers, and related species) are the largest butterfly family, comprising the majority of all butterfly species (Pierce 1987). This diversity of species has made it possible to observe some of the evolutionary transitions that may have led to the evolution of carnivory (feeding on ant broods within ant nests) in these insects. In some species larvae are not only tended by ants, which feed on secretions produced by the larvae, but are shepherded between the ants' nest and the lycaenid's food plant. In a few cases, however, the lycaenid larvae have become specialized to feed directly on the ant larvae within the nests. The five species of *Maculinea* in Europe, each of which associates with a different ant species, all feed initially on a host plant but attain only 1% of their final mass on this host before moving into the ants' nest for the following ten months. While in the nests, three of the *Maculinea* species prey on ant larvae, and the other two on ant regurgitates (Thomas et al. 1989, 1991).

Other new kinds of interaction develop from opportunistic encounters rather than from preexisting relationships. Most remain rare events with little or no effect on the evolution of either species: a population of the tree frog *Hyla truncata* in Brazil sometimes includes fruits in its otherwise arthropod diet (Da Silva, Britto-Pereira, and Carmaschi 1989); the free-living amoeba *Negleria fowleri,* which lives in freshwater and moist soil, occasionally invades the brain and meninges of humans swimming in warm freshwater lakes, ponds, or streams (Yaeger 1985). Only a tiny proportion of new interactions ever evolve to become more common through natural selection on one or both species. But over evolutionary time some of these rare encounters between species develop into new and highly specialized interactions, which can in turn open up yet other new possibilities for specialization.

The ever-changing composition of biological communities itself regularly creates opportunities for new interactions. The rapid dismantling of natural communities during the past several centuries has even escalated the rate at which new interactions develop between species. It is a fascinating process to observe in a macabre sort of way. As species are introduced into new areas, either intentionally or not, they are stripped of their old interactions and acquire new ones. Introduced crops, weeds, and pests, along with game fish, birds, and mammals are the most obvious of these introductions. Even species that are not being moved around by humans must often cope with invaders. For instance, some North American, South American, and Australian flowers are now visited and pollinated primarily by introduced honeybees.

The observations that some highly specialized interactions are quite old whereas others are both new and already highly specialized suggest two conclusions about the evolution of specialization that will be explored in more detail in later chapters: specialization in interactions does not necessarily lead to extinction, and specialization is commonly an evolutionarily dynamic state capable of change—sometimes rapid change—rather than an evolutionarily static dead end. These two observations will become part of the argument I develop about the process of coevolutionary change: long-term reciprocal evolution between two or more species may often be a geographic process in which populations are molded and remolded in their degrees of specialization and adaptations to other species. The evolution of extreme specialization and reciprocal change within local populations is only part of the raw material for the coevolutionary process.

The Lexicon of Evolving Interactions

In constructing arguments on specialization and the process of coevolution, all of us must draw on a terminology that is still inadequate for dealing with the complexity of ways in which species specialize and evolve with one another. The descriptive vocabulary of interspecific interactions, patterns of specialization, and coevolution is a product mostly of the past one hundred years. Darwin used some form of the word 'competition' sixty-eight times within the *Origin of Species,* sometimes referring to intraspecific interactions and at other times to interspecific interactions. In describing interactions in general, he often used the phrases 'mutual relation(s)' and 'mutually

adapted', and in describing reciprocal evolution between species, he sometimes used the word 'coadaptations'. He used 'prey' often (thirty-nine times),
but never the words 'predation' or 'predator'. Instead he talked about birds
of prey, etc. He used 'parasite' or 'parasitic' and occasionally the word
'browse', but never 'grazer' or 'grazing'. 'Symbiosis' was not coined until
1879, when de Bary used it to refer to intimate associations between species,
whether parasitic, commensalistic, or mutualistic. Since then it continues to
be used in that way by many biologists but has become synonymous with
'mutualism' when used by other biologists and in popular literature. The
words 'mutualism' and 'commensalism' made their way into the lexicon of
evolving interactions in the 1870s through the writings of the Belgian zoologist Pierre Van Beneden (Boucher 1985).

Much of the terminology on evolving interactions is even more recent,
paralleling the rise of population biology and evolutionary ecology. 'Coevolution', 'genetic feedback', and 'character displacement' are all terms that
were coined in the 1950s and 1960s to refer to processes of evolving interactions. The study of these processes influenced the terminology that developed
in the 1970s and 1980s. Parasites were no longer just the cestodes, trematodes, and nematodes of traditional parasitologists and the bacteria and fungi
of pathologists. Price (1977, 1980) showed that the parasitic mode of life
spans many phyla, producing a lifestyle in which individuals are adapted to
living on a single host individual throughout development or during an entire
stage of a complex life history. The evolutionary consequences of parasitism
transcend taxonomic boundaries and differ from those of other modes of life
and forms of interaction. At the same time as Price was showing the similarities in the parasitic lifestyle regardless of kingdom or phylum, Anderson and
May (1979) highlighted some of the differences. They subdivided parasites
into microparasites and macroparasites based upon differences in population
dynamics. By their definition, microparasites are those that reproduce directly within a host, have generation times much shorter than their host, and
induce acquired immunity in recovered hosts; macroparasites are usually
larger, reproduce outside the host, have generation times more similar to their
host, and generally induce at best only short-term immunity based upon the
number of parasites present.

From an evolutionary viewpoint, parasitism, grazing, and predation differ
in the kinds of selection pressures they place on species (Thompson 1982,
1986c, 1990). The evolution of specialization to particular prey or host species, the evolution of defense, the evolutionary transition from antagonism
to mutualism, and the modes of coevolution all vary among these different
forms of interaction. No longer is it sufficient for plant ecologists interested
in interactions to use predation to refer to all kinds of herbivory, because we
now know that different forms of herbivory can differ in the selection pres-

sures they exert on plant evolution. Some herbivores are parasites, some are grazers, and some are predators.

There continues to be the inevitable problem of how precisely to fine-tune the terminology on interactions in order to make meaningful generalizations about selection pressures and population biology. For example, as a counterbalance to studies on the similarities in interactions between parasites and hosts, Janzen (1985) tabulated differences in the nature of the interactions between plants and their ectoparasites (e.g., insect larvae that live on a single plant throughout development) as compared with animals and their ectoparasites (e.g., insects, such as lice, that live on a single animal host throughout their lifetimes). Similar problems have occurred in describing other kinds of interactions. Lawton and Hassell (1981) argued that many of the cited cases of competition between phytophagous insects were in fact amensalism, because they provide no evidence for reciprocal negative effects on fitness. This distinction is useful, because the predictions for both community structure and evolution differ between these modes of interaction.

The infusion of a more interaction-based terminology into evolutionary studies is still occurring. Phrases such as 'evolutionary arms races', 'enemy-free space', and 'tritrophic interactions' are heuristic tools for describing process or pattern, although each is a catchall phrase for a variety of processes and patterns. An evolutionary arms race between competitors is very different from one between species at different trophic levels. All typological descriptions of interactions lump together relationships that can differ fundamentally in the mechanisms that shape their outcome (Abrams 1987; Thompson 1988d). Moreover, all groupings of interactions mask to some extent the variation in pattern and process that leads to different degrees of specialization and coevolution. That, however, is the perennial problem of science: to move beyond individual observations and reach generalizations, yet to understand the limits of each generalization.

I have avoided coining new terms in discussing different modes of interaction because I wish to emphasize in this book patterns in the evolution of specialization and in the coevolutionary process rather than the classification of interactions. But, frankly, we will ultimately need some new terms. In discussing the evolution of antagonistic interactions between trophic levels, I will use the general terms 'parasitism', 'grazing', and 'predation' to indicate common evolutionary patterns of specialization and coevolution that result from particular lifestyles regardless of taxon. Hence, to emphasize similarities in selection pressures, I will sometimes group taxa as diverse as trematodes and gall-making insects when discussing parasites, bison and vampire bats and some grasshoppers when discussing grazers that move between two or more victims without killing them quickly and directly, and lions and seed predators when discussing predators. We will need terms that make finer distinctions. Social parasites such as cuckoos and cowbirds are similar to true parasites in some of the ways that natural selection acts on specialization, but they also differ from true parasites in some other respects. We need to draw distinctions in modes of interaction wherever it is important ecologically and evolutionarily to do so, but we also need to see how far we can generalize about the effects of particular general modes of interaction such as parasitism, grazing, and predation.

In some cases, we simply lack parallel terms. Parasites attack hosts, predators attack prey, but we have no general term for what grazers attack (if we use the term 'grazer' for taxa other than just vertebrates that chew on parts of plants). Hence, I have opted for the word 'victim' for now. We also lack a separate terminology for the partners in nonsymbiotic mutualisms. I have used 'visitors' and 'hosts' here and elsewhere (e.g., Thompson 1982), and some others have adopted that usage (e.g., Cushman and Beattie 1991), but we need a better word for the recipients of these short-term visits just as we need a better word for what grazers attack.

Terms for interactions are the generalizations—simply tools for helping to organize our thoughts about pattern and process—and the terms change as our understanding changes. Most of the refining of terms on interactions in recent decades has not simply been pedantry. Instead, it is the inevitable result of the shift in any science from simple description to the study of processes and the search for patterns. It has occurred repeatedly in all biological subdisciplines as they have become more evolutionary in their approach. The word 'species' has been reformulated in the past several hundred years from a typological concept that reinforced belief in the unchanging nature of organic life to one that implies a lineage of individuals cleaved from other groups by evolution. Similarly, terms for interactions are changing from vehicles for typological description to aids in organizing our ideas about patterns and processes in how interactions evolve. The following two chapters examine how this shift from description to the study of pattern and process has occurred in the study of specialization and coevolution, creating a progressively more dynamic view of evolving interactions.

CHAPTER TWO

From the Entangled Bank to
the Evolutionary Synthesis

By the time Darwin's *Origin* was published, naturalists had fairly well outlined the range of diversity among living species. Ever since the time of Linnaeus, naturalists had traveled widely, collecting specimens and shipping them back to the academic centers of Europe to be stamped with Latin binomials. Linnaeus had provided the system that allowed this diversity to be cataloged, and the very process of arranging that diversity of forms showed to naturalists the plan of the Creator. Interactions also showed that plan, and it was a view held not only by naturalists but by other scientists as well. Joseph Priestley, now known principally as the discoverer of oxygen, wrote: "Are not all plants likewise suited to the various kinds of animals which feed upon them? . . . The various kinds of animals are, again, in a thousand ways adapted to, and formed for, the use of one another. Beasts of a fiercer nature prey upon the tamer cattle: fishes of a larger size live almost wholly upon those of a less: and there are some birds which prey upon land animals, others upon fishes, and others upon creatures of their own species" (Priestley 1764, in Boorstin 1948, 43).

After publication of the *Origin,* the study of the diversity of species was changed forever. It took longer, however, for interactions to get the broad, comparative evolutionary attention that species received. One cannot preserve interactions and ship them back home to be cataloged and displayed in museums. Interactions have no morphologies that can be studied side by side as an aid to how they relate evolutionarily to one another. Moreover, they rarely leave a direct fossil record. Nevertheless, as for the study of species, the publication of the *Origin* and Darwin's study of orchids three years later inspired naturalists to confront the diversity and intricacies of interactions and their adaptive meaning. The study of evolving interactions became the purview of a small number of academic biologists and of the amateur naturalists who spent their time tromping through the entangled bank.

It was the naturalists, especially, who undertook the cataloging of interactions and provided the first adaptive interpretations. These careful observers showed the intricacies of the entangled bank to be much greater than anyone had imagined. The discovery of mimicry, ant gardens, and the many different forms of pollination all came about through the work of naturalists in the several decades immediately following publication of the *Origin*. Meanwhile, other biologists began to study how parasitism and symbiotic mutualism arise from other forms of interaction. In the 1860s and 1870s, de Bary, Van Beneden, and others coined new words such as 'symbiosis', 'mutualism', and 'commensalism' to categorize the new forms of interaction they were finding. Biologists began to construct phylogenies and develop general arguments about how parasitic taxa were related to saprophytic and free-living forms in everything from bacteria to lichens, mycorrhizae, and annelid worms, and they looked for patterns in how interactions differ in the selection pressures they place on populations (e.g., Smith 1887; Andrews 1891; Pound 1893). In the *American Naturalist,* Smith (1887) wrote, "Parasites, whether they be animal or vegetable, have certain characters in common which are due to their relation to their host rather than to their own intrinsic organization." These were the kinds of thoughts that would eventually lead to the rigorous search for pattern and process in evolving interactions.

Pollination Biology and the Origins of Coevolutionary Studies

It was the study of pollination and mimicry, however, that provided the impetus for evolutionary thinking about the origins of specialization in interactions and the conditions that favor coevolution. Ever since Darwin, biologists have understood that species may evolve specialized interactions with one another and that some of these interactions may result in reciprocal evolution. In describing coadaptation between bees and flowers in the *Origin,* Darwin wrote the first account of the process of coevolution. He began with two assumptions and an observation. The assumptions were that natural selection favors bees that are quicker than others at obtaining food, and it favors plants whose flowers are constructed in ways that maximize fertilization by pollinators. The observation was that honeybees ("hive-bees") could reach the nectar of incarnate clover, but only bumblebees ("humblebees") could reach the nectar of common red clover. Whole fields of red clover offered nectar that was inaccessible to honeybees.

Darwin then imagined how coevolution might occur between local populations of honeybees and common red clover. He imagined the size, body form, or length and curvature of the proboscis of honeybees evolving to better match the shape and length of the corolla tubes of the local population of red clover, so that they could take advantage of this resource. He then imagined what would happen if the bumblebees became rare. Natural selection might then favor common red clover plants that had shorter or more deeply divided corollas, thereby allowing yet better pollination by honeybees. He wrote,

"Thus I can understand how a flower and a bee might slowly become, either simultaneously or one after the other, modified and adapted in the most perfect manner to each other, by continued preservation of individuals presenting mutual and slightly favourable deviations of structure" (Darwin 1859, 94–95).

Building upon this theme, Darwin ([1862] 1979) devoted the first book that he published after the *Origin* to *The Various Contrivances by Which British and Foreign Orchids Are Fertilised by Insects, and on the Good Effects of Intercrossing.* It was a detailed study of descent with modification: how old floral parts are molded and remolded over evolutionary time into new shapes and functions. It explores the ways in which orchids have diverged and become specialized to different kinds of pollinators and mechanisms of cross-fertilization. Its painstaking descriptions of floral morphology and anatomy make it perhaps the most difficult of Darwin's books to read from cover to cover. Yet it is precisely those descriptions that make the book such a marvelous achievement. By focusing on those details, Darwin showed how floral structures have been modified and convoluted to take advantage of their visitors: "The use of each trifling detail of structure is far from a barren search to those who believe in natural selection" (p. 351). Those trifling details show us that related species have similar structures modified in different ways. In concluding his analysis of the homology of orchid structures, Darwin writes, "Can we, in truth, feel satisfied by saying that each Orchid was created, exactly as we now see it, on a certain 'ideal type' . . . ? Is it not a more simple and intelligible view that all Orchids owe what they have in common to descent from some monocotyledonous plant . . . each modification having been preserved which was useful to each plant, during the incessant changes to which the organic and the inorganic world has been exposed?" (pp. 306–7).

Darwin saw specialization to particular pollinators (or, more often, groups of pollinators) and the associated increased efficiency of pollination as the basis for the evolution of floral morphology (fig. 2.1). His book prompted others to study specialization to pollinators, and he was able to write in the second edition that he thought it

a safe generalization that species with a short and not very narrow nectary are fertilised by bees and flies; whilst those with a much elongated nectary, or one having a very narrow entrance, are fertilised by butterflies or moths, these being provided with long and thin proboscides. We thus see that the structure of the flowers of Orchids and that of the insects which habitually visit them, are correlated in an interesting manner,—a fact which has been amply proved by D. H. Müller to hold good with many of the Orchideae and other kinds of plants. (1877, 30)

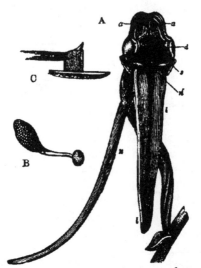

HABENARIA CHLORANTHA, OR BUTTERFLY ORCHIS.

a a. anther-cells. *d.* disc of pollinium. *s.* stigma. *n.* nectary. *n'.* orifice of nectary. *l.* labellum. A. Flower viewed in front, with all the sepals and petals removed except the labellum with its nectary, which is turned to one side.	B. A pollinium. (This has hardly a sufficiently elongated appear- ance.) The drum-like pedicel is hidden behind the disc. C. Diagram, giving a section through the viscid disc, the drum-like pedicel, and the attached end of the caudicle. The viscid disc is formed of an upper membrane with a layer of viscid matter beneath.

Fig. 2.1. Drawing of a *Habenaria* orchid from Darwin's *The Various Contrivances by Which British and Foreign Orchids Are Fertilised by Insects*. Darwin used the long corolla of *Habenaria* and the long proboscises of lepidopterans as an example of how "the structure of the flowers of Orchids and that of the insects which habitually visit them, are correlated in an interesting manner."

Darwin predicted that the Madagascan orchid *Angraecum sesquipedale,* with its "nectaries eleven and a half inches long," would be found to be polli- nated solely by moths with correspondingly long proboscises. He argued that even if the nectary were to fill sufficiently so that smaller insects could reach the rewards, the structure and positioning of the pollinia were such that only a large moth, with its head pushed hard against the flower, could effectively remove the pollinia: "The pollinia would not be withdrawn until some huge moth, with a wonderfully long proboscis, tried to drain the last drop" ([1862] 1979, 201). When writing the second edition he returned to this yet unful- filled prediction, perhaps a bit testy that "[t]his belief of mine has been ridi- culed by some entomologists," and noted that Fritz Müller had seen a sphinx moth in Brazil with "a proboscis of nearly sufficient length" (1877, 163).

Darwin viewed the evolution of these extremely specialized corolla tubes and proboscises as the result of a coevolutionary race: "there has been a race in gaining length between the nectary of Angraecum and the proboscis of certain moths" ([1862] 1979, 202–3). But what Darwin saw as fine-tuning of adaptation and specialization of species through reciprocal evolution, others saw as compelling evidence for special creation. In his 1867 book, *The Reign of Law*, the duke of Argyll used the extreme specialization of this orchid as evidence for the existence of God (Kritsky 1991; Smith 1991). This extremely specialized orchid and its predicted extremely specialized pollinator could not possibly have come about one before the other. They must have originated at the same time. The duke's arguments prompted Wallace to respond in turn (Wallace 1867). In an article entitled "Creation by Law," he patiently and clearly described how natural selection working on variation within orchid and moth populations could lead to reciprocal changes and the extreme specialization found in these species. Following Darwin's logic that fertilization in this species would occur only by moths that had to fully insert their proboscises into the flowers to reach the nectar, he described how through natural selection "there would in each generation be on the average an increase in the length of the nectaries, and also an increase in the length of the proboscis of the moths."

Wallace, however, went further. He knew that his argument taken alone would predict that all interactions between pollinators and plants should result in coevolution of pairs of species and extreme specialization. Yet not all orchids and pollinators are so specialized and tightly coevolved. This, in fact, was one of the duke's points: the existence of less-specialized orchid species indicated that there was nothing about orchids that required such extreme specialization. Consequently, the exact adjustment of this one orchid species to one pollinator must be evidence of God's work. To rebut this part of the duke's argument, Wallace noted that "there are a hundred causes that might have checked this process" and such checks have undoubtedly happened in other orchids. For example, if variation in the quantity of nectar was greater than variation in the length of the nectary, fertilization by smaller moths would have been possible. His implication was that under these conditions natural selection would not have favored the evolution of longer nectaries. Alternatively, if the numbers of the moth species with the longest proboscis had been diminished by an enemy, then natural selection again would have favored flowers with shorter nectaries pollinated by smaller moths.

This interchange of arguments shows that both Darwin and Wallace had, right from the start, understood how coevolution could result sometimes, but not always, in extreme specialization. Ghiselin (1984) has argued that all work on coevolution traces directly or indirectly to Darwin's orchid book. That work is also the origin of all subsequent work on the evolution of ex-

treme specialization. Darwin's study of orchid flowers resulted almost immediately in a cottage industry of similar analyses by other naturalists, who
began to study specialization and, in some cases, coevolution in other taxa.
In short order, Delpino produced a classification of flower types according to
the kinds of pollinators they attracted, and Hermann Müller began to write
inexhaustibly on specialization in both flowers and their insect visitors (review in Müller 1883). Müller was especially concerned with the sequence of
evolutionary events by which flowers and insects have adapted to each other.
As had Darwin, he began with the premise that adaptations in floral structure
were generally a means of ensuring cross-fertilization. To this he added the
premise that, in general, floral-visiting insects "are not limited by hereditary
instinct to certain flowers" (i.e., to one plant species). From his studies of
flower visitors in Europe he had concluded that such extreme specialists "do
not form 1 per cent" (Müller 1883, 570).

From these two premises, Müller developed a view of how specialization
evolved in these interactions. Because flower-visiting insects are generally
not specific to any one plant species, natural selection has favored plants with
conspicuous flowers that enhance their attractiveness to potential pollinators.
This conspicuousness, however, also attracts other, nonpollinating insects.
Consequently, floral evolution sails between the Scylla and Charybdis of
making floral rewards too available, thereby attracting nonpollinators, and
making them too difficult to reach, thereby discouraging potential pollinators. He argued that there are a number of routes a plant can take in solving
this problem, but once it adopts a particular route it is likely to continue in
that direction, sometimes but not always becoming more specialized to yet
fewer pollinators. Meanwhile, floral visitors evolve to cope with the changing positions of floral rewards within the flowers, sometimes causing species
to become specialized at least to particular floral types if not to particular
plant species.

Müller's book was originally (1873) published only in German. Fearing
that the "book is in the hands of only very few Englishmen," he published
between 1873 and 1877 seventeen short articles in *Nature* under the general
title "On the Fertilisation of Flowers by Insects and on the Reciprocal Adaptations of Both." The first half dozen of these articles described floral adaptations for cross-fertilization and insect adaptations for collecting nectar and
pollen. But during his summer vacation in the Alps in 1874, Müller noticed
that, in the Alpine, butterflies were more frequent visitors to flowers and bees
less frequent visitors than at lower elevations in the mountains. Moreover,
a number of Alpine plants had floral characteristics apparently adapted for
pollination by butterflies. He had time to make only a few observations that
first summer, but over the next several years he returned to the Alps and studied the frequency of insect visits to a wide range of flowers. Most of the

subsequent papers in the *Nature* series develop the theme of how plants at different elevations in the Alps have become specialized for different pollinators.

Müller approached the evolution of specialization by comparing closely related species or genera that differ in their elevational distribution. Wherever possible he chose what he considered to be sister species. For each pair of species, he compared floral morphologies and recorded the frequency of visits by lepidopterans and bees. He concluded that among such pairs, pollination by lepidopterans was more common in the Alpine and pollination by bees was more common at lower montane elevations. This was a strikingly powerful analysis. Müller did not simply note the relative frequency of butterfly-pollinated flowers and bee-pollinated flowers at different elevations, which could have come from differential sorting of plant families according to elevation. He carefully compared closely related species that he considered to have diverged in floral morphology from a shared ancestral form, and he tested his designations of "butterfly-type" and "bee-type" by recording the frequency of visits to these flowers. Although rudimentary, it was the first attempt to combine phylogenetic and ecological approaches as a way of understanding patterns in how interactions evolve and how specialization develops. His brother, Fritz Müller, used the same two-pronged approach in developing in the 1870s his own ideas about the evolution of mimicry. The brothers Müller both embedded their ideas about evolving interactions within a phylogenetic framework, and they used the relative abundance of species and the relative frequency of interactions as major ecological conditions shaping the direction of specialization.

Mimicry and the Relationship between Specialization and Coevolution

Even more so than interactions between pollinators and plants, the study of mimicry became a focal point for the early development of ideas on specialization and coevolution. Although the theory of mimicry did not develop explicitly from a concern over the evolution of specialization, the crux of the problem was exactly that: under what ecological conditions should natural selection favor a species tying itself evolutionarily quite tightly to the color pattern found in a coexisting species? Henry Walter Bates addressed the problem in a paper to the Linnean Society in 1861, suggesting that the great similarity in bright colors among some coexisting species may be due to the convergence of palatable species on unpalatable models. What made this such a brilliant idea was the fact that there is no direct interaction between the model and the mimic. The evolution of the mimic is tied to a specific model through the actions of a third, predatory species that confuses the palatable mimic with the unpalatable model. Darwin heard Bates read the paper and wrote to him two weeks later saying, "I think you have solved one of the most perplexing problems which could be given to solve" (December 3, 1861, in Darwin 1896).

Bates's paper was published the following year (Bates 1862) and became one of the major initial applications of Darwin's theory of natural selection. After reading the paper, Darwin became even more impressed with the power of Bates's concept of mimicry. He wrote again to Bates: "In my opinion it is one of the most remarkable and admirable papers I ever read in my life." Later in the letter, he added, "Your paper is too good to be largely appreciated by the mob of naturalists without souls; but, rely on it, that it will have *lasting* value" (November 20, 1862, in Darwin 1896). Both Darwin and Wallace highlighted the concept of mimicry in their subsequent writings on interactions between species. Darwin wrote a laudatory review of Bates's paper (Darwin [1863] 1977), and incorporated the arguments later in *The Descent of Man, and Selection in Relation to Sex* (Darwin [1871] 1981). Wallace (1870) devoted a full chapter to it in his book *Natural Selection* and mentioned it repeatedly in subsequent essays and books (e.g., Wallace [1889] 1905, 1895).

Bates's arguments did not invoke coevolution between the model and its mimics. In fact, he assumed that the mimic was generally much rarer than the model, and that the model was unaffected by the mimic. The idea of reciprocal change in mimicry came from another of Darwin's long-standing correspondents, Fritz Müller, who was studying natural history in Brazil while his brother crisscrossed the Alps studying flowers and their pollinators. It was Francis Darwin's (1896, II, 221) impression that of all his father's "unseen friends Fritz Müller was the one for whom he had the strongest regard." As early as 1867, Müller and Darwin were corresponding about mimicry, and Müller was considering the possibility that mimicry might also occur in plants (July 31, 1867, in Darwin 1896). By the 1870s Müller had conceived of a different form of mimicry in butterflies in which both or several taxa are distasteful and may converge on nearly identical color patterns through natural selection driven by predators. He published these ideas in 1879 in a short paper in the journal *Kosmos,* which Darwin pointed out to Professor Meldola, another of his correspondents. Meldola translated the paper into English and had it published the same year in the *Proceedings of the Entomological Society of London* (Poulton 1896).

Müller's idea of convergence of distasteful species arose from his attempt to understand the origin of similar color patterns in two genera of butterflies, *Ituna* and *Thyridia.* According to the logic of Batesian mimicry, the mimic should be rarer than the model, it should have diverged much more from related species than the model, and, unlike the model, it should be palatable.

In studying *Ituna* and *Thyridia* as well as some other species groups of models and mimics, Müller concluded that "all these characters sometimes leave us in the lurch." He therefore argued that some cases of resemblance among species arise from convergence among distasteful forms, and that if the species are of near equal abundance, both would evolve and it would not be possible to specify which was the model.

The importance of Müller's paper, besides its broadening of the conditions favoring mimicry, was its specific argument for the conditions under which reciprocal evolution may occur. Gilbert (1983) has noted that the first quantitative statement of the conditions favoring coevolution is Müller's argument stating that when several distasteful species are equally common, then "resemblance brings them a nearly equal advantage, and each step which the other takes in this direction is preserved by natural selection."

During the decades following the appearance of Fritz Müller's paper, other naturalists provided additional examples of mimicry and wrestled with the problem of how it originates within populations and spreads through natural selection (e.g., Wallace [1889] 1905; Poulton 1890; Beddard 1892). Romanes argued in the first of his three-volume treatise on Darwinism that "it is impossible to imagine stronger evidence in favour of natural selection" than mimicry of another species so as to deceive natural enemies (1892, 327). During these years books describing the natural history of different regions of the world routinely included examples or ideas about mimicry. Belt ([1874] 1928) dedicated his book *The Naturalist in Nicaragua* to Bates and suggested a number of possible additional examples of models and mimics in insects. Forbes (1885), writing in *A Naturalist's Wanderings in the Eastern Archipelago,* acknowledged his debt as a naturalist to Bates and Wallace and suggested cases of mimicry in birds. Beccari ([1904] 1989), whose view of evolution was strongly Lamarckian, suggested in his *Wanderings in the Great Forests of Borneo* that mimicry was somehow due to the action of the environment. Shelford ([1916] 1985) adopted a more strongly Darwinian view in *A Naturalist in Borneo* and devoted a long chapter to questions on mimicry. In two earlier detailed and beautifully illustrated papers, he also suggested dozens of possible cases of mimicry in brightly colored insects from Australia, Southeast Asia, and the Pacific islands, especially Borneo (Shelford 1902, 1912). These included remarkably close resemblance between brightly colored cockroaches and particular species of beetles, between flies and either bees or wasps, and between a moth and a true bug. This ever-enlarging list of new examples sustained interest in mimicry among naturalists and evolutionary biologists, and it fueled continued speculation on how this kind of close resemblance between two or more species could have evolved repeatedly.

The issue of reciprocal evolutionary change in groups of mimetic species became a focus of arguments beginning in 1894. In a series of papers, Frederick Dixey (1894, 1896, 1897, 1906) argued that although Batesian mimicry relied upon the relative rarity of the mimic and was unlikely to result in reciprocal evolutionary change, Müllerian mimicry often involved "give-and-take changes." Most reciprocal change would occur when two or more species are equally abundant and distasteful to birds. Dixey referred to this give-and-take as reciprocal mimicry, noting that previous writers had not stressed the point that mutual change is involved.

The foil to Dixey was G. A. K. Marshall (1908), who maintained that even in equally unpalatable species, one is likely to be rarer than the other and it is only the rarer one that will undergo evolutionary change, converging on the pattern of the more abundant species. He viewed reciprocal mimicry, or diaposematism as it was also called, as "merely a complication of Müller's theory," and one that would occur rarely in nature. He did allow, however, that under rare circumstances, two unpalatable species might fluctuate in relative abundance such that each serves as the model at different points in time. He referred to this as "alternating resemblance" to set it off from reciprocal change, in which species converge simultaneously on some intermediate pattern. Dixey (1909) replied that alternate mimicry was no different from what he had been calling reciprocal mimicry because "[n]o one could suppose that every step from one side is exactly in point of time coincident with a step from the other; nature works on successive individuals, and whether or not at any given moment the general trend is in one direction rather than another is immaterial." (This interchange over simultaneous versus alternate change, and evolutionary versus coevolutionary change, is interesting for the subsequent development of coevolutionary theory, because the same arguments have been repeated in different guises throughout the twentieth century.)

After the rise of Mendelian genetics at the turn of the century, Punnett (1915) used Marshall's arguments and findings on discontinuous inheritance to argue against the likelihood that species could gradually and mutually converge on a color pattern. Punnett's book appeared during the years in which evolutionary biologists were polarized into different schools of thought: the Naturalists and Biometricians on the one hand, who argued for the importance of natural selection in shaping adaptations and the gradual nature of those changes, and the Mendelians on the other, who argued for discontinuous change (saltations) as the driving force in evolution (Provine 1971). The whole problem of the evolution of mimicry became, in the words of R. A. Fisher, "the greatest post-Darwinian application of Natural Selection, [playing] an especially important part towards the end of the nineteenth and the beginning of the twentieth century" ([1930] 1958, 163). Perhaps to emphasize that point, Fisher chose a picture of models and mimics in Australian and tropical American insects as the frontispiece for his book. The evolution of mimicry had become a part of all the major arguments on evolutionary processes.

Besides its importance in debates over coevolution, mimicry had been invoked in arguments over sexual selection (e.g., Darwin [1871] 1981; Poulton 1909) and in arguments over whether evolution proceeded by discontinuous steps or by small, gradual, continuous changes (Poulton 1909). Fisher ([1930] 1958) devoted an entire chapter to mimicry in *The Genetical Theory of Natural Selection,* and it was the only kind of interspecific interaction he considered in depth in his book. In discussing and dismissing Marshall's arguments, he wrote, in typically Fisherian fashion, that the argument "is based upon insecure reasoning, and on examination is found to be wholly imaginary" (p. 188). He countered Punnett just as bluntly by arguing that gradual and mutual change was possible given Mendelian inheritance.

Pollination and mimicry, then, were the major interactions through which post-Darwinian biologists explored the ways in which species evolve to be tightly coupled to each other. They confronted the problem of how specialization evolves, they understood that reciprocal change was at least sometimes part of the evolutionary process, and they tried to analyze how specialization and reciprocal evolutionary change shaped interactions. From the solid base established in the late decades of the nineteenth century, the study of evolving interactions had the potential to advance rapidly in the early decades of the twentieth century. Instead it almost stopped dead.

The Rift between Evolution and Ecology

The impression that one develops in reading the literature of the first hundred years of evolutionary biology is of a science with deep roots in natural history that became less ecological as it branched out and incorporated other biological subdisciplines during the early decades of the twentieth century. The appreciation of the importance of Mendel's work following 1900 made the genetics of adaptation and speciation a major focus of evolutionary biology. The opportunity for linking ecology and evolution through the study of genetics was certainly there. As early as 1905 Biffen had shown that wheat resistance to yellow rust (*Puccinia striiformis*) followed Mendel's laws, and other similar studies appeared in subsequent years through agricultural research (de Wit 1992). These results could have been used to show how the outcomes of ecological interactions between species had a genetic, and hence evolutionary, basis. But the study of interactions during the early years of the twentieth century was becoming more ecological and less evolutionary. The development of both community ecology in the early 1900s and population ecology in the decades until the 1960s was mostly nonevolutionary in approach.

Although the concept of niche appeared early in the history of ecology (Grinnell 1917; Elton 1927; Gaffney 1975), the niche of a species was static and ecologists used the concept mostly as a way of understanding how communities are organized. Kingsland (1985) wrote in her insightful history of population ecology that ecological research in the early decades of the twentieth century was devoted to two types of study: the relationships of organisms to their environment (what was sometimes called autecology) and the natural associations of species in space (biogeography) and time (succession). Some ecologists certainly viewed their work as showing the processes that composed the struggle for existence (McIntosh 1980; Kingsland 1985), which had been a part of the tradition of German physiological plant ecology during the latter decades of the nineteenth century (McIntosh 1985; Cittadino 1990). Clements's (1905) text *Research Methods in Ecology* included a long section on "experimental evolution." The primary purpose of such studies, however, was to understand the origin of new forms in nature. One reads these pages in vain for any inklings of thoughts on the ecological relationships of species in an evolutionary context. To be sure, there were exceptions, such as Brues's (1920, 1924) studies of the evolution of host specialization in phytophagous insects. Nevertheless, for the most part ecology in these decades developed outside the realm of evolutionary biology (Baker 1983; Colwell 1985).

The rift was so nearly complete that, as Kingsland (1985) notes, Elton felt compelled to begin the chapter on evolution in his 1927 book *Animal Ecology* with the statement, "It may at first sight seem out of place to devote one chapter of a book on ecology to the subject of evolution." Even the subsequent research by Elton and others at his Bureau of Animal Population from the 1930s through the 1960s was on the proximal causes of population regulation and dynamics rather than the interplay between ecology and evolution (Crowcroft 1991). Similarly, Gause's models of competition, which eventually formed part of the foundation of evolutionary ecology, assumed that the intensity and nature of competition between species did not change. Although he did consider how the interactions between predators and prey could change over evolutionary time, these ideas had little immediate effect on biological thought (Kingsland 1985, 1986). The fields of evolution and ecology were then simply too far apart for the evolutionary aspects of Gause's ideas to have much of an impact.

When Clements and Shelford (1939) wrote their text *Bio-ecology* in an attempt "to correlate the fields of plant and animal ecology," their treatment was almost completely devoid of any evolutionary perspective. They were concerned with the current structure and organization of populations, interactions, and communities. When ecologists studied interactions, they did so to understand the distribution and abundance of species or the composition of communities. When evolutionary biologists in these decades invoked interactions, they did so almost invariably only as illustrations of the kinds of selection pressures shaping adaptation in populations. The interactions themselves were generally unchanging types (parasitism, predation, competition), just as species were types before the *Origin*. Things would change after the 'evolutionary synthesis' of the 1930s and 1940s, but they would change slowly.

CHAPTER THREE

Specialization and Coevolution since the Evolutionary Synthesis

The evolutionary synthesis in the 1930s and 1940s was initially more a victory for the study of the process of evolutionary change within species than for the study of the evolving interactions among species. With the rift between ecology and evolution still wide, much of the debate over the evolution of specialization occurred among paleontologists and systematists, who were interested primarily in the evolution of morphology and patterns in the radiation of species, rather than in the evolution of interactions.

Nevertheless, in these and the subsequent decades four major developments occurred in evolutionary thought that opened the way to a more dynamic view of the evolution of specialization and the coevolutionary process. First, the view that specialization is irreversible and an evolutionary dead end was abandoned or at least saddled with so many qualifiers that it allowed for other evolutionary outcomes. Second, the group-selectionist view that interactions evolve toward reduced antagonism was replaced by an individual-selectionist view that allowed for a diversity of evolutionary outcomes under different ecological conditions. Third, ecological studies of interspecific interactions gradually adopted a more evolutionary approach, resulting in the field of evolutionary ecology. Fourth, multiple approaches to the coevolutionary process developed, and these different approaches showed that the word 'coevolution' was an umbrella for a variety of mechanisms and outcomes of reciprocal evolutionary change. This chapter considers how each of these four developments in evolutionary thought occurred, allowing for a more ecologically dynamic view of specialization and coevolution.

Prelude to Development 1: Specialization and the Pathways of Evolution

In 1896 Cope had made two linked arguments about patterns in the evolution of specialization. One was the "law of the unspecialized": new evolutionary groups arise from relatively unspecialized species, because specialized species are generally incapable of adapting to new conditions. The other was that definite progress has occurred through geologic time, with body size increasing within phyletic lineages (Cope's Rule). Taken together these two arguments would seem to suggest that body size would tend to increase over evolutionary time because larger species are less specialized. Cope, however, developed a very different argument. He suggested that as species become larger they also become more specialized, especially in morphology. The largest species within lineages are therefore the most specialized and are evolutionary blind alleys. The point of departure for subsequent progressive evolution is neither these largest and most-specialized species nor necessarily the most-primitive species. Instead, it is through intermediate species with "a combination of effective structure and plasticity"—that is, species with the "golden mean of character"—that progressive evolution occurs. According to Cope, both size and specialization increase over geologic time through the differential survival of these intermediate species. Such progressive evolution, however, is not inevitable. Citing parasites as an example, he argued that degeneration also occurs, but it too produces extreme specialization and decreased chance of future survival of a lineage.

The mechanisms that Cope invoked to explain trends in specialization, progression, and degeneration were neo-Lamarckian rather than Darwinian. When he used ecological arguments, he did so to explain how the environment and neo-Lamarckian evolution produced definite directions in evolutionary history. Cope's proposed evolutionary mechanisms had a short half-life, as a result of the rise of Mendelism only four years later. But his major theme—linking evolutionary specialization primarily to morphology and size—appeared repeatedly over the next half century as the evolutionary synthesis developed and spread among the subdisciplines of biology. Cope had also argued in his law of the unspecialized that lack of specialization in some kinds of ecological characters, such as omnivory, created a greater chance of survival than more-specialized feeding habits, but morphological specialization and the link to body size were the core of his argument. Evolutionary writers seemed to relish describing these allegedly terminal branches of morphological evolution within phyletic lineages. Osborn (1917) wrote of the "*cul-de-sac* of structure" suffered by organisms finely specialized to their physical and living environments, leading them ultimately to extinction as relatively unspecialized taxa become the origins of new adaptive radiations. Others wrote of evolutionary dead ends, dead-end streets, or blind alleys.

The evolution of specialization was routinely intertwined with ideas about evolutionary progress. From the writings of Herbert Spencer onward, some evolutionary biologists searched fossils and extant taxa for examples of progress, variously defined. Cope's linking of progress to a specific relationship between size and specialization was one way of explaining some patterns in geological history. Even after the evolutionary synthesis, the relationship between specialization and progress continued to be an important issue. Julian Huxley (1942) made it a prominent part of his summary of evolutionary theory in his book *Evolution: The Modern Synthesis*. He considered evolutionary progress to be an improvement in the efficiency of life in general, and specialization an improvement in the efficiency of adaptation for a particular mode of life. He allowed, however, that neither is inevitable, and both can lead to evolutionary dead ends. For Huxley, tapeworms are specialized, degenerative blind alleys without any potential for further advance. Jellyfish are specialized primitive types left behind long ago by more successful forms. More than anything else, however, the reason for Huxley's ideas on specialization and progress came from his overall view of what constituted progress, which was evolution toward conscious and conceptual thought. Consequently, he argued that human evolution is the only line of evolutionary progress still continuing. Evolution in other phylogenetic branches had been simply a series of blind alleys, leading often to specialization but not progress.

Schmalhausen ([1949] 1986) also made the relationship between specialization and progress a major part of his thinking about patterns in evolution, but unlike others before him, he used interactions within and between species as the basis for his arguments. He suggested that intraspecific competition favored decreased fecundity, loss of evolutionary plasticity due to intense selection, and increased specialization in feeding, which he did not define further. In his view this tendency for progressive specialization to narrower parts of the environment is particularly common in predators at the tops of food chains. The reason is that these top predators have no enemies and hence their population structure is governed mostly by intraspecific competition and selection for increased efficiency. "Specialization in nutrition" and increase in body size evolve in tandem in these species, leading eventually to evolutionary blind alleys.

As with many other evolutionists of that time, the specialization that concerned Schmalhausen was not to particular prey species but to highly specialized morphologies that allowed individuals to capture prey more efficiently. Whenever intense intraspecific competition eliminated most of the variation in morphology, species could no longer respond rapidly to natural selection for morphological change and became doomed to extinction. In Schmalhausen's view, the ultimate victory in the struggle for existence went to relatively unspecialized organisms that retained large reserves of variation in morphology, which could be molded and remolded by natural selection.

Besides progressive, adaptive specialization, Schmalhausen envisaged several other kinds of evolutionary change. Among these was adaptive but nonprogressive specialization. It is to this last form of evolution that he relegated organisms such as free-living bacteria, diatoms, and insects. Schmalhausen argued that taxa of small organisms such as these are defenseless against predators and parasites. This defenselessness has left them no prospect for progressive evolution, which depends upon freedom from enemy attack and selection driven mostly by intraspecific competition. These defenseless taxa may rapidly diverge through adaptation to use different foods, but they are incapable of fundamentally altering their organization. From Schmalhausen's developmental and morphological perspective on evolution, these taxa, with few exceptions, "have almost stopped evolving." Those that have managed to undergo some evolution beyond what he called their modest sphere of existence are those that have acquired some means of passive defense such as the thick chitinous armor of beetles or the vigorous flight abilities of flies, bees and wasps, and butterflies and moths.

Schmalhausen's overall view of evolution, then, relied upon his assumptions about the ways in which interactions within and between species, together with the evolution of defense, influenced morphological specialization. Neither Schmalhausen nor Huxley mentioned Cope directly in their major treatises on evolution, but Rensch (1959) drew explicitly on Cope's views in his major treatise on evolution above the species level. He attempted to take the patterns in the evolution of size and specialization proposed by Cope and provide a Darwinian interpretation to them rather than Cope's neo-Lamarckian view (Rensch 1959, 1980). Even with his Darwinian view, however, Rensch accepted Cope's links between specialization, morphology, and size, and the eventual outcome of extinction. He argued that escape from the blind alley of specialization was possible only occasionally, when a reversal of specialization is brought about by prolonging the juvenile phase of development. Rensch, however, made even fewer links between specialization and ecology than other writers of the evolutionary synthesis. He wrote his book without knowledge of Huxley's (1942), Mayr's (1942), or Simpson's (1944) books, which he saw only after he had corrected the proofs of his own book (Rensch 1980). He seemed unsure about how to deal with the many life histories and evolutionary outcomes that fell outside the patterns in the evolution of morphology that he expected. To Rensch, the major features of evolution were the links between increased complexity, morphological specialization, and size—and their combined effects on the history of progressive evolution. Evolution did not invariably move in these directions, but it did so commonly enough, and they produced many of the major patterns in evolutionary history.

Parasitism and other kinds of intimate association between species fell outside this view. Like Huxley, Rensch saw progress in evolution as increasing independence from and control over an organism's environment. He argued that parasites are the result of "regressive" evolution. Because mutation is random, regressive evolution could occur at any time within a phyletic lineage. He noted that parasites are common, almost embarrassingly so—accounting, he said, for 25% of German species—and they have radiated and specialized onto different hosts. Nevertheless, he devoted only several paragraphs to their evolution, leaving the evolution of specialization in these taxa undiscussed.

These views—from Cope to Huxley and Schmalhausen and Rensch—did not deal directly with the relationships between specialization and coevolution. Nevertheless, their general arguments about the evolution of specialization have had a lasting impact on our perceptions about how interactions evolve between species. Over the decades, they have become rearranged into arguments over whether highly host-specific or prey-specific species evolve from more-generalized taxa, whether highly coevolved pair-wise mutualisms evolve from more-generalized interactions among species, and whether host-specific parasites and mutualists are terminal ends on evolutionary branches.

Development 1: A Change of View on Specialization

During these middle decades of the century, some other evolutionary biologists were more uncertain about the outcomes of specialization in whatever form it took. As more biologists began to question the idea that specialization was a dead end and offered evolutionary ways out of the blind alleys, a change of view gradually took hold. The writings of Mayr and Romer in the 1940s show the trends in thinking that led eventually to a change of view on specialization. Mayr (1942, 294–95) argued that the "principle of specialization"—that every phylogenetic sequence leads from a generalized, primitive form through more-specialized forms and finally to overspecialization and extinction—was "true only to a limited degree." Most major taxa include both "primitive" forms and more-specialized forms, and specialization is not always an advantage. He asked rhetorically which was better adapted, an insect species adapted to spending its entire life cycle on one plant species or another insect species that can feed on a hundred? Increased specialization may sometimes be favored and lead to "a new 'adaptive plateau,' with unsuspected evolutionary possibilities." It may also, however, lead to extinction, because the specialists are adapted to only one specific set of environmental conditions. Consequently, "unspecialized lines often outlive their specialized offshoots or descendants."

Romer (1949), too, accepted that there was "at least a modicum of truth" to the idea that extinction is a result of specialization, preventing the escape of species from evolutionary blind alleys. But, like Mayr, he viewed any direct link between specialization and extinction as "facile verbal explanation." He argued that the fossil record appears to show that generalized forms survive better than specialized forms "because extinct lines are *ipso facto* characterized as specialized; those which survived are labeled generalized. These terms have little meaning otherwise."

Some years later, Mayr took a much stronger view against the law of the unspecialized, writing that the "generalization is certainly not supported by the known evolutionary facts" (Mayr 1959, 370). He argued further that it was a vestige of archetypal thinking in that it assumed that ancestors are "in every respect generalized and unspecialized." In fact, the more he thought about it, the sillier it all seemed. Using Osborn's phylogenetic tree of the proboscidians and their putative ancestors as an example, he insisted that the generalized ancestors often posited by these students of evolution could never have existed in nature and that this method of phylogenetic analysis made all fossils aberrant sidelines.

As Mayr and Romer questioned the law of the unspecialized by attacking the way in which species are characterized as generalized and unspecialized, Amadon (1943) and Hardy (1954) questioned the premise that specialization leads to evolutionary dead ends and offered ways out. In an article entitled "Specialization and Evolution," Amadon agreed that specialization has sometimes led to dead ends, but he maintained that it also has led to "many of the most important advances in evolution" (p. 133). He supported his stance with examples from the fossil record, but he also drew on ideas and results from the developing field of population genetics. Referring to Dobzhansky's (1941) then recent book *Genetics and the Origin of Species,* he suggested that current knowledge of genetics gave "no support to the belief that specialization decreases a species' potentialities for further change" (p. 138). Rather, in his view, the theory of population genetics supported the idea that specialization has played a major role in evolution. He began with Wright's concept of adaptive peaks and argued that specialized species often have small populations and experience genetic drift. Although these conditions could result in high probabilities of extinction, they could sometimes produce a novel genotype preadapted to a new, unoccupied adaptive peak.

Hardy (1954) used a different tack in suggesting ways out of the apparent blind alleys caused by specialization. Marshaling arguments from developmental biology, he suggested that retention of juvenile characters through processes such as neoteny provided an evolutionary pathway of escape. His article, entitled "Escape from Specialization," in a book that he coedited with Julian Huxley, was a direct rebuttal of Huxley's arguments that all evolutionary pathways except that toward humans were too specialized to allow for any significant evolutionary progress. Like a boxer shaking hands with his opponent before a fight, he began by writing, "I would indeed seem ungracious if my contribution to this volume were thought to be put forward in any spirit of attack upon Dr. Huxley's suggestion." You just knew it was going to get good after that. Drawing heavily on the arguments of his father-in-law,

Walter Garstang, he suggested that retention of juvenile characters, when less specialized than adult characters, could provide the opportunity for new radiations. This suggestion was not new, as Hardy acknowledged, but he felt that arguments such as Huxley's had ignored this interrelationship between specialization, ontogeny, and phylogeny. Some years later, Gould's (1977) masterful history of the study of the relationship between ontogeny and phylogeny showed how paedomorphosis, achieved through either neoteny or progenesis, has been commonly used throughout the nineteenth and twentieth centuries to explain escape from extinction in various taxonomic groups.

Most of the arguments about the evolution of specialization developed from work on animals, but Stebbins (1950) addressed the problem for plants and concluded that progressive specialization into dead ends was probably uncommon in plants. He worried specifically about the occurrence of both relatively generalized and highly specialized forms within a lineage. He argued that a broad range of specialization is more likely to occur within lineages in plants than in animals for three reasons. The number of different solutions to the problem of cross-pollination and seed dispersal in plants is greater than that available for reproduction in at least vertebrate animals, so there are more potential lines of adaptive radiation. Each of these lines of plant radiation, however, is relatively short, because plant structure is relatively simple compared with that of animals. Moreover, adaptive radiation in plants affects only reproductive characters, and selection for specialization in these characters will vary among habitats. Hence, plant lineages are often a combination of relatively generalized forms and short spurs of more-specialized forms. In Stebbins's view, then, the arguments developed for patterns in the evolution of specialization within lineages of animals simply do not apply to plants.

Although biologists working during the middle decades of the century all provided their own Darwinian reasons why specialization may not be an evolutionary dead end, it was George Gaylord Simpson, with his paleontological perspective, who more than anyone placed both morphological and ecological specialization in a broad Darwinian framework. The evolution of specialization became one of the central points in Simpson's (1953) paleontological views on the radiation of species. When writing his 1944 book, *Tempo and Mode in Evolution,* he was not yet ready to confront the quagmire of concepts, beliefs, and misconceptions that were continuing to bog down evolutionary thinking about specialization. Simpson's purpose in writing *Tempo and Mode* was to place paleontology within a Darwinian framework. As Gould (1980a) has written, Simpson's book lay completely outside the traditions of his profession. Rather than a book documenting patterns and laws, such as Cope's Rule, Simpson produced a treatise on Darwinism as applied to the fossil record, and he adopted a statistical approach to the study of tempo and mode in evolution.

In discussing specialization, he noted that within a phyletic lineage, it is common, although not universal, for descendent lineages to occupy progressively narrower adaptive zones. Moreover, it is also common, although again not universal, for relatively specialized taxa to become extinct before the less specialized: "the rule of the survival of the unspecialized." But he noted that specialization did not imply distance from a defined primitive condition or point of departure, nor did it imply something about the intensity of selection. Rather, it indicated "specificity of adaptation or inverse width of the zone of tolerance" (Simpson 1944, 143). He briefly considered how some conditions could favor the survival of relatively specialized forms rather than unspecialized forms, but then left it there, saying that the concept of specialization, "which is both complex and vague," was beyond the scope of the book.

Simpson returned to the problem of specialization in his 1953 book, mixing the morphological specialization that concerned other paleontologists and systematists with ecological specialization on particular resources. He developed further his concept of adaptive zones, which was based upon the idea that radiations of species occur by organisms adopting a novel way of life that allows them to use previously unexploited resources. Hence, the evolution of flying insects opened up a potentially new adaptive zone for flying predators. Radiation of flying predators occurred as populations specialized on different prey taxa or capture techniques. He argued that although an island could itself be an adaptive zone for radiation, as it has been for Galápagos finches and Hawaiian plants and animals, the more evolutionarily important adaptive zones were generally a change in the way of life rather than a change in geography. As he said, "The horses that became grazers did not have to go anywhere to do so" (Simpson 1953, 207).

Simpson viewed each of these adaptive zones as divided into subzones. The adaptive zones and subzones have discontinuities between them, which represent intermediate forms unlikely to last long in evolutionary time. The specialized forms in the new zones and subzones play a role in creating these discontinuities. He argued, for example, that the evolution of specialized grazing horses placed less-specialized grazers and facultative grazer/browsers at a selective disadvantage with respect to both the specialized horses and other taxa such as artiodactyls with whom they overlapped in their way of life.

In Simpson's view specialization might evolve in any of three ways (fig. 3.1). One possibility was sequential occupation of narrower and narrower zones (fig. 3.1A). Another possibility was speciation into more-specialized

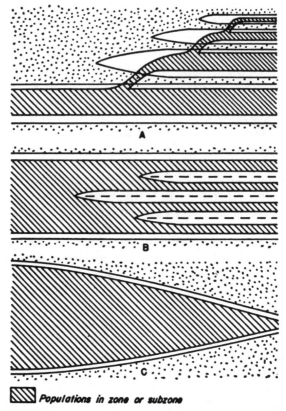

Fig. 3.1. Simpson's (1953) view of how specialization, which he defined generally as specificity of adaptation or as narrowness in width of the zone of tolerance, may evolve in any of three ways. Reprinted, by permission, from G.G. Simpson, *The Major Features of Evolution,* ©1953 by Columbia University Press.

forms within an adaptive zone (fig. 3.1*B*). This is how he viewed, for instance, radiation of the Galápagos finches. A third possibility was a narrowing of the adaptive zone itself such that the same species fills an ever narrowing zone (fig. 3.1*C*). He suggested that this third possibility may have occurred in some host-specific parasites and monophagous animals, but he thought it much rarer than specialization through speciation.

Simpson argued against the view that specialization is an evolutionary dead end:

> Extinction in which specialization (of any sort) is involved is usually said to be due to overspecialization, but this way of putting things

confuses the issue. It seems to imply that overspecialization is a definable extra degree of specialization or that it is something quite different, even moralistically so, 'bad' as opposed to 'good' specialization. Of course it is nothing of the sort. Overspecialization is just specialization that has become disadvantageous because the environment has changed. (1953, 299)

He argued further that nothing about the evolutionary outcome of specialization was assured: "[S]urvival of the less specialized . . . has been dignified by recognition as an evolutionary principle. Such a 'principle,' however, is merely a description of what sometimes happens and this should be accompanied by all possible alternatives . . . all of which also frequently occur" (Simpson 1953, 214). The evolutionary problem was not to understand which outcome was the most common but rather what conditions favored each outcome. Competition and broad changes in the availability of resources were the kinds of environmental conditions that Simpson employed in evaluating when more-specialized, rather than less-specialized, forms are more likely to survive. A less-specialized form (in what he called a lower adaptive zone) may become extinct when the zones are not different enough to prevent competition with the more-specialized form (fig. 3.2). When the problem is not competition, however, but rather a general deterioration of conditions for all populations, then the less-specialized forms may be favored (fig. 3.2). He used as an example a specialist species that feeds only on one food as compared with a less-specialized species that has four foods. If the food of the more-specialized form disappears or if all four foods decrease in availability, the more-specialized form is more likely to become extinct than the less-specialized form.

Throughout the 1940s and 1950s the arguments in paleontology and systematics mixed many different kinds of specialization. An example of the evolution of a specialized morphology would sometimes be followed by discussion of specialization to a particular habitat or specialization to another species. Similarly, arguments on speciation invoked, and sometimes mixed, various forms of specialization in discussing how it may be involved in the evolution of reproductive isolation among populations. In *Animal Species and Evolution,* Mayr (1963) grappled briefly with the problem of how different kinds of organisms specialize in different ways. Leaving morphological specialization aside, he argued that there were at least five kinds of ecological specialization: specialization for a very narrow niche, broad tolerance of individuals to a variety of ecological conditions (i.e., broad phenotypic plasticity), genetic polymorphism within populations producing individuals adapted to particular subniches, ecotypic variation among populations within geographic areas, and geographic races covering major parts of a species'

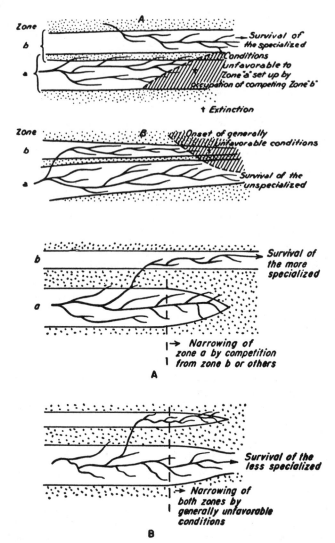

Fig. 3.2. Simpson's (*top*) 1944 and (*bottom*) 1953 views of the conditions favoring survival of (*a*) more-specialized forms and (*b*) less-specialized forms within a lineage. Reprinted, by permission, from G.G. Simpson, *Tempo and Mode in Evolution*, ©1944 by Columbia University Press, and *The Major Features of Evolution*, ©1953 by Columbia University Press.

range. Mayr's first two categories are actually the two ends of the spectrum of specialization. The other three categories are ways in which groups of individuals within a species can become specialized to exploit different habitats or resources. Mayr argued that extreme specialization was common in insects but virtually unknown in plants. By this statement, it is unclear where he would place figs, yuccas, and other plants that are adapted to pollination by a single pollinator species. Mayr's concern, however, in constructing these categories was not specifically to analyze the evolution of specialization but rather to understand how different forms of specialization might be involved in speciation. Hence, his suggestion that extreme specialization is virtually unknown in plants may reflect more a concern with specialization to particular microhabitats than with specialization in interactions, as occurs in pollination or defense against parasites and pathogens.

Development 2: Selection on Individuals and the Outcomes of Interactions

While paleontologists and systematists explored the evolutionary consequences of morphological specialization, and other evolutionary biologists investigated how different forms of specialization influence speciation, some ecologists and population biologists began to consider more closely how species coevolve. Unlike our current views on specialization and coevolution, which include evolutionary arms races and a variety of other possible outcomes, a common view during these middle decades was of an intrinsic directionality in evolution of interactions toward cooperation or, at least, reduced antagonism. Predators take the weak and old among their prey, and parasites evolve to become less virulent. That view would have to change before the study of coevolution could progress.

The Chicago "school" of ecology in the 1930s and 1940s was strongly evolutionary in its approach. Their multiauthored text *Principles of Animal Ecology* (Allee et al. 1949) used evolution to provide a theoretical underpinning for patterns in ecology. They complained that although ecology had been an outgrowth of nineteenth-century natural history, which strongly emphasized evolution, "modern ecologists have been somewhat reticent in developing evolutionary principles" (p. 598). They accepted the common view of the time that specialization "results in limitations toward further evolution" and evolutionary blind alleys. At the same time, they argued that the probability of survival of populations increases with the degree to which they become harmoniously adapted to other organisms and to the physical environment. In the view of Allee and colleagues, then, the balance of nature is a result of differential survival of harmoniously adapted species, which, through their specialized adaptations to one another and the physical environment, are limited in further evolution.

This harmonious adaptation of species, in their view, has come about because evolution has favored toleration over exploitation. They believed that species coevolved—what they called "reciprocal adjustment within an interspecies system"—toward cooperation. The arguments they used to support their view were firmly group selectionist and proceeded from their overall belief in the balance of nature (Mitman 1988). In England, Wynne-Edwards (1962) adopted a similar view in his thinking on how natural selection molded the behavior of animals and the dynamics of populations.

The Chicago school and Wynne-Edwards were not alone. Prior to the 1960s, this group-selectionist view of evolution influenced a good deal of work on the evolution of behavior within populations, the ways in which population levels are regulated, and, most importantly for the relationship between specialization and coevolution, how interactions evolve between species. For example, the major evolutionary paradigm in parasitology during these decades was a broadly group-selectionist view that expected that interactions between parasites and hosts almost invariably evolve toward commensalism over evolutionary time. A high degree of virulence in a parasite was often used as evidence that an interaction between a parasite and a host must be relatively new (Allison 1982). In contrast to this group-selectionist thinking, Haldane (1949) suggested that the evolution of interactions between parasites and hosts may favor the development of polymorphisms for defenses in host populations. This suggestion, however, had little immediate effect on the thinking of parasitologists.

The reaction to the group-selectionist view came during the 1950s and especially the 1960s (Futuyma 1986). Lack (1954), Orians (1962), Hutchinson (1965), and MacArthur (1965), among others, all emphasized the role of selection at the level of the individual in considering how evolution influenced the structure of populations and communities. Similarly, Hamilton (1964), Maynard Smith (1964), and Williams (1966) argued forcefully that the evolution of behavior occurred through natural selection among individuals rather than among groups. These views opened the way for studying how different ecological conditions favored different degrees of specialization and different outcomes of coevolution.

The infusion of ideas from population and evolutionary genetics also gradually forced some ecologists to think more precisely about the ways in which genetic variation among individuals could influence the structure of populations and interactions among species over short time periods. This change in thinking in ecology is still under way, and is far from complete, but it had its beginnings in these decades. Bradshaw's (1952) studies of ecotypic differentiation in plant populations, Kettlewell's (1955) studies of industrial melanism in moths, Birch's (1960) arguments on the role of genetics in population ecology, Ford's (1964) views of ecological genetics, and Levins's (1968) work on fitness sets all contributed, along with other studies, to an increasingly evolutionary viewpoint in ecology. Thomas Park and colleagues showed that the outcomes of competition between species were highly variable, and that the variation in outcome that they observed had a partially genetic basis. Park's work on two species of flour beetle, *Tribolium confusum* and *T. castaneum,* showed that the outcome of competition between these species depended upon initial genotypes of the beetles, temperature, humidity, and presence of a protozoan parasite (Park 1948, 1954, 1962; Park, Leslie, and Mertz 1964). Park's results were important in that they helped to show that competition coefficients depend upon both the mix of genotypes and the environment. They suggested that the outcome of competition is subject to natural selection, and that the mean outcome can evolve in different ways in different environments.

The gradual integration of studies in ecology, population genetics, and evolution slowly changed how evolutionary biologists viewed specialization and evolving interactions. Views on how predators acted for the good of prey populations by removing weak and old individuals were replaced with ideas on evolutionary arms races between species. The battle against the view that parasites automatically evolve toward commensalism was eventually replaced by an understanding that all evolutionary outcomes, from decreased to increased antagonism, are possible. The fields of parasitology and pathology remained until recent decades the last holdout of a directional view of evolving interactions, retaining longer a remnant of the "progressive" views of evolution popular earlier in the century.

Just a little over a decade ago, when Anderson and May (1982) constructed mathematical models of coevolution between parasites and hosts, they thought it so important to impress upon parasitologists that coevolution toward commensalism was not inevitable that they placed their major conclusion in italics (removed here): "In general, we conclude that the complicated interplay between virulence and transmissibility of parasites leaves room for many coevolutionary pathways to be followed, with many endpoints" (p. 411). Eight years later, they still found it necessary to make this point one of three major themes in another paper on coevolution, noting that group-selectionist views of these interactions continue to be written in textbooks in biology, medicine, and veterinary science (May and Anderson 1990). Meanwhile, Price (1977, 1980) had shown that parasites of widely divergent taxa share many common ecological and evolutionary features, including extreme specialization to particular hosts. The work of Price, Anderson, and May began to push parasitology into the realm of evolutionary ecology.

Development 3: Evolutionary Ecology and Evolving Interactions

Although parasitology was a holdout, the view of most other interactions had already changed with the development of evolutionary ecology. By viewing the evolution of interactions between species as a major problem in both evolution and ecology, the field of evolutionary ecology provided the impetus for a renewed interest in how specialization develops and how species coevolve. Although evolutionary ecology arose from the amalgamation of a variety of approaches to the study of animals and plants (Collins 1986), it was primarily the work of David Lack and G. Evelyn Hutchinson and their academic progeny and associates that fostered the rigorous study of how interactions evolve in natural populations. Orians's (1962) paper entitled "Natural Selection and Ecological Theory" heralded this new fusion of ecology and evolution, which was being spearheaded by animal ecologists. Meanwhile, plant ecology was being left behind. To redress this problem, Harper (1967) entitled his Presidential Address to the British Ecological Society "A Darwinian Approach to Plant Ecology" and ended it by saying that plant ecologists could find enough interesting questions in the *Origin of Species* to keep them busy for the next hundred years.

With the rise of evolutionary ecology, the evolution of adaptation within species became intertwined with the evolution of interactions. Part of the reason was that an understanding of species *and* their interactions was needed to develop a theoretical basis for community structure and organization. From his studies of Galápagos finches, David Lack ([1947] 1983) suggested that adaptive differences in populations may arise while they are isolated from one another. These differences would then increase through competition if the populations come back into contact, thereby permitting coexistence. This is the view that Hutchinson (1959) used later in developing the arguments in his paper "Homage to Santa Rosalia *or* Why Are There So Many Kinds of Animals?" He wrote, "Before developing my ideas I should like to say that I subscribe to the view that the process of natural selection, coupled with isolation and later mutual invasion of ranges leads to the evolution of sympatric species, which at equilibrium occupy distinct niches, according to the Volterra-Gause principle" (p. 146). Brown and Wilson (1956) referred to this process of divergence in competing species as character displacement. It suggested one way in which interactions between two species could result in increased specialization in use of resources and, in the process, produce reciprocal evolutionary changes between species.

Just how similar species could be and yet coexist became the subject of intensive research, beginning with Hutchinson's (1959) observation that coexisting species often differ in morphology by a particular ratio. Hutchinson's work was soon followed with mathematical models on the limits to similarity between competitors (MacArthur and Levins 1967) and with field studies documenting the degree of similarity among coexisting species. The primary concern in many of these studies was not directly with the evolution of specialization or the process of coevolution in interactions but rather with how species are packed into communities and how similar they could be in diet and life history yet invade a community and coexist. Empirical studies recording differences among species in sympatry became so routine that, in frustration, Schoener complained that "some investigators are still content mainly to document differences between species, a procedure of only limited interest" (1974, 37), then went on to list a series of more-profitable questions that researchers might ask about the actual process of resource partitioning. Nevertheless, these studies helped to set in motion a kind of thinking that asked how specialization evolves in natural communities and how reciprocal evolutionary change can affect the organization of biological communities. Concern over the role of competition in structuring communities made the evolution of niche width one of the major topics that spurred the development of evolutionary ecology (MacArthur and Levins 1967; Roughgarden 1972).

Ecological studies of interactions between trophic levels were also becoming more evolutionary, using a variety of approaches. Optimal foraging theory developed from inquiries into animal behavior. It used the currencies of time and energy (and occasionally nutrient levels) to predict how animals should search for food, what they should search for, and how many different kinds of food items they should include in their diets (Emlen 1966; MacArthur and Pianka 1966; Schoener 1971). The assumption in the various versions of optimal foraging theory is that natural selection favors those individuals that are the most efficient in their foraging behaviors, with efficiency being defined variously as a combination of maximizing the mean or minimizing the variance in energy intake during the time spent foraging, minimizing the total time spent foraging, and minimizing the risk of predation or other antagonistic interactions while foraging (Real and Caraco 1986; Stephens and Krebs 1986; Schoener 1987; Mangel and Clark 1988). The theory has been very successful at predicting foraging patterns for behaviorally flexible animals that choose among nutritively similar foods, such as in bees taking nectar from an array of plant species or oystercatchers choosing among mollusc species of different sizes. Nevertheless, as Futuyma and Moreno (1988) pointed out, optimal foraging theory has been used more as an ecological theory of how behaviorally flexible animals should choose among foods given current encounter rates, energy rewards, and handling times for each food item than as a theory of the evolution of specialization. It is not a theory of how behavioral flexibility, along with morphology and physiology, might evolve to favor individuals that are genetically narrower or broader in the spectrum of host or prey species they choose among. Rather, as Crawley and Krebs (1992) note, it is a theory for exploring "the products of selection, the design features of animals."

Yet other approaches to specialization and coevolution developed from ecological, physiological, and biochemical studies of the evolution of arms races between animals and plants. The presence of more than 10,000 different kinds of secondary metabolites in plants, the nutritional inadequacy of diets of single plant species for many herbivores, the mixing of plant species in the diets of some insect and vertebrate herbivores, yet the extreme specialization of others to one or a few plant species, all demanded approaches to the evolution of specialization that involved more than time and energy as the currencies of natural selection. In addition, Ehrlich and Raven's (1964) paper on coevolution of butterflies and plants showed strong phylogenetic constraints on which plant species were used as larval hosts and how those constraints could shape the coevolutionary process. Their paper was part of a tradition of studying how the phylogeny of parasites tracks the phylogeny of hosts. Their analysis, however, differed fundamentally from these past studies in suggesting that patterns of diversity in the plants and the butterflies, and the patterns of specialization in the butterflies were a result of a particular form of coevolution. Together these studies of interactions between animals and plants focused a great deal of research on the problem of how the evolution of defense and counterdefense produces patterns in the evolution of specialization. Smith's (1970) study of coevolution between conifers and red squirrels was one of the first of this new generation of ecological studies to try to analyze how antagonistic interactions between herbivores and plants actually led to coevolution in natural populations, and Gilbert and Raven's (1975) edited volume on *Coevolution of Animals and Plants* highlighted many of the evolutionary problems for both animals and plants in mutualistic, as well as antagonistic, relationships.

Development 4: Expanding Approaches to the Study of Coevolution

Ehrlich and Raven's (1964) paper on patterns of specialization in butterflies and coevolution with host plants was one of five approaches to the study of coevolution that developed during the late 1940s through 1960s. Two of the other approaches were based upon the genetics of resistance to parasites and pathogens. Flor (1942, 1955) had shown that the variation in outcome of interactions between flax and flax rust could be interpreted as a gene-for-gene relationship, with each gene for resistance to rust in flax matched by the evolution of a specific gene for virulence in flax rust. Flor's analyses eventually transformed the field of plant pathology, making gene-for-gene relationships the paradigm for work in phytopathology. The first mathematical model of coevolution, in Mode's 1958 paper in *Evolution,* was explicitly a model of gene-for-gene interactions.

The theory of gene-for-gene interactions was based initially on frequencies of alleles, with little attention to the population ecology of species. In contrast, Pimentel (1961, 1988; Pimentel et al. 1965; Pimentel, Levin, and Olson 1978; Pimentel and Bellotti 1976) took a different approach to coevolution by linking the genetics of interactions with population dynamics. Pimentel's hypothesis of genetic feedback involves a multistep process. Populations of hosts and parasites initially fluctuate wildly as the susceptible host population evolves resistance to the parasites. Evolution of new defenses in the host, in turn, produces new mutant parasites that are able to overcome the resistance. According to Pimentel's hypothesis, as he continued to develop it over subsequent decades, the host populations will accumulate defenses over many generations, which will eventually lead to a damping in the amplitude of the oscillations in the populations. Pimentel's hypothesis has received a series of partial tests in which population fluctuations have been monitored in cultures of parasitoids and either susceptible or resistant host flies. No experiment has been run long enough, however, to actually see the evolution of damped oscillations in both hosts and parasites.

The fourth approach to the study of coevolution developed during these decades came from theoretical studies designed to analyze the dynamics of change under different ecological conditions in antagonistic interactions. The theory of coevolution of competitors made its appearance from an amalgam of approaches, including Brown and Wilson's (1956) study of character displacement, papers on limiting similarity in coexisting species and on patterns in niche shifts (Hutchinson 1959; MacArthur and Levins 1967; Roughgarden 1972, 1974, 1976; Schoener 1974), evaluation of life histories from the perspectives of r- and K-selection (MacArthur and Wilson 1967), and studies of the evolution of competitive ability (Park 1962; Futuyma 1970; Gill 1972, 1974). Together these studies suggested that associations between competitors could result in rapid reciprocal evolutionary change, at least under some conditions. Meanwhile, models of coevolution between predators and prey were developed, based mostly upon the idea of an evolutionary arms race in which one species is evolutionarily pursuing another in a race of defenses and counterdefenses (Van Valen 1973). These early models (e.g., Schaffer and Rosenzweig 1978; Hofbauer, Schuster, and Sigmund 1979), along with later approaches (e.g., Abrams 1986; Rosenzweig, Brown, and Vincent 1987; Brown and Vincent 1992) suggested that small changes in the initial conditions of an interaction could lead to very different evolutionary outcomes and that, under some conditions, the strategies of predators and prey may cycle over evolutionary time. Together, most of the evolutionary models of predators and prey, and those between competitors, suggested a highly dynamic view of the coevolutionary process in which many outcomes were possible depending upon the initial conditions.

These first four approaches to coevolution were for antagonistic interactions either within or between trophic levels. Janzen's (1966, 1967a,b) studies on the ecology and evolution of interactions between ants and acacias provided a fifth approach, but for mutualistic interactions. His seminal papers on mutualism between ants and acacias provided the same kind of impetus for studies of mutualistic interactions between animals and plants that Ehrlich and Raven's paper had done for antagonistic interactions, but with an added twist. Ehrlich and Raven's paper suggested how phylogenetic studies might be used to search for patterns in the evolution of interactions between phytophagous insects and plants, whereas Janzen's observations and experiments showed how evolutionary ecologists could study directly some aspects of the process of coevolution in natural populations by linking ecological studies and systematics. At about the same time, the study of interactions between pollinators and plants was being pulled from decades of description by botanists into careful analyses of the processes shaping these interactions (review in Baker 1983). Feinsinger (1987) attributes this change in emphasis to three landmark reviews: Grant and Grant's (1965) analysis of pollination in the family Polemoniaceae, which considered the evolutionary viewpoints of both the plants and the pollinators; Baker and Hurd's (1968) review of the evolution of traits important to the interaction in both plants and pollinators; and Heinrich and Raven's (1972) emphasis on the role of energetics in shaping these interactions.

This diversity of approaches to coevolution led to an enthusiastic search for reciprocal evolution as well as to some confusion as to just what constituted coevolutionary change. Coevolution can involve the evolution of polymorphisms as well as directional selection, speciation as well as adaptation, and patterns in phylogeny as well as patterns in the evolution of host specificity or specialization on prey. During the 1960s and 1970s, coevolutionary studies sometimes concentrated on one of these aspects of reciprocal evolutionary change and made pronouncements on the whole concept of coevolution based upon those results. Even worse, as more biologists began to study coevolution the word itself slowly started to become synonymous with anything having to do with interactions between species, whether it involved reciprocal change or not. In sheer desperation, Janzen (1980) wrote a short paper entitled "When Is It Coevolution?" which pleaded for a more careful use of this important concept. Janzen's paper and the arguments in several books in the early 1980s (Thompson 1982; Futuyma and Slatkin 1983; Nitecki 1983) attempted to direct the focus on coevolution back onto patterns in reciprocal evolutionary change. The result has been studies much more clearly directed to understanding the processes by which reciprocal change occurs. In a questionnaire filled out several years ago by members of the British Ecological Society (Cherrett 1989), ecologists ranked coevolution and animal-plant coevolution among the fifty most important concepts in ecology (ranked 24 and 34 respectively). It is likely to become more important as our ideas on the processes and outcomes of coevolution become more refined.

Although studies on the evolution of interactions, specialization, and coevolution increased quickly from the 1960s onward, it took until the 1970s for textbooks in ecology and evolution to incorporate discussions on the evolution of interactions based upon natural selection acting on individuals. Ecology texts began to mention these issues in the early 1970s, and Roughgarden's (1979) text on evolutionary ecology provided a full treatment of the range of theory on evolving antagonistic and mutualistic interactions. At the same time, Futuyma's (1979) *Evolutionary Biology* became the first textbook on evolution to devote a chapter specifically to conceptual issues on the evolution of interactions. Until then generations of students in courses in evolutionary biology were trained without realizing that they were studying only half of the problem of the evolution of adaptation and diversity. Or they studied evolving interactions as parts of other courses, giving the impression that the evolution of interactions was a specialized topic outside the scope of the fundamental questions in evolution. Only now are texts beginning to reflect the mix of ecology, genetics, systematics, physiology, molecular biology, developmental biology, and paleobiology composing the study of evolving interactions in general, and specialization and coevolution in particular. The study of evolving interactions demands the mixing of so many subdisciplines of evolutionary biology that it is not really surprising it has taken so long.

Even now the major questions on evolving interactions and coevolution are still being refined, and new ones are being posed. Almost all the major questions, however, stem in one way or another from the general observation that species differ in their degree of specialization to other species. It is to the causes of the range in specialization found among species that the following chapters now turn.

PART II
THE EVOLUTION OF
SPECIALIZATION

CHAPTER FOUR

Phylogeny of Specialization

Interactions between species are shaped in part by the phylogenetic baggage of structure, physiology, and behavior that organisms inherit from their ancestors. The evolution of specialization proceeds by testing the boundaries of past history as new mutations, genetic combinations, and ecological opportunities present themselves. It is distressingly easy to observe an association between a pair of species and concoct adaptive stories—emphasizing current natural selection and ignoring phylogeny—about how specialization has evolved and how coevolution may have shaped the interaction. The problem, as Janzen (1980) said, is that a species invading a new community will fit in where it can. A parasite will attack those species whose defenses it can circumvent, a predator will kill the prey species it can catch, and a competitor will fit wherever it can avoid being outcompeted. Skutch (1980) has written that the bananas on his farm in Costa Rica seem perfectly adapted for pollination by the hummingbirds that visit them, yet these are new, fortuitous interactions, formed only since human introduction of bananas into Central and South America. Similarly, merlins (*Falco columbarius*) in Saskatoon, Saskatchewan, give the impression of being specialists on introduced house sparrows and of being highly adapted for catching these abundant birds, which now make up to 70% numerically of this falcon's diet (Oliphant and McTaggart 1977; Sodhi 1992). Yet this, too, is a new interaction for this population of merlins.

These examples illustrate how difficult, and in some cases impossible, it is to take a snapshot of a community at one moment in evolutionary time and distinguish new interactions from those that have been fine-tuned through specialization and coevolution over long periods of time. The evolutionary origins of specialization, and the process of coevolution, can be unraveled only through a combined understanding of the ecology, genetics, and phylogeny of interactions.

Darwin, the Müller brothers, and other early evolutionary biologists understood fully the need for an intimate knowledge of the natural history of interactions, selection pressures, and a phylogenetic perspective in studying the evolution of specialization. Darwin's orchid studies, Hermann Müller's work

on specialization to different pollinators, and Fritz Müller's ideas on the evolution of mimicry all relied upon a combined understanding of natural history and a phylogenetic perspective to provide the context for the action of
natural selection. The links between ecology and phylogeny, however, became weaker during the first two-thirds of the twentieth century as different
traditions developed in ecology and systematics.

Those links have been reforged in the last quarter of this century through
a complex mingling of influences. Ehrlich and Raven's (1964) paper on coevolution of butterflies and plants was a mix of phylogenetic and ecological
perspectives on, among other things, patterns in the evolution of specialization, as were Janzen's (1967b, 1974b) papers on the ecology and phylogeny
of associations between ants and acacias. Gould and Lewontin's (1979) critique of the "adaptationist program" argued convincingly for the need to separate historical and structural constraints from adaptations in understanding
how organisms evolve. In a different way, the critiques of competition theory
by Simberloff, Strong, and colleagues (Strong, Szyska, and Simberloff 1979;
Simberloff 1983) influenced how ecologists subsequently used comparisons
among species to study evolving interactions. Price's (1977, 1980) studies on
patterns in adaptive radiation and specialization in parasites were built upon
comparisons among related taxa. Finally, the development of cladistic and
molecular methods in systematics, with their potential for evaluating not just
the relatedness of species but also the phylogenetic sequence from ancestral
to descendent taxa, has allowed for a more powerful alliance between systematics and ecology (e.g., Brooks and McLennan 1991, Armbruster 1991).
The result of this mingling of approaches and critiques has been an increasing number of studies designed to analyze the phylogeny of particular interactions and address major hypotheses on the phylogeny of specialization.

In this chapter I examine patterns in the phylogeny of specialization between interacting taxa. I am particularly interested in evaluating two long-
standing ideas on the phylogeny of species that influence how we think about
the evolutionary processes driving specialization and their implications for
coevolution: Is specialization generally a derived condition and an evolutionary dead end within lineages? Does specialization in parasites lead to phylogenetic tracking of speciation in their hosts?

Natural Selection in a Phylogenetic Context

Closely related species share many life history, behavioral, morphological,
and physiological traits and constraints as a result of their common ancestry.
A certain degree of 'phylogenetic inertia' can maintain this similarity over
long periods of evolutionary time, and as a consequence, much of evolution
occurs as minor changes within the overall structural body plan (the

Bauplan) of a lineage. Evolution proceeds, as Jacob (1977) wrote, by tinkering with current structures rather than by engineering new ones from scratch. Species diverge within the architectural and historical constraints of their body plans (Gould and Lewontin 1979).

Two consequences of phylogenetic history are important for our understanding of the evolution of specialization and its relationship to coevolution. First, the adaptations carried forward from ancestors influence how new interactions are formed. A species colonizing a new community, such as a parasite from Europe introduced into North America, is likely to interact with species in the new community in ways that are similar to its interactions with species with which it has evolved. If the colonizing species can fit in, survive, and reproduce successfully, it becomes part of its adopted assemblage of species. Fitting in for a monogenean parasite, phytophagous insect, or rust fungus often, although not always, means feeding on a host species that is phylogenetically closely related to the host in its parent population. This phylogenetic bias in how a species enters into a new community can convey the false impression, to an observer arriving after the colonization event, that it is highly adapted and specialized to the species with which it now interacts (Janzen 1980). The large number of host shifts by insects over the past hundred years onto introduced crops, ornamental plants, and weeds (e.g., Strong 1974, 1979; Tabashnik 1983; Thomas et al. 1987; Feder, Chilcote, and Bush 1988) cautions against adaptive interpretations of the evolution of specialization in the absence of both a phylogenetic and a geographic perspective.

Second, because closely related species are often simply minor variations on an evolutionary theme, adaptation to one species may preadapt a population to interactions with still other, related species within the same lineage. That is, specialization can fortuitously open the door to limited despecialization. If a parasite or predator population evolves the ability to attack one host or prey species successfully, those same adaptations may also allow it to attack, to some degree, one or more other, closely related species that share similar life histories and defenses. Similarly, if a plant population becomes adapted for pollination by one bee species, it may become fortuitously available for visitation and pollination by other, similar bee species. Expansion to the use of multiple closely related species may itself then trigger a new round of selection for specialization as populations become adapted to one or another of these closely related species in different parts of the geographic range. The result can be a pattern of specialization that varies geographically as populations evolve and change in their geographic distributions.

Consequently, even when it is evident that an entire lineage interacts with only one other lineage, it is often not possible to know with certainty that the associations between particular pairs of species have been intact over long periods of evolutionary time. For example, most moths in the oecophorid genus *Depressaria* feed on plants in the Umbelliferae, and eleven of the twelve species in the *D. douglasella* group feed upon umbellifers in the genera *Lomatium* and *Cymopterus*. Moreover, most of these species are specific to one *Lomatium* or *Cymopterus* species (Clarke 1952; Hodges 1974; Thompson 1983a,b,c). These facts suggest that the interaction between *Depressaria* and these two plant genera is old enough to have produced a radiation of sibling *Depressaria* species. These observations alone, however, are insufficient for determining if associations between pairs of *Depressaria* and *Lomatium* species have been stable over a long period of evolutionary time. Such a conclusion requires a phylogenetic analysis of the moths, coupled with concomitant ecological studies on geographic variation in host use and oviposition preference, to determine just how strong the genetic preference is for one plant species.

The potential for interacting with several closely related species can create different degrees of specialization among populations and species. For instance, there are more than a score of willow-feeding chrysomelid beetles in Europe. Among these species, *Phratora vitellinae* is commonly restricted within local populations to one host. Although the host species used by these beetles differ geographically, the attacked willows always contain high levels of salicin and related salicylates, which this beetle uses to manufacture its own chemical defense against predators (Rowell-Rahier 1984a,b; Tahvanainen, Julkunen-Tiitto, and Kettunen 1985; Pasteels, Rowell-Rahier, and Raupp 1988; Denno, Larsson, and Olmstead 1990). Sympatric with *P. vitellinae* in some communities is another species, *Galerucella lineola,* which in some populations uses two or three willow species, although it may show a preference for one of the willows (Tahvanainen, Julkunen-Tiitto, and Kettunen 1985; Denno, Larsson, and Olmstead 1990). *Galerucella* does not sequester salicin and, in fact, appears to avoid plants rich in this and related compounds. The impression left by the feeding patterns of these two species is that *P. vitellinae* is an extreme specialist, whereas *G. lineola* is more of a willow generalist.

Nevertheless, the differences in number of local hosts used by these beetles may say more about the number of coexisting salicin-poor willows as compared with salicin-rich willows than it says about differences in specialization in these beetles. Both beetle species show a hierarchy of preference among willow species in oviposition preference and feeding trials (Rowell-Rahier 1984b; Tahvanainen, Julkunen-Tiitto, and Ketunen 1985; Denno, Larsson, and Olmstead 1990), and the degree of preference corresponds to the concentrations and composition of salicin and related compounds in the plants. It is therefore possible that if there were more salicin-rich than salicin-poor willow species within these communities, then *P. vitellinae* would appear to be the greater generalist. All these beetles feed on a small range of related willows in a genus renowned for hybridization and introgression among its recognized species. Our view of the relative degrees of specialization in these beetles may reflect the differing past histories of hybridization and introgression among salicin-rich, compared with salicin-poor, willow species and the similarity of some of these species as hosts for the beetles.

These influences of phylogenetic history on evolving interactions almost guarantee that a range in specialization will be found within any lineage. Moreover, they would seem to preempt any possible long-term direction in the phylogeny of specialization: as populations shift geographically over evolutionary time, they encounter new species, some of which are related, phylogenetically or ecologically, to species with which they already interact. The specializations evolved in one environment become opportunities for new interactions in other environments. Nevertheless, two major ideas about the phylogeny of specialization have been put forward repeatedly in different forms since the late 1800s that incorporate the limitations imposed by specialization while ignoring opportunity. The first is that specialists evolve from generalists and that specialization of any sort is an evolutionary dead end, a slippery slope that cannot be rescaled and instead careers a species down into the pit of extinction. The second is that parasites (actually, parasites, intimate commensals, and mutualistic symbionts) speciate along with their hosts, such that the branching patterns in the phylogenetic trees of hosts and parasites look essentially the same. The following sections take up these hypotheses in turn.

Is Specialization Derived?

The available studies suggest that all possible phylogenetic sequences involving specialists and generalists occur—generalist to generalist, generalist to specialist, specialist to specialist, and specialist to generalist—and there is no overall direction to specialization within lineages. The hypothesis that extreme specialists evolve from generalists, however, requires a more unambiguous phylogeny of species than is currently available for most taxa. That is not a very encouraging way to begin an analysis of the phylogeny of specialization, but it is the current reality of the available data. Most phylogenies that we have now are still based upon morphological and ecological data that are important in interactions. If one uses, for example, host associations in parasites as characters in a phylogeny (e.g., specialization on primates as compared with specialization on rodents), then, depending upon the assumptions built into construction of the phylogeny, the resultant phylogenetic tree may simply reflect the assumption that parasites attacking primates and parasites attacking rodents are each a monophyletic group. Although there are a number of ways to control partially for this potential circularity (e.g., Armbruster 1992), the best is to use characters that are independent of the phenotypic traits important in interactions. This is one of the reasons molecular phylogenies offer such important promise for the study of evolving interactions.

The ideal analysis for understanding whether extreme specialization is generally a phylogenetically derived condition would be to take a group of fairly large monophyletic lineages and determine the proportion of times that specialization is the evolutionarily derived condition within each lineage. The available molecular phylogenies, however, are still too few. Moreover, many available phylogenies have one or more missing species that, for various reasons, were not included in the final analysis. These missing species affect the interpretation of the probability of shifts toward and from specialization.

With these cautions in mind about current phylogenies, some studies indicate that specialists are not necessarily terminal branches within lineages. The moth family Prodoxidae (yucca moths and related genera) shows most of the possible phylogenetic sequences between specialization and generalization and has been the focus of combined phylogenetic and ecological studies designed specifically to explore the evolution of specialization, the phylogeny of mutualism, and the process of coevolution. Intensive fieldwork on this family in recent years has allowed construction of morphology-based phylogenies of the prodoxid genera (Nielsen and Davis 1985; Wagner and Powell 1988; Davis, Pellmyr, and Thompson 1992), and molecular phylogenies are nearing completion (Brown, Harrison, Pellmyr, and Thompson, in preparation). Until recently only the hosts of the yucca moths were known, but work over the past decade has now revealed the host associations and basic biology of most of the species in the other prodoxid genera (e.g., Thompson 1987c; Davis, Pellmyr, and Thompson 1992; Pellmyr and Thompson 1992; Thompson and Pellmyr 1992). Since the interactions between yuccas and yucca moths are routinely cited as one of the classic cases of specialization and coevolution, the studies of these related genera are needed to provide the broader phylogenetic context in which to interpret specialization, mutualism, and coevolution in these associations.

Each of the monophyletic groups with two or more species within the subfamily Prodoxinae includes species known from only one or two hosts and other species known from at least several hosts (table 4.1). Hence, at least at this level there is no indication that primitive clades have more-generalized species and derived clades more-specialized species. The more-important questions, however, are whether species of extreme specialists can give rise to new clades (Futuyma and Moreno 1988) and whether specialists are able

Table 4.1. The range in the number of plant species used as hosts within each monophyletic group with two or more species within the moth subfamily Prodoxinae (yucca moths and allies)

Monophyletic group	No. of species	No. of species used per moth species
Greya politella group	2	2–7
Greya punctiferella group	4	1–4
Greya solenobiella group	5	1–5
Tegeticula spp.	3+	1–19[a]
Prodoxus spp.	10	1–11
Agavenema spp.	2	1

Sources: Data from Davis 1967; Thompson 1987c; Davis, Pellmyr, and Thompson 1992; and Pellmyr and Thompson, unpublished.

[a]*Tegeticula yuccasella,* which has been reported from at least nineteen *Yucca* species, is now considered to be a complex of sibling species, each with a much narrower range of hosts (Miles 1983; Addicott et al. 1990).

to survive even if the species they specialize on becomes extinct. The species-level phylogenies within most clades are not yet sufficiently resolved to answer these questions convincingly, but recent work on the phylogeny of other taxa, and on the evolutionary genetics of specialization (discussed in the next chapter), suggests that specialists are not evolutionary dead ends.

A finer-scale analysis of the phylogeny of specialization is possible using the chrysomelid beetle genus *Ophraella,* which is a group of leaf-feeders found in North America and Mexico (fig. 4.1). All thirteen beetle species feed on plants within the Asteraceae. Futuyma and McCafferty's (1990) phylogeny of the beetles, based upon morphological characters and allozymes, does not support the idea that evolution within a lineage is toward greater host specialization. Most species have only one known host plant, and those that feed upon more than one host include both primitive and derived species (fig. 4.2).

A closer look at the distribution of host specialization in *Ophraella* and *Greya* (one of the largest prodoxid moth genera), however, does show several common patterns in the distribution of the number of hosts used by insects and other taxa that feed parasitically. Most species are recorded from only one or a few hosts, but a few species stand out as much less host specific (fig. 4.3). That is, the distributions are highly skewed. The skewing, however, has a pattern: the species recorded from the most hosts (*Greya politella; Ophraella communa*) are those with the broadest geographic ranges within these genera, and populations of these species tend to use different hosts in different parts of their ranges. Hence, the list of recorded hosts gives an inaccurate view of the degree of specialization found within populations of

Fig. 4.1. *Ophraella sexvittata,* one of thirteen leaf-feeding beetles that compose the genus *Ophraella.* Reprinted, by permission, from D.J. Futuyma, in P.W. Price et al., eds., *Plant-Animal Interactions,* pp. 431–54, copyright ©1991 by John Wiley & Sons and D.J. Futuyma.

these widely distributed species. Moreover, most of the species with relatively long host lists use exclusively, or mostly, a group of related hosts in a single host genus. *Greya subalba* (five hosts) feeds on five *Lomatium* species, and *O. cibrata* (five hosts) feeds upon five *Solidago* species.

This interplay of different geographic distributions among the insect species, different geographic distributions in available hosts, and geographic variation in host use points to potential pitfalls common to all analyses of the phylogeny of specialization. Both *Greya* and *Ophraella* include endemic species and widespread species, and both groups include species feeding on species-rich plant genera and species-poor genera. For example, the genus *Lomatium,* which is the host for *Greya subalba* (five hosts), contains about eighty species distributed throughout western North America (Mathias 1965), whereas the genus *Bowlesia,* which includes the sole host for *G. powelli,* contains only two species in North America, the remaining twelve occurring in South America (Mathias and Constance 1965). Hence, the opportunities for colonization of closely related plant species are greater for the *Lomatium*-feeder than for the *Bowlesia*-feeder. Add to this the incomplete

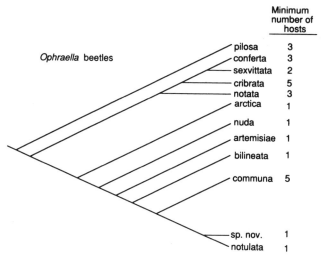

Fig. 4.2. Phylogeny of specialization in the beetle genus *Ophraella*. Data from Futuyma 1991 and Futuyma and McCafferty 1990.

reproductive isolation of some host species and the greater commonness of hybridization in some plant genera than in others, and the differences between one, three, or five hosts blur even more. Several studies have shown that some parasites exhibit higher rates of parasitism within hybrid zones of their plant or animal hosts (e.g., Sage et al. 1986; Whitham 1989; Le Brun et al. 1992), although hybridization itself may not always be the direct cause of higher parasitism rates in such hybrid zones (Le Brun et al. 1992; Paige and Capman 1993).

This mix, then, of different geographic ranges in insects and plants, different numbers of species in the host genera colonized by insects, population differentiation in the insects and plants, and different degrees of hybridization among hosts almost guarantees a distribution from one upward in the number of hosts used by any reasonably large genus of insects—or by any parasitic group. A new endemic species attacking a single, geographically restricted host in a small genus may bud off from a widely distributed species with a long list of known hosts. Elsewhere, a new widespread species may arise from an endemic species as one of its populations happens to shift to a large genus of chemically similar hosts with indistinct species limits. In between these two extremes are all other possible combinations and outcomes. Each evolutionary shift onto a new host comes with its own problems and its own opportunities for specialization and subsequent radiation among closely related hosts or prey.

Recent studies of other taxa also indicate no tendency toward increased

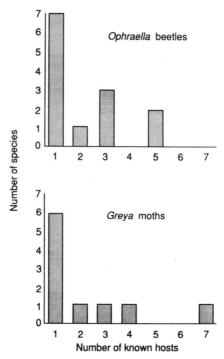

Fig. 4.3. Distribution of the number of host species used by the beetle genus *Ophraella* and the moth genus *Greya*. Data from Davis, Pellmyr, and Thompson 1992; Futuyma 1991; and Thompson and Pellmyr, unpublished.

specialization during phylogeny. Parasitic cowbirds, which lay their eggs in the nests of other avian species, provide no evidence that specialized species derive from more-generalized species. The six cowbird species include one nonparasitic and five parasitic species, which range in specialization from a single host to the use of more than 200 hosts. Lanyon (1992) constructed a phylogeny of the parasitic cowbird species based upon 852 base pairs of the mitochondrial cytochrome-b gene. He compared these sequences with twenty additional blackbird species and two other species from related passerine families. The analysis produced a single most-parsimonious tree that suggested two conclusions (fig. 4.4). First, the current designation of genera in parasitic cowbirds does not reflect monophyletic groupings: *Scaphidura* falls within *Molothrus* in the molecular phylogeny. Second, primitive parasitic cowbirds are much more host specific than the more-derived species. Based upon this cytochrome-b phylogeny, there is a clear trend toward use of more hosts by derived species. Even if additional analyses show that one or

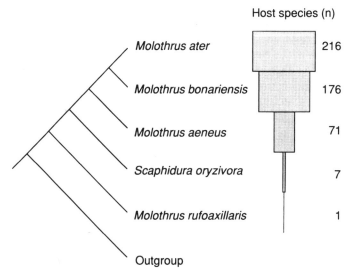

Fig. 4.4. Phylogeny of specialization in parasitic cowbirds, indicating evolution of host generalists from host specialists. Redrawn from S.M. Lanyon, "Interspecific Brood Parasitism in Blackbirds (Icterinae)," *Science* 255:77–79, copyright 1992 by the AAAS.

two of these cowbirds are misplaced in this phylogeny, it seems unlikely that a revised phylogeny would completely turn this pattern on its head and indicate a trend toward increasing specialization. At most, any revision of this phylogeny would likely create a view of no general pattern in the evolution of specialization.

The mix of generalists and specialists within phylogenetic lineages suggests that there is little reason to regard specialization in interactions as an evolutionary dead end. Mutation, natural selection, and drift in an extreme specialist would seem just as likely to give a new radiation of descendent species as the same evolutionary processes acting on a generalist. An insect fully dependent upon one host plant throughout its geographic range, or a plant fully dependent wherever it grows on one pollinator species, may become extinct if the other species became extinct. But it seems just as likely that, within a specialist species, one or more populations may give rise to individuals less fully dependent on that one other species.

We can address the question of what happens to a specialist in the absence of the other species by asking how species with different degrees of specialization respond either when placed into a new environment or when new species are added into their environment. There are thousands of examples of these two kinds of 'natural experiment', which have occurred as species have been moved around the world either purposely or accidentally. All possible results seem to occur, showing that specialists often have some flexibility in their reliance on particular species. The entire field of biological control is based upon the assumption and hope that species showing extreme specialization in their interactions in their native habitats will show the same degree of specialization when introduced into new habitats. Fortunately, that has often turned out to be so, but each failure of a specialist biological control agent to remain on its intended host is an example of the evolutionary flexibility found in species that are normally extreme specialists.

Does Specialization Lead to Phylogenetic Tracking?

The second major question on the phylogeny of specialization is whether interacting taxa speciate in tandem. Fahrenholz's Rule is one of the oldest ideas in evolutionary parasitology. As formulated by Eichler in the 1940s, it states that "in the case of permanent parasites . . . the relationship of the host can usually be inferred directly from the systematics of the parasite" (translation from Hennig 1966). This hypothesis has sometimes mistakenly been spliced to Ehrlich and Raven's (1964) view of coevolution, whose hypothesis, in fact, generates different expectations (discussed in chapter 16). According to the rule, primitive hosts within a taxon have primitive parasites, and the more-derived host species harbor more-derived parasites. The result is variously called phylogenetic tracking, parallel cladogenesis, or Fahrenholz's Rule (or sometimes coevolution by biologists who disregard the central point that coevolution involves reciprocal evolutionary change).

Like most biological rules of this sort, the degree to which the rule holds true varies with the taxonomic group being studied and the taxonomic scale at which the study occurs. Fahrenholz's Rule is not so much a rule as it is an amalgam of three components of the evolution of interactions: the degree to which parasites (and other symbionts) are specific to a single host, the extent to which host use in parasites is governed by phylogenetic inertia, which prevents shifts onto unrelated hosts, and the extent to which speciation in hosts leads directly to speciation in parasites. The best use of the rule is as a null hypothesis against which to evaluate the combined influences of these three factors on evolving interactions.

Almost any reasonably sized genus of parasites or mutualistic symbionts will have a few species that live on a group of closely related hosts and one or more species that feed on aberrant—or, at least, not so closely related—hosts. Geographic subdivision of host populations, followed by speciation in either the host, the parasite, or both, causes some of the cases of phylogenetic tracking. As a counterbalance, however, some parasite populations will have colonized unrelated hosts. Hence, the moth genus *Depressaria* feeds mostly upon plants in the Umbelliferae, but *D. artemisiae* has colonized the phylogenetically unrelated, but chemically related, *Artemisia dracunculus* in the Asteraceae. The same shift has occurred in the *Papilio machaon* group of swallowtail butterflies: *P. oregonius* feeds upon *A. dracunculus,* whereas most of the other species feed on umbellifers. The conclusion of whether a group follows Fahrenholz's Rule therefore depends upon how many exceptions one allows.

We know enough now from comparisons of cladograms of interacting taxa to conclude that Fahrenholz's Rule is simply a statement of one of the many possible outcomes of evolving interactions. Fahrenholz was concerned primarily with the pattern of speciation found in well-defined groups of very specialized parasites. Any collection of studies on parallel speciation between interacting taxa provides evidence for all possible combinations of phylogenetic tracking and nontracking of one lineage on another (e.g., Stone and Hawksworth 1986; Brooks and McLennan 1991). Differential speciation rates of interacting taxa, differential extinction rates, differences in geographic ranges among interacting species, novel mutations, and new ecological opportunities together prevent complete concordance in almost all comparisons. A run of parallel speciation is soon broken by a shift in one or more parasite populations onto a phylogenetically unrelated host. The larger the number of species in the group, the lower the chance of sustained phylogenetic tracking.

For example, after thoroughly reviewing the worldwide pattern of specialization in agromyzid flies, whose 2500 known species are mostly stem and leaf miners in plants, Spencer (1990) concluded that "there is really little logic in the hosts which have been colonized by the majority of genera" and that there have been "countless switches between unrelated families." The result is a "remarkable degree of randomness" in the phylogenetic patterns of specialization in these insects. The 1190 agromyzid species with known hosts have colonized everything from bryophytes, horsetails, and ferns to gymnosperms and angiosperms. There are clear runs of speciation among related host taxa, as in the 91 species of *Phytomyza* on Ranunculaceae, but a broader look throughout that genus of 318 species shows that it has also produced a radiation of 64 species on the very distantly related Umbelliferae and incidental colonizations and minor radiations on twenty-six other plant families. The same pattern of radiation in species on a few plant families and a smattering of colonizations by other species over many other plant families occurs also in many of the other genera of agromyzid flies.

Similarly, Sprent (1982) has argued that radiation and specialization in ascaridoid nematodes attacking vertebrate carnivores have not been the result of gradual adaptive radiation among related host taxa but instead a series of "haphazard excursions." Powell (1980) reached essentially the same conclusion in his evaluation of patterns of specialization in microlepidoptera, of which almost every family has radiated to include a wide array of plant families as larval hosts. At least thirteen of the ditrysian microlepidopteran families have radiated onto twenty or more plant families, and many of the others have been recorded from ten or more families.

The randomness in patterns of specialization that has struck Spencer, Sprent, Powell, and other systematists is the observed lack of sustained phylogenetic tracking. But no one would argue that the shifts onto new host taxa are truly random. Phylogenetic history certainly imposes constraints on how parasites are distributed among hosts. Nevertheless, new mutations and ecological opportunity now appear to make phylogeny less of a direct long-term constraint on patterns of specialization than some systematists in previous decades believed. The role of ecological opportunity in breaking through phylogenetic constraints seems apparent, for instance, in ascaridoids that alternate hosts. Radiations of these parasites, although phylogenetically haphazard, are ecologically linked. Some genera, such as *Phocanema* and *Anisakis,* alternate between fish and the predatory fish, reptiles, birds, and mammals that eat those fish. These trophic links appear to have opened up new adaptive zones, allowing some genera of ascaridoids to colonize even more phylogenetically unrelated hosts after moving, for example, from freshwater to terrestrial environments (Sprent 1982).

The problem in using ascaridoids, agromyzid flies, or any other single taxonomic group as examples of the lack of phylogenetic tracking is that, by picking and choosing carefully among parasites, one can reach any conclusion one wishes about the phylogeny of specialization. One way of addressing the problem in a more systematic manner is to take one taxonomic group of hosts and evaluate phylogenetic tracking in all the parasite groups attacking it. This approach eliminates the criticism that one has specifically chosen parasite groups that fit one's views. (The approach does not, however, eliminate the possibility that some host taxa may be more susceptible to phylogenetic tracking by parasites than other host taxa.) The plant family Umbelliferae is a convenient group for such an analysis, because a diverse array of insects attacks the approximately 3000 known species of umbellifers, and many of these umbellifer-feeding species are specialized to attack only one or a few closely related host species (Berenbaum 1983, 1990; Thompson 1986c). At least seven insect genera distributed over five families in three orders have radiated into five or more species on Umbelliferae in North America and Europe (table 4.2). There are closely related species within some of these insect genera that have speciated onto closely related hosts, as has occurred in the radiation of at least eight species in the *Depressaria douglasella* group onto the umbellifer genus *Lomatium* (Clarke 1952; Thompson 1983b,c,d). Nevertheless, none of these insect genera shows evidence of sustained phylogenetic tracking of an umbellifer genus. All have

Table 4.2. The major insect genera that have radiated in species onto Umbelliferae in North America and Eurasia

Taxon	No. of species	No. of umbellifer-feeding species	No. of umbellifer genera known as hosts	No. of plant families known as hosts worldwide	References
Homoptera					
Aphididae					
Aphis	659	31	28	125	Patch 1938; Richards 1976
Diptera					
Agromyzidae					
Melanagromyzidae	125	11	11	26	Spencer 1990
Phytomyza	318	64	48	28	Spencer 1990
Lepidoptera					
Oecophoridae					
Agonopterix	125	34	32	12	Hannemann 1953; Hodges 1974
Depressaria	100	41	22	3	Hannemann 1953; Hodges 1974; Thompson, personal observation
Papilionidae					
Papilio	220	10	27+	38	Scriber 1984; Miller 1987a; Sperling 1991; Wehling and Thompson, personal observation
Prodoxidae					
Greya	16	5	4	2	Davis, Pellmyr, and Thompson 1992

Note: The genera listed are those in which at least five species are known to feed on umbellifers.

Table 4.3. The plant genera in three families shared as hosts by the four major lepidopteran genera that include specialists on the Umbelliferae

Insect genus	Umbelliferae	Rutaceae	Compositae
Papilio (Papilionidae)	*Lomatium*	*Ruta*	*Artemisia*
Agonopterix (Oecophoridae)	*Lomatium*	*Ruta*	*Artemisia*
Depressaria (Oecophoridae)	*Lomatium*	*Ruta*	*Artemisia*
Greya (Prodoxidae)	*Lomatium*	—	—

Note: In addition to *Lomatium*, the genera *Papilio*, *Agonopterix*, and *Depressaria* have several other host genera in common within the Umbelliferae (e.g., *Angelica, Cnidium, Daucus, Heracleum*).

colonized four or more plant genera within the Umbelliferae and two or more plant families overall.

In some cases the shifts of these umbellifer-feeding insects onto unrelated hosts are explicable *post hoc* through the chemical similarity of the plants. For example, it has long been argued that the use of Umbelliferae, Rutaceae, and Compositae by the *Papilio machaon* group of swallowtail butterflies is due to chemical similarities among these plant families (Dethier 1941). That conclusion can be reinforced by the observation that two other insect genera attacking umbellifers—*Depressaria* and *Agonopterix* in the moth family Oecophoridae—have colonized some of the same plant genera in two of the other plant families colonized by *Papilio* (table 4.3). All three of these insect genera have species that use *Ruta* (Rutaceae) and *Artemisia* (Compositae) as well as some of the same genera within the Umbelliferae.

Nevertheless, as with phylogenetic relatedness, chemical similarity explains only part of the phylogeny of specialization among hosts. The other lepidopteran genus that has radiated onto umbellifers—*Greya*—feeds on neither Rutaceae nor Compositae. Instead, all the species that do not feed upon Umbelliferae feed on plants in the Saxifragaceae (Davis, Pellmyr, and Thompson 1992), which none of the other six umbellifer-specialist insect genera uses as a host. The lack of use of at least Compositae by *Greya* is not simply a result of lack of ecological opportunity. Some *Greya* species occur in the same habitats as the *Papilio* and *Depressaria* species that feed upon *Artemisia* (personal observation). In contrast, both of the agromyzid fly genera that have radiated in species on the Umbelliferae have colonized Compositae—just like *Papilio*, *Depressaria*, and *Agonopterix*—but, unlike these three lepidopteran genera, neither includes Rutaceae among the host taxa. Instead, these genera have undergone extensive colonizations of more than twenty-five other plant families. The 318 known species of *Phytomyza* alone include 64 on Umbelliferae, another 91 on Ranunculaceae, and additional species distributed among all the other subclasses of dicotyledons (Spencer 1990).

The impression left by the distribution of specialization in all these insect genera feeding on umbellifers is one of an intricate tangle of phylogenetic inertia, ecological opportunities, and selection pressures that have together produced a pattern of specialization unique to each genus. The distributions are not random, but each pattern results from a unique interrelationship of history, ecology, genetics, and chemistry. As a result, the evolutionary trajectory of specialization in these, or any, parasite lineages and, conversely, the set of specialists with which any host species must contend over evolutionary time are unpredictable, or only partially reconstructable, results of past history. As Gould (1989) has said about the history of life in general, if we replayed the tape of the history of life, the results would almost assuredly be very different. Hence, as life has evolved, the Northern Hemisphere Umbelliferae has attracted at least seven genera of insects that have radiated onto its species. During that time, it has also attracted the odd colonization from other parasite taxa: at least two of the seventy-seven species of the gall midge genus *Contarinia,* the others of which are distributed over twenty-eight plant families (AnanthaKrishnan 1984; Gagné 1989; Thompson, unpublished data), and at least one species each from the weevils *Apion* and *Smicronyx* (Ellison and Thompson 1987). It seems unlikely that a replay of the history of life would produce the same suite of interactions.

Conclusions

There is no reason to suspect that extreme specialization is necessarily the more derived condition in the phylogeny of interactions. Nor is there any reason to expect that specialization is an evolutionary dead end from which species cannot escape, or that specialists, as a rule, show sustained phylogenetic tracking of host taxa. Although it has become popular to think about the various specializations found in species as constraints on future evolution, these same specializations are sometimes the fortuitous preadaptations that allow success in new interactions. Adaptation to one species or group of species can have correlated, fortuitous effects that open up the possibility of interactions with other, unrelated taxa as species expand their geographic ranges. The result is that any lineage is likely to have a mix of species with different degrees of specialization in their interactions with other species. These combined effects of phylogenetic inertia and new ecological opportunities are part of the reason why the entangled web has become so entangled, sometimes producing unexpected combinations of interacting and coevolving species.

If there is no intrinsic direction to the evolution of specialization, and if specialists are not necessarily evolutionary dead ends, then specialization also has the potential to evolve in different directions in different populations of a species. The result will be a dynamic geographic mosaic created by the degree to which species are specialized to one another, without any general tendency of populations to evolve necessarily to greater specialization and then extinction. This evolutionary flexibility in specialization becomes one part of a geographic view of coevolution.

CHAPTER FIVE

Evolutionary Genetics of Specialization

If specialization is often evolutionarily dynamic rather than a dead end, then we should find the origins and mechanisms of that flexibility in the genetics of species. We should be able to see how the genetics of specialization, combined with ecological opportunity, allow new interactions to form, creating a geographic mosaic in the pattern of specialization within species and a patchwork pattern of specialization among species within most phylogenetic lineages.

In this chapter I analyze the evolutionary genetics of specialization in order to provide a genetic grounding for a geographic view of specialization and the coevolutionary process. I begin with the molecular genetics of specialization between root-nodule bacteria and their host plants for the insights that these detailed studies provide about the genetic complexity of the processes leading to specialization. I then turn to the genetics of host and prey choice in actively searching animals. The latter part of the chapter dwells in detail on the evolutionary genetics of host choice and use in swallowtail butterflies as a case study of the evolution and geographic structure of specialization. Later chapters on coevolution will incorporate yet other aspects of geographic structure in the genetics of specialization (e.g., gene-for-gene relationships between plants and pathogens; the genetics of mimicry).

What we know about the genetics of specialization in interactions comes mostly from detailed studies of a small number of species. For specialization on hosts and prey these include mostly a few bacteriophages, *Rhizobium* and several other bacteria, a smattering of rust fungi and related pathogens attacking plants and invertebrate parasites attacking animals, a small sampling of phytophagous insects, and an even smaller number of vertebrates. For specialization in defense against enemies or attraction of mutualists, the studies include an equally tiny range of microbial, plant, and animal taxa. Nevertheless, studies of this small subset of species have already shown that there is a rich variety of genetic mechanisms influencing the evolution of specialization. Major genes, modifier genes, polygenes, and epistatic effects have all been found to influence patterns of specialization. In addition, plasmids mediate interactions between some bacteria and their hosts (Young and Johnston

1989), and maternal effects, either genetic or nongenetic, influence the outcome of yet other interactions (Mousseau and Dingle 1991; Rossiter 1991a,b). Genes for specific defenses against two or more enemies are sometimes clustered together on a small region on one chromosome (Dickinson, Jones, and Jones 1993), yet genes for specialization to a small group of hosts are sometimes distributed over two or more chromosomes (e.g., Thompson, Wehling, and Podolsky 1990).

As more studies accumulate, it will become possible to mine them in a search for patterns in how different modes of inheritance influence the evolution of specialization in different ways. One component of such an analysis will be to ask if different life histories and modes of interaction favor different genetic bases for the evolution of specialization. Complex life histories, for example, may involve genetic mechanisms of specialization different from those of simpler life histories.

Genetics of Specialization in the Symbiosis between Legumes and Rhizobia

Genetic studies of the symbiosis between legumes and root-nodule bacteria have revealed a remarkable sequence of steps, molded by coevolution, through which these species specialize in their responses to one another. As in the few other interactions in which the molecular genetics have been studied in detail—for example, that between bacteria and the relatively simple phage *fl* (Bull and Molineux 1992) or *Agrobacterium* and its host plants—these detailed studies are showing how very complex the genetics of specialization and coevolution can become. Despite the genetic complexity of these interactions, the results of these studies are showing how small genetic and associated biochemical changes can influence patterns of specialization, leading sometimes to geographic differences in associations between species.

Three nonrelated genera of soil bacteria (*Rhizobium, Bradyrhyzobium,* and *Azorhizobium*), collectively called rhizobia, have evolved complex nitrogen-fixing symbioses with plants. The bacteria invade plant tissues, usually the roots, and induce formation of nodules. Within the nodules, the bacteria reduce nitrogen to ammonia, which is used by the host plant. With one exception—several species of *Parasponia* (Ulmaceae)—these symbioses are restricted to legumes, but they are pervasive among legumes and occur in most of the 15,000 legume species (Young and Johnston 1989).

Rhizobia vary tremendously in their degree of specificity to particular host plants. One known strain of *Rhizobium* is capable of nodulating thirty-five different legume genera as well as *Parasponia,* whereas, at the other extreme, some other rhizobia are restricted to one plant species or even to varieties within a crop-plant species (Dénarié, Debellé, and Rosenberg 1992). Studies of the molecular genetics of specificity in rhizobia have now shown that there is no single, simple genetic basis for the observed range of specificity found in rhizobia. Successful nodulation and nitrogen fixation require a sequence of steps controlled by genes in both the rhizobia and the plants. Differences in specificity among rhizobia strains can arise from differences in genes in each step of the sequence.

It is worthwhile to follow the major sequence of events that leads to successful nodule formation (reviews in Young and Johnston 1989 and in Dénarié, Debellé, and Rosenberg 1992) because the sequence suggests how the complex genetics of this interaction can bring about differences in host specificity among closely related species and even populations within species. The symbiosis begins as plant flavonoids are exuded from the roots and induce the expression of bacterial genes called *nod* (nodulation) genes. In effect, these otherwise free-living bacteria use the flavonoids as cues to availability of a host and activate the genes for the symbiosis only in response to the presence of the appropriate flavonoids. The regulatory gene *nodD* produces constitutively NodD proteins, which control transcription of the common and specific *nod* genes (Schlaman, Okker, and Lugtenberg 1992). In the presence of the appropriate flavonoids, these NodD proteins are activated, and transcription of the common and specific *nod* genes begins. The *nodD* genes differ among bacterial strains, and the NodD proteins from different bacteria respond to different sets of flavonoids. Hence, the first determinant of specificity in rhizobia is the range of flavonoids to which the NodD proteins produced by a *nodD* gene respond.

Variation in specificity among rhizobia is further increased at this first step because the number of *nodD* genes differs among strains. Some rhizobia have only one *nodD* gene, whereas others have up to three, and each of these genes may respond to a different set of flavonoids (Honma and Ausubel 1987). Consequently, the number of *nodD* genes harbored by a strain can influence the range of hosts it attacks, but there is no simple relationship between the number of *nodD* genes and the range of hosts that a bacterial strain can nodulate (Dénarié, Debellé, and Rosenberg 1992). In some rhizobia with multiple *nodD* genes, small changes in any one gene may have little effect on the number of plant species they can nodulate. In other rhizobia, however, even a point mutation can influence specificity. For example, a point mutation in the single *nodD* gene of *Rhizobium leguminosarum* bv. *trifolii* allows this strain not only to respond to the flavonoids of clover, its normal host, but also to undergo at least the early stages of nodulation on *Parasponia* (McIver et al. 1989).

The next step in the formation of this symbiosis is the activation of the common and host-specific structural *nod* genes by the NodD proteins. These

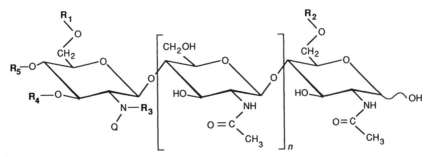

Fig. 5.1. Structure of the Nod factors showing the basic core structure and acyl moiety (Q) present in all Nod factors and the sites (R_1–R_5) that are modified by the host-specific *nod* genes. Redrawn from I. Vijn et al., "Nod Factors and Nodulation in Plants," *Science* 260:1764–65, copyright 1993 by the AAAS.

genes code for the synthesis of another group of molecules, called the Nod factors, which activate the transcription of genes in the plant. The common *nodABC* genes occur in all the rhizobia that have been studied, and these genes appear to be responsible for producing the basic structure (an oligosaccharide) of the Nod factors. The host-specific *nod* genes (e.g., *nodFE, nodH, nodPQ*) modify the basic structure by decorating it with side groups (fig. 5.1). Each of these decorations affects specialization by influencing the range of hosts that can be activated by the Nod factors (Roche et al. 1991; Vijn et al. 1993).

Upon contact with the host, the Nod factors elicit formation of nodule meristems and associated tissues (Dénarié, Debellé, and Rosenberg 1992). This process involves a series of nodule-specific (nodulin) genes, some of which are expressed before nitrogen fixation begins whereas others are expressed later (Govers et al. 1987; Scheres et al. 1990; Yang et al. 1993). Once nodules have been formed, a complex series of regulatory and structural genes determines whether nitrogen fixation takes places (Merrick 1992). The wrong combination of genes can result in nodule formation but no nitrogen fixation (Young and Johnston 1989).

Hence, genes determining specificity occur at every stage in the formation of this symbiosis. With so many genes involved in the process, it is easy to understand why there is such a broad range in host specificity among rhizobia. It is more difficult to understand why rhizobia are generally not either more host specific or less host specific. On the one hand, it is a marvel that any rhizobia are capable of forming symbioses with more than one or a few legume species, given the number of genes involved and the number of signals that must be sent back and forth between the bacteria and the plant in the process of nodule formation and nitrogen fixation. On the other hand, if the association between rhizobia and legumes is nearly always mutualistic, then the question arises as to exactly why such a long and complex series of signals determining specificity in these associations is maintained over evolutionary time.

A partial answer to why there is no universal *Rhizobium* strain capable of nodulating all legumes may lie in the associations between plants and their pathogens (Young and Johnston 1989). Plant roots are continually challenged by pathogens attempting to invade their tissues. Over evolutionary time, different plant species will evolve different ways of defending themselves against invasion by pathogens. To form a symbiosis, rhizobia must circumvent these defenses and invade the plant tissues. Natural selection will therefore favor specialization in rhizobia as they evolve solutions for invading one or another plant species.

Overall, the complexity of the molecular genetics of host specificity in rhizobia reinforces the conclusion that a range of specialization is almost bound to occur within any group of related species. Moreover, geographic differences in specialization are likely to develop among populations as they coevolve. Some results from rhizobia support this view. Unlike commercial peas, the primitive pea line Afghanistan carries the gene *sym2*, which allows it to form nodules with strains of *R. leguminosarum* bv. *viciae* containing an extra *nod* gene, *nodX*. When commercial pea lines are crossed and then introgressed so that they carry the *sym2* gene, they become capable of forming nodules only with *R. leguminosarum* bv. *viciae* strains carrying the *nodX* gene (Vijn et al. 1993). Hence, this particular host gene requires a specific symbiont gene for nodule formation to occur. Similar results, producing geographic differences in specialization and coevolution due to pairs of host and symbiont genes, must surely occur in natural populations as well.

Evolutionary Genetics of Preference Hierarchies in Animals

No example of the genetics of specialization and coevolution between actively searching animals and their hosts, victims, or prey has been worked out to the same depth in molecular biology as that between legumes and rhizobia. Good progress, however, is being made on the population and evolutionary genetics of some interactions involving actively searching animals. These ecological results complement the molecular results available for legumes and rhizobia by showing, in some cases, the relationship between genetics and the geographic structure of specialization.

The paradigm that has driven much of the research on the genetics and evolution of specialization in actively searching animals is that individuals exhibit hierarchies of preference. Given several equally accessible hosts or prey, an individual will be more likely to visit or attack one species than another. It may prefer one species to such a degree that given the choice, it always interacts only with that most-preferred species. Alternatively, an individual may rank two or more species as equal in preference and interact with all of them at least initially. Through experience, it may become more efficient at extracting resources from one species than another and may come to specialize on it.

There are now hundreds of papers, with viewpoints ranging from optimality theory to physiological-state models, suggesting how behaviorally flexible animals should adjust their degree of specialization to particular ecological conditions (reviews in Stephens and Krebs 1986; Schoener 1987; and Futuyma and Moreno 1988). Many of these are purely ecological, rather than evolutionary, in point of view. They take the current preference hierarchy, range of behavioral flexibility, or relationship of physiological state to diet breadth and make predictions about how an individual should adjust its degree of specialization throughout its lifetime within the specified limits. Studies ranging from insects to vertebrates to sea slugs have shown that individuals within populations do in fact differ significantly in preference or degree of specialization to particular hosts or prey (e.g., Arnold 1981a,b; Wiklund 1981, 1982; Ng 1988; Thompson 1988e; Courtney, Chen, and Gardner 1989; Trowbridge 1991). Some of these differences may reflect differences in experience or maternal effects, whereas others may reflect differences in genetics.

The evolutionary problem, however, is to understand how preference ranking, degree of preference for one species over another (specificity), range of behavioral flexibility in choice, and ability to use a range of species (e.g., measures of performance such as growth rates on different hosts) evolve. Arnold's (1981a,b) study of slug-feeding in garter snakes (*Thamnophis elegans*) is one of the few that has demonstrated direct genetic control of prey selection in a vertebrate predator. Slug-feeding is common in the wet, mollusc-rich environments of coastal northern California. Here slugs constitute over 90% of the diet of these garter snakes. Farther inland, however, where slugs are less common, the snakes feed mostly on frogs and fish. Coastal and inland populations are genetically polymorphic for slug-feeding, with the slug-eating morph the more common on the coast and the slug-refusing morph the more common inland (fig. 5.2). Naive, newborn hybrids between the two morphs are somewhat intermediate in their tendency to feed on slugs (Arnold 1981a,b). The number of genes involved has not been determined. Arnold's studies show that there is a strong genetic basis to the geographic pattern of prey choice in garter snakes. The genetics of preference in this case creates a range of specialization within populations and large differences in the distribution of preference between populations.

Many traits associated with food selection in other vertebrates also vary

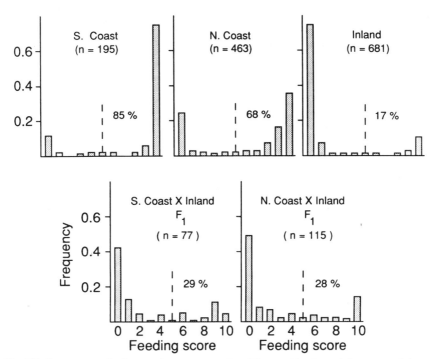

Fig. 5.2. Percentages of naive, newborn garter snakes (*Thamnophis elegans*) from coastal, inland, and laboratory hybrid populations showing preference for slugs (scores above 5) or avoidance of slugs (scores below 5). Redrawn from Arnold 1981b.

genetically within and among populations. Some of these traits, such as bill size and shape in Darwin's finches (Grant and Grant 1989), are governed by a number of genes and evolve as quantitative traits. Studies of some vertebrate predators that are polymorphic for jaw morphology have indicated that at least some of these polymorphisms, and the geographic differences in their occurrence, may be genetically based. Crosses between wide- and narrow-mouth morphs of the goodeid fish *Ilyodon* yield both morphs among the hybrid offspring, but nothing more about the genetics is known (Grudzien and Turner 1984). Crosses between three sympatric morphs of arctic char (*Salvelinus alpinus*) that differ in jaw morphology produce hybrids that initially resemble the maternal morph, but whether these maternal effects carry over into adulthood is unknown (Skúlason, Noakes, and Snorrason 1989). Since jaw polymorphisms in some fish are induced by diet (Meyer 1987), genetic studies of trophic polymorphism in these taxa must include analyses of interactions between genotype and diet.

Some anecdotal results on the genetics of specialization in other species can be gleaned from the feeding habits of hybrids found in nature, indicating geographic patterns in specialization influenced by hybridization and introgression among populations. In Arizona, two broadly sympatric species of catostomid fish, the Sonoran sucker *Catostomus insignis* and the desert sucker *Pantosteus clarki,* have been commonly reported to hybridize. These two species differ markedly in diet. The Sonoran sucker is primarily carnivorous, whereas the desert sucker has cartilaginous sheaths on its jaws that it uses to scrape diatoms, algae, and other organic matter from solid surfaces. In two sites where both fish and their presumed hybrids co-occur, Sonoran sucker diets are 55–60% animal matter (by volume), desert sucker diets are 10–18% animal matter, and hybrids are somewhat intermediate, with 28–53% (Clarkson and Minckley 1988). These differences, however, may have little to do directly with specialization on prey species. The amount of animal matter in the hybrid diet may result from other causes, such as differences in microhabitats searched for food or differences in the efficiency of prey capture associated with jaw structure.

Evolutionary Genetics of Preference in Insects

Much more comparative work has been done on the evolutionary genetics of preference and specialization in insects than in vertebrates. These studies are unraveling some of the ways in which the genetics of preference hierarchies can shape local and geographic patterns of specialization. The following paragraphs summarize what is currently known about the ways in which the genetics of preference shapes the geographic structure of specialization in swallowtail butterflies, and then contrast these results with those for other insects.

The *Papilio machaon* group of swallowtails is composed of six to ten species with a Holarctic distribution (fig. 5.3). These insects appear to be in the process of differentiating and speciating geographically, and hybrid zones occur among some of the recognized species and subspecies. Some groups of populations recognized as subspecies by some taxonomists are considered full species by others (Collins and Morris 1985) and vice versa (Sperling 1987, 1991). In North America, current taxonomy recognizes from four to about six species (Tyler 1975; Collins and Morris 1985; Sperling 1987). The widely recognized species include *P. indra, P. machaon, P. polyxenes,* and *P. zelicaon.* In addition, *P. oregonius* has generally been recognized as a separate species, but some evidence suggests that it and two other recognized species, *P. bairdii* and *P. brevicauda,* may best be regarded as subspecies of *P. machaon* (Sperling 1987, 1991). The problems of taxonomy aside, the close relationships among subspecies and species in the *P. machaon* group have made them useful for studies of the evolutionary genetics of specializa-

Fig. 5.3. Swallowtail butterflies within the *Papilio machaon* group in North America: *P. zelicaon* (*upper left*), *P. oregonius* (*upper right*), *P. indra(lower left), P. polyxenes (lower right).*

tion, because even populations that do not normally hybridize in nature can be forced to do so in the laboratory by hand-pairing individuals. Viable F_1 hybrids can generally be obtained, and sometimes, although much more rarely, F_2 hybrids.

The *P. machaon* group uses hosts from three plant families: Umbelliferae, Rutaceae, and Compositae. The two species that have been studied in most detail in North America are the anise swallowtail, *P. zelicaon,* and the Oregon swallowtail, *P. oregonius.* These two species occur sympatrically or parapatrically over part of their ranges in western North America. *Papilio zelicaon* feeds primarily on the Umbelliferae, and populations differ in their umbelliferous hosts (Emmel and Shields 1978; Sims 1980; Thompson 1988b; Tiritilli and Thompson 1988). Some populations in California, however, have shifted in recent decades from the Umbelliferae onto *Citrus* in the Rutaceae (Shapiro and Masuda 1980).

In the Tucannon River drainage of the Blue Mountains of eastern Washington State and Oregon, *P. zelicaon* lays its eggs on two umbellifer species, *Lomatium grayi* and *Cymopterus terebinthinus.* Other species of *Lomatium* occur within the habitats visited by ovipositing females but are rarely used as hosts. When the two normal hosts are offered to *P. zelicaon* along with similar amounts of fennel (*Foeniculum vulgare*), which is another umbellifer used as a host in other populations, and *Artemisia dracunculus,* a composite

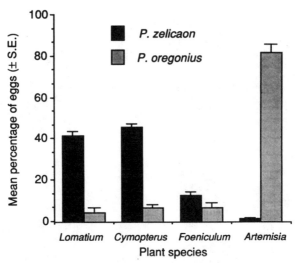

Fig. 5.4. Oviposition preference in *Papilio zelicaon* in the Blue Mountains of eastern Washington State as compared with *P. oregonius* in the surrounding steppe communities. The hosts are *Lomatium grayi*, *Cymopterus terebinthinus*, *Foeniculum vulgare*, and *Artemisia dracunculus*. Redrawn from Thompson 1988e.

that is the host of *P. oregonius,* females show a characteristic preference hierarchy (fig. 5.4). Females from most families prefer their normal two hosts and appear to treat them as one species, showing almost no discrimination between them. A small proportion of eggs are often laid on fennel, and few or no eggs are laid on *Artemisia.* The preference for the normal hosts is relative rather than absolute. In the absence of the normal hosts, females will readily lay their eggs on fennel.

The preference hierarchy for *P. oregonius* is completely different (fig. 5.4). In the steppe communities to the west of the Blue Mountains, *P. oregonius* has available the same umbellifer hosts as *P. zelicaon* but oviposits exclusively instead on *Artemisia dracunculus.* Given the same four plant species as *P. zelicaon* under the same experimental conditions, *P. oregonius* females from a population near Palouse Falls characteristically lay almost all their eggs on their normal host *Artemisia,* generally ignoring the three umbellifer species (Thompson 1988e).

In both *P. zelicaon* and *P. oregonius,* females from different families differ in the degree to which they are specific to their normal hosts (Thompson 1988e). Females of both species from the Blue Mountains area always prefer their normal host(s), but females from some families of both these species consistently lay a few eggs on plants low in the preference hierarchy (fig.

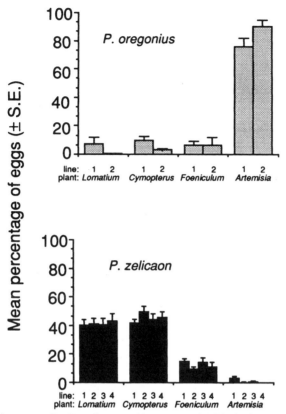

Fig. 5.5. Variation among *Papilio zelicaon* and *P. oregonius* families in the tendency to lay a few eggs on plant species low in the preference hierarchy and not used as hosts in natural populations. The plant species names are given in the legend to fig. 5. 4. Redrawn from Thompson 1988e.

5.5). This variation is evolutionarily interesting in that it suggests that even extreme specialization such as occurs in *P. oregonius* is not an evolutionary dead end. At least some individuals in these populations would be capable of shifting onto an appropriate novel plant species in the absence of the normal host.

The exact cues that females of these two species use in ranking hosts are unknown. Mixtures of chemical compounds probably play a role at least once a female has landed on a plant and the chemoreceptors in her tarsi have come in contact with the plant surface. In three other *Papilio* species (*P. polyxenes*, another North American member of the *P. machaon* group; and *P. xuthus* and *P. protenor*, which are Japanese species more distantly related to the *P.*

machaon group) the oviposition response of females after contact with im-
pregnated filter paper depends upon the relative proportions of several plant
compounds (Ohsugi, Nishida, and Fukami 1985; Honda 1986; Nishida et al.
1987; Feeny et al. 1988; Feeny 1991; Baur, Feeny, and Städler 1993). Single
compounds elicit less of a response than combinations of compounds, and
some combinations produce a better response than others. Moreover, at least
in *P. polyxenes* a combination of these contact stimulants and volatile compounds more commonly elicits an oviposition response than contact stimulants alone (Feeny et al. 1989). It is easy to imagine how these short-distance
responses together with longer-distance visual cues could produce a hierarchy in oviposition preference among a range of available plant species.

Interspecific crosses between *P. zelicaon* and *P. oregonius* have begun to
reveal the genetic basis for the preference hierarchies in these species. The
major differences in oviposition preference are determined primarily by one
or more loci on the X chromosome (Thompson 1988c). This is evident in
the differences in preference in reciprocal crosses (fig. 5.6). Females are the
heterogametic sex in Lepidoptera, so X linkage is indicated by traits in hybrids that are similar to the paternal species. Although it is not yet clear exactly how many loci on this chromosome affect preference, the results so far
indicate that this one chromosome is primarily responsible for controlling
how these butterflies rank potential host plants.

The effect of the X chromosome, however, is not symmetrical in the
crosses (fig. 5.6). Although the genetic basis of this lack of symmetry is not
yet clear, crosses between *P. oregonius* females and *P. zelicaon* males generally produce offspring that have oviposition preferences almost identical to
those of *P. zelicaon* females. Crosses between *P. zelicaon* females and *P. oregonius* males, however, produce offspring that vary in preference from a pattern similar to that of *P. oregonius* to patterns that are more intermediate
between the two species. The variation found among crosses appears to have
two genetic sources. First, there appears to be variation at the preference
locus (loci) on the X chromosome (Thompson 1990). The current results,
however, cannot distinguish whether this variation arises from variation at
one locus or among several loci. Second, one or more loci not on the X chromosome appear to affect preference, thereby modifying the effect of the X
(Thompson 1988c). Consequently, the evolution of preference in these swallowtails may involve loci on at least two chromosomes, although the X chromosome has the largest effect, at least on the differences in preference between *P. zelicaon* and *P. oregonius*.

Preliminary experiments on another group of swallowtail species, the tiger
swallowtails in North America (Scriber, Giebink, and Snider 1991), and a
heliothine moth, *Heliothis virescens* (Waldvogel and Gould 1990), have indicated the X chromosome may also be involved in the evolution of oviposition

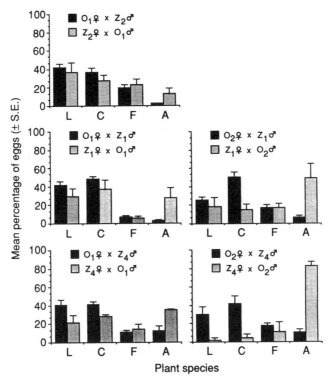

Fig. 5.6. Reciprocal crosses between the swallowtail butterflies *Papilio zelicaon* and *P. oregonius* showing X linkage of oviposition preference. O_1, O_2, Z_1, Z_2, and Z_4 are the first-generation progeny of individual females. The plant species names are given in the legend to fig. 5.4. Redrawn from Thompson 1988c.

preference in other lepidopteran species. In five interspecific crosses between tiger swallowtail species, *Papilio canadensis* females and *Papilio glaucus* males, F_1 hybrids all showed a strong preference for the host of *P. glaucus* (Scriber, Giebink, and Snider 1991). Unfortunately, the reciprocal crosses did not produce hybrid offspring. Similarly, crosses between two strains of *Heliothis virescens* preferring different hosts show differences between reciprocal hybrids consistent with an interpretation of X-linked inheritance of oviposition preference. Although Waldvogel and Gould (1990) did not give a specific genetic interpretation to their results, they did note that the preferences of the reciprocal hybrids were similar to those of the paternal strain and different from those of the maternal strain. This is the expected result for X-linked traits in Lepidoptera. The results for this species, however, must be interpreted cautiously because, unlike in the tests of the other species, groups of females, rather than individual females, were placed in test cages. Consequently, the final distribution of eggs in each cage was a composite of the preferences of females. This kind of design masks variation among females within strains and crosses, making statistical analysis and interpretation more difficult (Thompson and Pellmyr 1991). Nevertheless, the trend in the results is the same as that found in the other lepidopteran species that have been tested.

It is not clear why the X chromosome has become the site for localization of genes for host preference in at least some Lepidoptera. Under some conditions, evolution on the X chromosome can be faster than on the autosomes and allow the buildup of coadapted gene complexes (Charlesworth, Coyne, and Barton 1987; Jaenike 1989b; Hagen and Scriber 1994). Although it is unknown for any of these lepidopteran species exactly how many genes are controlling oviposition preference, the strong influence of one chromosome in a group such as butterflies, which characteristically have about thirty pairs of chromosomes, suggests that changes at relatively few genetic loci could have large effects on the preference hierarchy of these species.

The results for these few species, however, do not imply that all preference hierarchies in insects have a simple genetic basis. *Papilio* is currently the only insect group in which the genes for oviposition preference have been localized with some certainty to one chromosome. But then, only a few other insect taxa have begun to be evaluated for the mode of inheritance of preference. Few other insects outside the Lepidoptera have a sex-determination system in which females are the heterogametic sex. Consequently, even if the X chromosome turns out to be important in determination of preference in other insects, the evolutionary implications would be different from those of Lepidoptera.

The most detailed studies of mode of inheritance for another insect taxon have been Jaenike's experiments on *Drosophila tripunctata*. Oviposition preference in this species is inherited polygenically through autosomal loci with significant dominance and interaction effects (Jaenike 1987). The number of loci involved, however, remains unknown. The results of crosses show nonadditive effects and an approximately normal distribution of preferences among families, which, as Jaenike (1987) noted, could have resulted from either a few loci or many loci with small effects.

Using another species of *Drosophila*, Lofdahl (1987) analyzed acceptance of a novel host as a quantitative trait, using a half-sib design in which males were mated to 2–113 females. *Drosophila mojavensis* breeds in rotting necroses of five or six cacti in the southwestern United States and in Mexico. Individual populations, however, are restricted to one or two cactus species. The necroses of each of these cactus species have a unique community of bacteria and yeast that is the actual food of the *Drosophila* larvae and adults. Lofdahl tested whether there was heritable variation in a monophagous *Opuntia*-feeding population for a novel cactus (agria cactus). Agria cactus is a host of *D. mojavensis* elsewhere, but this population never encounters it. She constructed simulated agria cactus necroses and tested 1100 females for acceptance of this 'host' in no-choice trials. She also tested for heritable variation in the number of eggs laid by each female on this host. Heritable variation was low for both characters, accounting for about 11–18% of the total phenotypic variance for acceptance and 14% of the phenotypic variance in number of eggs laid. This low result may mean that there is very little genetic variation in this population for accepting this novel host. Alternatively, it may mean that very little of the variation occurs as additive genetic variance (i.e., variation contributed by several to many genes contributing equally to the trait and unaffected by dominance or epistasis).

Very little is known about the evolutionary genetics of specialization in insects other than phytophagous species and *Drosophila*. The only genetic crosses to test for the genetics of specialization in carnivorous insects have been on two species of green lacewings. Both species feed as larvae on aphids, but *Chrysopa quadripunctata* has been recorded as feeding on a half dozen or more species throughout its range (although whether local populations are more specialized is apparently unknown), whereas the sibling species *C. slossonae* has been found feeding only on the woolly alder aphid (*Prociphilus tesselatus*). Tauber and Tauber (1987) crossed *C. quadripunctata* from California with *C. slossonae* from New York and tested the parental species and their hybrids on either green peach aphids (*Myzus persicae*) or a combination of green peach aphids and woolly alder aphids. All females of both species laid eggs in cages with the green peach aphid/woolly alder aphid combination. Most *C. quadripunctata* females laid eggs in the cages with only green peach aphids, whereas no *C. slossonae* females laid eggs in those cages. Almost all the reciprocal hybrids in the F_1 generation laid in cages with green peach aphids alone and in the cages with both aphid species. Although only a small number of hybrids were tested, the results suggest a genetic basis for oviposition preference for particular prey in at least the more prey-specific of these two lacewings.

Preference Hierarchies and the Geographic Structure of Specialization

It is tempting at first inspection to view all this work on preference hierarchies as tautological: individuals attack what they prefer. But a closer look at geographical patterns in host use by the several insect species that have been studied in detail shows that the relationship between preference and specialization is more subtle, and it gives a glimpse at how geographic patterns of specialization may develop in part from the underlying genetics of preference and the metapopulation structure of species. There are now several polyphagous insect taxa that have been studied in enough detail to understand how variation in oviposition preference is partitioned both within and among populations: these include *Drosophila tripunctata* and two butterfly species, *Euphydryas editha* and *Papilio zelicaon*. This small group of species spans three of the possible ways in which genetic variation in preference can be partitioned within and among populations of a species: low to moderate within and between populations (e.g., *P. zelicaon*), high within and between populations (e.g., *E. editha*), and high within but low between populations (e.g., *D. tripunctata*).

Of these three species, *P. zelicaon* is the most evolutionarily conservative in preference ranking. That conservatism is evident both in shifts onto novel hosts and in the use of low-ranking hosts in populations outside the range of preferred hosts for hundreds of generations or more. During the past few hundred years, *P. zelicaon* has colonized populations of introduced fennel throughout the west coast of North America, and in some places, where the native hosts have been nearly eliminated, it feeds on nothing else. For one population studied intensively near Sacramento, California, this shift to fennel has not resulted in a major reorganization of the preference hierarchy. When offered the same four plant species used in the earlier trials for populations feeding on native hosts, these fennel-feeding butterflies lay a higher proportion of their eggs on fennel than do those from populations feeding on native hosts, but they show no clear preference for fennel over these other hosts (fig. 5.7). That is, there has been only a moderate change in the degree of preference, so that now all three umbellifer hosts are about equal to ovipositing females.

One possible interpretation of these results is that the population has been caught in the process of transition and that over time natural selection will favor those individuals with a strong preference for fennel. But that need not be so. It will depend upon how the genes influencing preference for fennel affect the genes influencing preference for species of *Lomatium* and *Cymopterus*. Study of another *P. zelicaon* population suggests that continuing evolution toward a strong preference for the local host may not be an inevitable outcome of long-term monophagy on that host, implying again that local ecological specialization is not an evolutionary dead end. In the northwestern United States, some coastal populations of *P. zelicaon* feed on *Angelica lucida* (Tiritilli and Thompson 1988). At Leadbetter Point in southwestern Washington, there is a *P. zelicaon* population isolated by hundreds of miles from the nearest potential *Lomatium* and *Cymopterus* hosts (fig. 5.8). This population may have been restricted to *Angelica* for hundreds or thousands of generations, although it is impossible to know for certain. The only other potential host is cow parsnip, *Heracleum lanatum,* which is rare nearby but

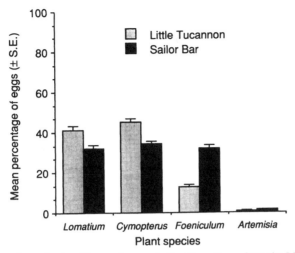

Fig. 5.7. Comparison of oviposition preference in *Papilio zelicaon* from the Little Tucannon population in Washington State, which feeds exclusively on native hosts (*Lomatium grayi* and *Cymopterus terebinthinus*), and the Sailor Bar population in California, which feeds on introduced fennel (*Foeniculum vulgare*). Plant species names are given in the legend to fig. 5.4. Redrawn from Thompson 1993.

serves as a more common host elsewhere along the coast. In choice trials, females from this population prefer *Lomatium* and *Cymopterus* to their normal host (fig. 5.9). The preference hierarchy of this population, in fact, is very similar to *Lomatium/Cymopterus*-feeding populations from Washington populations that never encounter *Angelica*. This geographic pattern of preference in *P. zelicaon* suggests that feeding exclusively on a local host for a fairly long period of evolutionary time does not necessarily lead to genetic specialization in preference for that one host.

This pattern of a highly conservative overall preference ranking among populations and little to moderate variation within *P. zelicaon* populations is at the other extreme from that found in *Euphydryas editha*. Like *P. zelicaon, E. editha* is a common butterfly in western North America, but it has a complex metapopulation structure with discrete, local demes (Ehrlich et al. 1980; Harrison, Murphy, and Ehrlich 1988). Populations of *E. editha* differ in the plants they use as hosts, and Singer, Ng, and Thomas (1988) found a high heritability ($h_2 = 0.9$) for oviposition preference in one population. Also like *P. zelicaon,* some *Euphydryas* populations have switched onto novel hosts without a major change in how they rank the hosts. Unlike *P. zelicaon,* however, individuals differ significantly within and between some populations in the order in which they rank hosts (Singer 1982, 1983; Thomas et al. 1987; Singer et al. 1992). The result is a complex geographic pattern of oviposition

Fig. 5.8. Sites of four *Papilio zelicaon* populations that feed on different plant species. The Little Tucannon population feeds on *Lomatium grayi* and *Cymopterus terebinthinus,* the Wawawai population on *Lomatium grayi,* the Leadbetter population on *Angelica lucida,* and the Sailor Bar population on *Foeniculum vulgare.* Reprinted, by permission, from Thompson 1993.

preference very different from the more conservative pattern found in *P. zelicaon.* The difference between the two species may be a consequence of the metapopulation structure of *E. editha.* Such patterns may even differ among closely related species. Like *E. editha,* the Baltimore checkerspot (*Euphydryas phaeton*) in the eastern United States has also switched in some populations from its native host, *Chelone glabra,* to plantain (*Plantago lanceolata*), an introduced plant. Although some populations now feed exclusively on plantain, in experimental trials they still prefer their native host and grow better on that host (Bowers, Stamp, and Collinge 1992).

The geography of preference in *Drosophila tripunctata* differs from both *P. zelicaon* and *E. editha.* There is so much variation in oviposition preference among *D. tripunctata* families within populations that there is no evi-

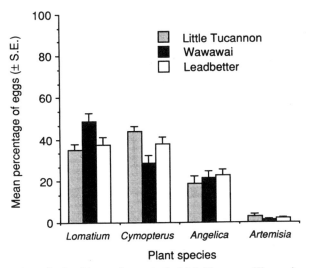

Fig. 5.9. Comparison of oviposition preference in the Little Tucannon, Wawawai, and Leadbetter populations of *Papilio zelicaon*. Redrawn from Thompson 1993.

dence of geographic differences among populations throughout the southern United States (Jaenike 1989a). Jaenike (1989a) has suggested that either disruptive selection or lax stabilizing selection within populations of this polyphagous species that feeds upon highly transient hosts (i.e., mushrooms, fruits) may be the cause of so much variation in oviposition preference maintained within populations. Geographic similarity in preference has also been found in the beetle *Callosobruchus maculatus* (Wasserman 1986) and in a cactophilic *Drosophila* in Australia. Oviposition trials for this *Drosophila* species, however, were performed using twenty females per cage, making it impossible to partition the within-population variance in this study (Barker 1992).

These insect species span all possible patterns in the geography of preference hierarchies except one: low variation within populations and high variation among populations. This one missing geographic pattern in specialization is probably rare and evolutionarily transient within species. Increasing gene flow among populations is likely to change it into one of the other patterns. The alternative—decreasing gene flow among populations that are evolving different preferences—is likely to result in speciation.

Some of the differences among these species may arise from differences in the life histories and metapopulation structure. In addition, some of the differences may arise from differences among taxa in how genetic correlations in preference evolve. In *E. editha,* preference ranking appears to be

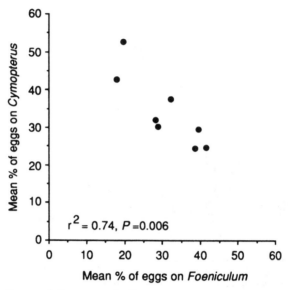

Fig. 5.10. Negative correlation among family means in oviposition preference for two plant species (among four) offered to *Papilio zelicaon* females from Sailor Bar, California. Each family mean is based upon 11–17 female progeny. Redrawn from Thompson 1993.

evolutionarily labile both within and between populations, and there is no evidence that preference for one plant species is genetically correlated, either positively or negatively, with preference for any other. In *P. zelicaon* there is some evidence of a correlation in preference among some plant species. Within the Sailor Bar population, which feeds upon fennel, there is a strong negative correlation between preference for fennel and preference for *Cymopterus* (fig. 5.10).

Studies of other species have suggested strong genetic or phenotypic correlations among preferences for hosts or prey, which could preempt or slow the evolution of specialization onto currently low-ranking species. In a study of oviposition preference in *Drosophila suboccidentalis,* Courtney, Chen, and Gardner (1989) found a positive phenotypic correlation in the percentages of females accepting two potential hosts. For southern cowpea weevils (*Callosobruchus maculatus*), Wasserman (1986) found variation among populations in the number of hosts used but not in how they ranked the hosts. And Arnold's (1981a,b) study of newborn garter snakes (*Thamnophis elegans*) showed that acceptance of leeches and acceptance of slugs evolve in tandem due to a positive genetic correlation. Hence, among the very small number of taxa that have been studied for correlations in preference hierarchies, there is an astonishingly broad range of potential correlations that could influence the degree of specialization found in species and the potential for evolutionary shifts in preference among potential hosts and prey.

Preference and Performance

Preference is only part of the problem in the evolution of specialization in actively searching animals. All else being equal, natural selection should favor those individuals that preferentially choose hosts that are nutritionally best for survival and growth. Nevertheless, the results of an increasing number of studies in recent decades have shown a much more complex relationship between preference and performance and how it affects the evolution of specialization. 'Performance' here is the sum total of survival, growth, and reproduction that composes Darwinian fitness.

In the *Papilio machaon* group of swallowtail butterflies, some of the genes affecting larval performance on hosts are different from the genes affecting oviposition preference. Unlike the genes for oviposition preference, there is no evidence that the X chromosome is involved in at least some aspects of larval performance in *P. zelicaon* and *P. oregonius* (Thompson, Wehling, and Podolsky 1990). The evolution of specialization in these species must therefore involve the coordination of different groups of genes expressed in the adults and larvae. Similar differences in the genetics of adult preference and larval performance are appearing in related studies of other groups of swallowtails (Scriber 1986a; Nitao et al. 1991; Scriber, Giebink, and Snider 1991).

These genetic results imply that in some organisms, the evolutionary relationship between preference and performance may remain somewhat variable within species. A close relationship between oviposition preference and particular components of larval performance has been shown for a few species of insects (e.g., Via 1986; Singer, Ng, and Thomas 1988), although the cause of these correlations has not been shown conclusively to have a genetic basis. In most other species in which such correlations have been sought—including butterflies, moths, aphids, beetles, flies, and mites—the results have been more variable (reviews in Thompson 1988b and in Jaenike 1990). This does not mean that close genetic relationships between adult preference and larval performance do not occur. To some extent they must if a lineage is to be successful. The results from *Papilio,* however, suggest that independent inheritance of preference and performance genes, and the different genetic bases among components of performance, may give rise to a whole spectrum of evolutionary relationships between the ranking of hosts and overall performance both within and among populations.

Trade-offs in Performance

Another genetic approach to the evolution of specialization has been to search for trade-offs in performance on different hosts. The assumption is that a jack-of-all-trades is a master of none. The problem, however, is to decide which components of performance are the most important for Darwinian fitness and most likely to be the focus of evolutionary trade-offs. As Rausher (1988) has pointed out, studies of performance often draw ecological and evolutionary conclusions from measurements of only one or a few components of performance.

Recent studies have shown that specialization in performance can evolve quickly in populations under some conditions, although it is not clear whether genetic trade-offs are directly involved. The nematode *Howardula aoronymphium* parasitizes two species groups of mushroom-breeding *Drosophila* in North America, including four species in the eastern United States (Montague and Jaenike 1985). Jaenike (1993) took one strain capable of attacking all four *Drosophila* species and cultured it on *D. falleni* exclusively for almost three years, during which time it lost the ability to attack the other three hosts. Jaenike interpreted the rapid evolution of specialization in this nematode as potentially a result of genetic drift on loci affecting infection of the other three hosts. One alternative explanation would be natural selection against alleles allowing attack of multiple hosts if those alleles result in slower growth rates, lower reproductive rates, or reductions in other components of fitness compared with alleles allowing attack of only one host. The natural selection hypothesis assumes that there are trade-offs in fitness in attacking two or more hosts rather than one, whereas the genetic drift hypothesis assumes that alleles allowing attack of two or more hosts are neutral in the laboratory culture.

When trade-offs are involved in the evolution of specialization, they could arise from any of a number of genetic causes. The simplest would be a single locus with alternative alleles allowing attack or use of either one species or another (e.g., single alleles responsible for initiating different specific detoxification pathways). Somewhat more complicated would be several loci all affecting one overriding aspect of performance. A trade-off would then be expressed (depending upon several conditions, including additivity of genes) as a negative correlation in performance on alternative hosts or prey.

Yet another possibility is what I will call the coadapted genes hypothesis, which I develop here as one way in which the interaction among different components of performance could maintain a range in the ability to use two or more host or prey species even within populations of extreme specialists. Assume that adaptation to two hosts or prey is at least partially genetically independent. At least some genetic loci affect an individual's ability to attack one host or prey but have no affect on performance on other hosts or prey. In addition, some other genes may affect performance on both hosts or prey. Some of those genes may assist individuals in attacking both species, whereas others may allow more successful attack of one species but at the expense of the ability to attack the other species. Individuals will differ in the combinations of independent and nonindependent performance genes. As a result, anything from a strongly positive to a negative correlation would occur within a population for any single component of performance, depending upon how these different genes are distributed among individuals.

If a population is encountering a novel host or prey, however, a negative correlation in performance on the novel, as compared to the normal, host should be uncommon for any single component of performance. The reason stems from the way in which variation in performance is likely to be distributed within the population. Natural selection for performance on the normal host or prey should have eliminated, over evolutionary time, at least the most deleterious genes. Nevertheless, a range in performance will still occur because some deleterious genes may have beneficial effects on traits unrelated to host use (pleiotropic effects). In addition, there may be a small number of individuals that perform poorly on their normal host as a result of new deleterious mutations or chance inbreeding effects, but overall most individuals will perform fairly well. As a result, there will be a range of outcomes in performance even on the normal host, although extremely poor performance should be rare.

In contrast, the population should show an initial distribution of performance on the novel host that reflects fortuitous genetic variation for the ability to use that host (fig. 5.11). Some genotypes may by chance perform poorly on both hosts due to inbreeding, new mutations, or incompatible combinations of loci, whereas others, also by chance, may perform relatively well, although probably not quite as well on the novel host as on the normal host. Hence, the range in outcome on both the normal and novel hosts may be broad, but very poor performance on the normal host and very good performance on the novel host should be rare.

This is the result that occurs when families of the extreme specialist *Papilio oregonius* are split into two groups, one reared on their normal host (*Artemisia dracunculus*) and the other on a novel host (*Lomatium grayi*) (fig. 5.11). Most families fare very well on the normal host but die on the novel host. (The cluster of lines with high survivorship on the normal host and low survivorship on the novel host hides the separate lines for some families.) The remaining families exhibit a range of abilities to survive on these two hosts.

Under these genetic conditions, specialization driven by trade-offs would result not from a strong negative correlation in any one component of perfor-

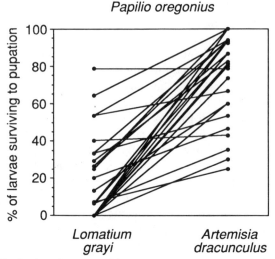

Fig. 5.11. The distribution of survivorship among *Papilio oregonius* families on a novel host, *Lomatium grayi* (Umbelliferae), compared with that on their normal host, *Artemisia dracunculus* (Compositae) (*n* = 27 families reared on both hosts). The cluster of lines with high survivorship on the normal host and low survivorship on the novel host hides the separate lines for some families. For each family, at least 10 larvae were fed separately on each plant species. In all, there were 400 larvae on *L. grayi* and 399 on *A. dracunculus*. All larvae were reared in a growth chamber set at 23° C and 16 hr. of light each day. Newly hatched first-instar larvae were placed in a 150 × 25 mm Petri dish with a vegetative sprig of one of the plant species. Each sprig had its base immersed in a florists' water pick. After the second instar, all larvae were placed in separate dishes. Each larva was checked daily, and food was changed every 2 days or more frequently, if needed. The positions of the Petri dishes within the environmental chamber were changed daily to minimize position effects. From Thompson, unpublished data.

mance but rather from the problem of favoring simultaneously the many different genes for performance needed to attack two different hosts. That is, trade-offs will occur when natural selection is not able to coordinate all the separate loci needed to maximize fitness on both hosts simultaneously. Any single component of performance may not show any evidence of a trade-off, as indicated by the results for survival of *P. oregonius* on a normal and on a novel host. The trade-off arises in the coordination of the many loci across many components of performance into a coadapted gene complex.

This view of the evolution of trade-offs suggests that the evolution of specialization is a combinatorial process. Some combinations of genes may allow less specialization to one host or prey, whereas other combinations of genes may demand more. The result could be differences in degrees of specialization even among populations and closely related species. Although this particular hypothetical view may not be correct, it highlights our need to evaluate more closely the actual genetics of trade-offs that create specialization and yet appear to allow the possibility of shifts onto new hosts.

Conclusions

The complex molecular genetics of symbiosis between legumes and rhizobia, the genetics of preference hierarchies in animals, and the genetic relationships between preference and performance all suggest that at least a narrow range of specialization will probably be common among groups of related populations and species, even when natural selection favors specialization. The sequence of steps in the genetics of symbiosis may not often allow a symbiont genotype to invade a wide range of hosts, but it does not necessarily limit it to one host either. One or a few mutations can sometimes allow an expansion in the number of host species that a symbiont can successfully invade, thereby opening the opportunity for geographic differences in the range of hosts involved in the symbiosis.

In a different way, the genetics of preference in animals often establishes a hierarchy of potential hosts or prey rather than absolute specialization to one species. Such hierarchies lead to geographic differences in the degree of specialization found within local populations. The imperfect genetic linkage between preference for hosts or prey and the suitability of those victims for survival, growth, and reproduction can maintain variation in how natural selection acts on specialization. The evolution of trade-offs in performance, which favors specialization, may itself sometimes result from the problem of coordinating many gene loci affecting different components of performance on different hosts (the coadapted genes hypothesis) rather than from trade-offs in any single component of performance. If so, then different combinations of genes will favor different degrees of specialization. Natural selection can act to narrow or broaden the distribution of specialization within and among species, but the complexity of the genetic processes involved will probably often maintain at least some range of specialization among individuals, populations, and closely related species.

CHAPTER SIX

Ontogeny of Specialization

The ontogeny of organisms provides a wide variety of ways in which the genetics of specialization can be expressed. The life histories of some organisms are partitioned into discrete stages, and genes for specialization to different species are expressed sequentially as individuals proceed through each stage. The life histories of other species favor developmental polymorphisms, in which individuals from a population differ in the species on which they specialize. The ability of the genetic structure of organisms to compartmentalize specialization through events during ontogeny increases the opportunities for one species to coevolve simultaneously with two or more species. It also increases the chance that different populations of species will evolve to specialize on different combinations of other species, adding to the geographic structure of evolving interactions.

In this chapter I examine a variety of ways in which organisms partition specialization in their interactions during ontogeny. These ontogenetic changes in specialization illustrate how even a single population can evolve to specialize on—and potentially coevolve with—more than one other species simultaneously.

We characterize the ontogeny of organisms according to a sequence of life history events: growth, timing and mode of reproduction, and senescence. But we can just as readily, and just as importantly, characterize the ontogeny of most organisms as a sequence of interactions with other species. At different stages of a life history, individuals of many species often specialize on different taxa. A butterfly larva may chew the leaves of one plant species, whereas the adult sucks the nectar of other plant species. A predatory fish may eat aquatic insects early in life and switch to feeding on other fish as it gets larger. A plant may produce thorns to protect itself against grazers when it is young and small but not when it is large (e.g., juvenile vs. adult tissues of black locust, *Robinia pseudoacacia*). This compartmentalization of interactions throughout development is so common that we take it for granted, yet it is a major part of how organisms evolve to specialize in their interactions with one another.

The Extreme: Host Alternation in Parasites

Parasites that alternate between different host species are the most extreme examples of how evolution can favor organisms that specialize on two or more species by partitioning that specialization to different times during ontogeny. A large portion of parasitology and pathology courses is taken up in describing the bizarre sequence of host alternation found in many parasitic species. But host alternation is not simply an esoteric side topic in the study of specialization and coevolution. Thousands, if not tens of thousands, of species have complex life cycles that link widely divergent taxa within most biological communities. There are about one thousand acanthocephalan species alone, all of which alternate between an arthropod intermediate host and a final vertebrate host (Conway Morris and Crompton 1982). About twenty-seven hundred aphid species alternate hosts, and several hundred of these alternate between hosts in different plant families (Eastop 1986). Among other invertebrates and fungi, host-alternating species include many of the most virulent known parasites of animals and plants (table 6.1), and the selection pressures that they place on both (or all) their host populations must often be quite intense.

Not all of these interactions necessarily involve coevolution between the parasite and its hosts. They illustrate, however, one potential way in which coevolution may involve one species coevolving simultaneously with two or more other species, sometimes even differing geographically in the combination of hosts used.

If the alternations were simply between closely related hosts, then the evolution of these complex life cycles would not seem so strange. But the hosts used by host-alternating species are often so distantly related that the most that can be said about their taxonomic similarity is that they are both animals or they are both green. For example, aphids in the subfamily Melaphidina alternate between sumac (*Rhus*) and mosses (Eastop 1986). The rust fungus *Puccinia graminis,* which has different races that attack different cereal crops, alternates between grasses and barberry (*Berberis*) or *Mahonia* (Luig 1983), and some primitive rusts alternate between firs (*Abies*) and ferns (Savile 1971). Some protozoa that cause human diseases (*Plasmodium, Trypanosoma, Leishmania*) have biting flies as intermediate hosts (Anderson and May 1991), and schistosome flukes have molluscs as intermediate hosts (Southgate and Rollinson 1987).

The multiple independent origins of these complex life cycles in organisms as phylogenetically distinct as protozoans, trematodes, pathogenic fungi, and phytophagous insects strongly indicate that the evolution of host alternation is not a fluke of the odd biology of one group of parasites. It is a

Table 6.1. Major parasite taxa in which at least some species alternate hosts

Example	Hosts	Transmission to definitive host	References
Viruses			
Yellow fever virus	Humans/*Aedes* mosquitoes	Inject	Anderson and May 1991
Maize mosaic virus	Maize/*Peregrinus maidis* planthopper	Inject	Greber 1984
Rickettsiae			
Q fever (*Coxiella burnetti*)	Humans/*Amblyomma* and *Dermacentor* ticks	Inject	Anderson and May 1991
Rocky Mountain spotted fever (*Rickettsia rickettsii*)	Humans/*Dermacentor andersoni* ticks	Inject	Burgdorfer 1984
Spiroplasmas and *Mycoplasma*-like organisms (MLOs)			
Aster yellows agent	(*Aster/Macrosteles fascifrons*)	Inject	Sinha 1984
Maize bushy stunt MLO	Maize/*Dalbulus* leafhoppers	Inject	Nault 1980
Rust fungi			
Wheat stem rust (*Puccinia graminis* f. sp. *tritici*)	Wheat/*Berberis* spp	Self	Luig 1983
Sweet fern rust (*Cronartium comptoniae*)	*Pinus* spp/*Comptonia peregrina* (sweet fern) and *Myrica gale*	Self	Hunt and Van Sickle 1984

Protozoa			
Malaria (*Plasmodium falciparum*)	Humans/*Anopheles* spp. mosquitoes	Inject	Yaeger 1985
Trypanosomiasis (*Trypanosoma brucei*)	Humans/*Glossina* spp. tsetse flies	Inject	D'Alessandro-Bacigalupo 1985
Leishmaniasis (*Leishmania donovani*)	Humans/*Phlebotomus* and *Lutzomyia* sand flies	Inject	Anderson and May 1991
Coccidiosis (*Sarcocystis suihominis*)	Humans/pigs	Ingest	Yaeger 1985
Flatworms			
Schistosomiasis (*Schistosoma mansoni*)	Humans/*Biomphalaria* snails	Self	Malek 1985
Bertiella obesa	Koalas/oribatid mites	Ingest	Beveridge 1982
Pork tapeworm (*Taenia solium*)	Humans/pigs	Ingest	Little 1985
Nematodes			
River blindness (*Onchocerca volvulus*)	Humans/*Simulium* blackflies	Inject	Orihel 1985
Hexametra angusticaecoides	Boas/chameleons	Ingest	Sprent 1982
Acanthocephalans			
Echinorhynchus salmonis	Yellow perch/*Pontoporeia affinis* and *Pallasea quadrispinosa* amphipods	Ingest	Tedla and Fernando 1970
Aphids			
Pemphigus betae	*Populus* spp./herbaceous roots	Self	Moran 1991

Note: Transmission to the definitive host is by injection by the vector host (inject), ingestion by the definitive host (ingest), or self-movement by wings, wind, or water (self).

common consequence of the parasitic lifestyle (Price 1992). Nevertheless, alternation of hosts does not occur in all groups of parasites. Among fungi, alternation between very different hosts is restricted almost entirely to rusts (Savile 1976), and among insects it is restricted to aphids (Moran 1991). Hence, it seems that there must be some combinations of life history and ecological conditions that have allowed some parasitic groups, but not others, to adopt alternation between different hosts. A few common patterns do in fact stand out. Many of the taxa that alternate hosts have the ability to undergo asexual reproduction on one or both hosts. Moreover, many of these taxa undergo asexual reproduction usually on one host (the intermediate host) and sexual reproduction on the other (the definitive host). Patterns of reproduction in some individual species are much more baroque than this simple division between intermediate and definitive host, but this general partitioning of hosts and modes of reproduction is common.

Another common aspect of host alternation is that the interactions seem always to be parasitic rather than mutualistic. I have found no cases in which a mutualistic symbiont regularly alternates between host species. This restriction of host alternation to parasitism may result from two causes. One is that host defenses or variation in host quality may be an important driving force for host alternation. In mutualistic interactions, natural selection on both the host and the symbiont should favor retention of the association as long as it is beneficial to both. The other implication is that host alternation may be partially linked to the partitioning of sites most appropriate for growth and reproduction. Unlike parasites, many mutualistic symbionts have forsaken sexual reproduction altogether (Law and Lewis 1983).

But why should any species alternate hosts, specializing on very different hosts during different phases of ontogeny? By comparing the life cycles of different parasite taxa and the ecological conditions in which they occur, three potential major routes to specialization on two or more hosts through host alternation become evident. All these routes provide the opportunity for geographic differences in hosts used within a species during development.

Host alternation via injection by a grazing vector. The first route is the use of an intermediate host to increase efficiency of transmission between final hosts. Hence, some viruses, rickettsiae, protozoans, and nematodes are transmitted between vertebrate hosts via biting flies, and other viruses, spiroplasmas, and *Mycoplasma*-like organisms are transmitted between plants by leafhoppers and other homopteran insects. In such cases, it is easy to understand how the use of an intermediate host could be selectively advantageous to the parasite: it is hitching a ride on a mobile species that is actively searching for the same host as the parasite. Consequently, it is not surprising that the intermediate vectors are generally species with piercing mouthparts that feed as grazers, moving between two or more hosts during their lifetimes and acting as mobile hypodermic syringes. The vectors used by these parasites sometimes vary geographically, as in the malarial parasite *Plasmodium faliciparum,* which is transported by different mosquito species in different geographic areas.

Host alternation via predation. The second evolutionary route occurs in parasitic species that pass directly from intermediate to definitive hosts up through a food chain, transferring, for example, from a herbivore to a predator of the herbivore and, sometimes, even to a predator of the predator. Some flatworms (platyhelminths), roundworms (nematodes), protozoa, and all acanthocephalans have adopted this kind of host alternation. It is an ingenious means by which parasites have repeatedly exploited the process of predation to their own ends: the ecological problem of having your host, and incidentally you, killed by a predator becomes instead an evolutionary opportunity. Attack by some parasites in fact alters the behavior of the host in ways that appear to enhance transmission to the alternate host (Dobson 1988). For example, mud snails *Ilyanassa obsoleta* infected with trematode parasites are more likely to remain in upper shore habitats during nighttime low tides than uninfected snails. In these upper shore habitats, the snails are prone to predation by the trematode's second intermediate hosts, which are semiterrestrial crustaceans such as amphipods and fiddler crabs (Curtis 1990).

It is not surprising, then, that in parasites that rely upon ingestion of their intermediate host, the final host is generally a vertebrate, and the particular host most commonly used can vary among populations. In acanthocephalans and ascaridoid nematodes, the definitive host is often a vertebrate predator at or near the top of a food web (Conway Morris and Crompton 1982; Sprent 1982). Moreover, these parasites have radiated in species throughout the vertebrates, infiltrating the upper reaches of a great many food webs and producing some wonderfully novel couplings of hosts. On Madagascar, for example, the ascaridoid *Hexametra angusticaecoides* occurs in both chameleons and their predators, Madagascar boas (Sprent 1982). In studying host alternation among ascaridoids, Sprent (1983) found four patterns, all of which involve movement up through two or three trophic levels before eggs are then deposited back into the environment, where the cycle begins again: invertebrate host to vertebrate host; vertebrate host to vertebrate host; aquatic invertebrate host to fish to fish-eater (e.g., crocodiles, pelicans); and terrestrial invertebrate host to small predator to large predator.

Although most ascaridoids alternate hosts, a smaller number, which are distributed over a variety of genera, do not (Sprent 1982). These exceptions provide a way of probing the ecological conditions that may favor the loss of complex life cycles and a change in the number of hosts on which populations specialize. Most of the species that do not alternate hosts attack herbivores that are probably often not susceptible to predation due to large size or inaccessibility: ox, hippopotamus, giant panda, land tortoise, and beavers, fruit bats, and mole rats. In these species, infection seems to occur either directly through ingestion of eggs or, in some species with mammalian hosts, through transmission from the host mother's placenta or milk to her offspring.

Independent host alternation. Host alternation via predation is evolutionarily similar to host alternation via a grazing vector in that in both cases the intermediate host is a transmission vector to the final host. There is a third route, however, in which the intermediate host is not a transmission vector. In these taxa, the parasites move via wings, wind, or water between their alternate hosts and, as in the other routes, the pattern of specialization can vary among populations. Aphids produce winged generations that fly between their summer and winter hosts, rusts often alternate hosts via the wind, and flukes travel between hosts through water. The evolutionary origins of this kind of host alternation are less clear than for the other two routes. For all these taxa, the alternate host has often been considered to be a way of coping with periods when the normal host is unavailable or less suitable for growth and reproduction. Genotypes that make use of an alternate host during those times, reproducing asexually while on that alternate host, will be favored over genotypes that remain on the normal host or become dormant. For aphids alone, at least five adaptive hypotheses have been proposed for alternation of hosts (Moran 1988), and all of them are versions of this general explanation. To some extent this overall hypothesis must be true, although as Savile (1976) wrote, referring to rust fungi, it is a "desperately dangerous device."

To argue, however, that alternation of hosts may originate as an adaptive response to the problem of either transmission between hosts or temporary unavailability/unsuitability of one host is not the same as arguing that host alternation is currently optimal and maintained by natural selection in every species that exhibits it. Sequential specialization on two or more hosts during development is often accompanied by the evolution of a separate morphology, physiology, and behavior to cope with each host. Once acquired, it may not be easy to lose. Taking an extreme view of such constraints, Moran (1988) has argued that host alternation in aphids occurs precisely because those species have not been able to abandon one of the morphs and switch completely from their ancestral woody hosts to herbaceous hosts. The morph emerging in the spring on woody plants from sexually produced eggs (the fundatrix) is highly specialized and quite different from the morph that develops through a series of asexual generations on herbaceous plants throughout the summer, ending with a sexual form that flies back to the woody host (fig. 6.1). The fundatrix is almost always wingless, has a specialized morphology that includes a large abdomen and short antennae and legs, and pro-

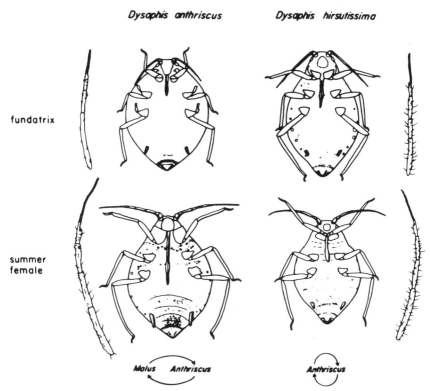

Fig. 6.1. The fundatrix morph and the summer female morph of a host-switching aphid species and a congeneric species that does not switch hosts. Reprinted, by permission, from Moran 1988.

duces more ovarioles than other morphs; it also often has specialized probing behaviors for initiating gall formation (Moran 1992). Moran (1988, 1992) has argued that dropping the woody host would demand two independent mutations: one altering oviposition preference for the woody host in sexual females, the other altering the fundatrix morph, which is specialized for asexual reproduction on the woody host in the spring. Those species that have succeeded in dropping the woody host have also lost the fundatrix morph.

Moran's argument, then, is that the fundatrix morph is a constraint that often blocks a complete shift to herbaceous hosts. She suggests that these life histories offer evidence for specialization as a dead end. The genetic changes, however, needed to drop the ancestral host and the fundatrix morph simultaneously may be less independent, and less unlikely, than Moran envisions. The study of heterochrony has shown that it is possible for species simultaneously to drop the second environment used by individuals during

development, abandon the specialized morph associated with that environment, and shift sexual reproduction to the environment in which it begins
development. That, for example, is what happens in some paedomorphic amphibians (Gould 1977, Semlitsch 1985). For aphids, the use of the woody
host and the associated morphology may be determined by a coadapted gene
complex capable of being turned off by a single genetic switch. That is probably an oversimplification, but it suggests one way in which the loss of one
host may not require the simultaneous appearance of two independent mutations suggested by Moran. Nonetheless, Moran's arguments are a construc-
tive counterbalance to the assumption that alternating hosts is currently adaptive in every species that currently exhibits it. Her arguments may partially
explain the patchwork distribution of host alternation within aphid taxa. Only
about 10% of aphid species, distributed among four of the sixteen subfamilies, alternate between hosts in different plant families; the remaining 90%
remain on one host species or a group of closely related host species throughout the year (Eastop 1986).

Inability to drop an ancestral host, however, is unlikely as a general explanation for host alternation in everything from aphids to rusts to schistosomes.
As in the evolution of all interspecific interactions, a combination of selection pressures and constraints causes the patchwork appearance of particular
forms of interaction in some taxa at a particular point in time. Neither selection nor constraint alone is likely to explain the whole pattern. Loss of one
host has happened repeatedly within taxa that usually alternate hosts, creating a patchwork geographic pattern in the number of hosts used by certain
taxa. These reduced life cycles (Istock 1967) indicate that species are not
always constrained to complex life histories once acquired and may respond
to selection against the use of two hosts. For instance, unlike populations
in more-temperate regions, Arctic populations of some rust fungi maintain
systemic infections within one host species rather than alternate hosts (Savile
1972; J. J. Burdon, personal communication). *Pemphigus betae* aphids alternate between *Populus* leaves and herbaceous roots where susceptible *Populus angustifolia* × *fremontii* hybrids occur but restrict feeding to herbaceous
roots at higher elevations where only the more resistant *P. angustifolia* occurs
(Moran and Whitham 1988; Whitham 1989; Moran 1991). The multitude of
complex life histories among parasite species appears to result from the combination of constraints on the kinds of life histories that permit alternation of
hosts, subsequent constraints on the elimination of a host once it is incorporated into a complex life cycle, the different uses of the alternate host (transmission vector vs. seasonal unsuitability of one host), and selection for retention of host alternation in some environments and for loss in others.

Host alternation is an extreme version of specialization to two or more
other species during a lifetime. These complex life cycles show clearly that

the genetic systems of organisms are capable of maintaining specialization (and potentially coevolving) simultaneously with two or more species. In some species, one of the hosts is dropped in some populations but not in others, thereby creating geographic differences among parasite populations in the number of hosts they use.

Complete Metamorphosis and the Ontogeny of Interactions

The other extreme way in which interactions (both interspecific and intraspecific) have been compartmentalized and specialized during ontogeny is through complete metamorphosis. During metamorphosis, individuals drastically change their morphologies, the species with which they interact, and even their mode of interaction (Wilbur 1980). Hence, amphibians often change from an aquatic tadpole form to a semiterrestrial adult form, some marine invertebrates have planktonic stages that differ greatly in morphology and lifestyle from adult stages, and holometabolous insects such as beetles, flies, hymenopterans, moths, and butterflies change from a larva into a morphologically distinct winged form. A host-specific leaf-chewing larva may become a host-specific nectar-sucking adult on a completely different plant species.

In some taxa, complete metamorphosis can be lost through the process of neoteny or progenesis, producing in some populations paedomorphic adults that are sexually reproductive yet juvenile in morphology (Gould 1977). In other taxa—for instance, some frogs in the genera *Liopelma, Arthroleptella,* and *Eleutherodactylus,* each of which is in a different family (Raff and Kaufman 1983)—the free-living larval morph has been completely abandoned and development proceeds directly to the adult morph. In yet other species, some populations may exhibit a mix of complex and simplified life histories. In some ambystomatid salamanders, paedomorphic and metamorphosed individuals co-occur, with the environment partially controlling the frequency of the morphs (Collins 1981; Semlitsch 1985; Semlitsch and Gibbons 1985). The coexistence of these different morphs within populations implies that, at least in some species, complete metamorphosis and the different interspecific interactions that accompany it are currently maintained by natural selection.

Complete metamorphosis, then, carries with it the possibility of genetic specialization on different species at different life history stages. Not all species that undergo complete metamorphosis are extreme specialists on different species at different stages in their lives, but this pattern of ontogeny allows the possibility. In some species specialization to different species can even be fine-tuned within an early stage of ontogeny. Remarkable examples occur in parasitoids in the wasp family Aphelinidae. At least nine genera of these wasps include species in which the ontogeny of males diverges in some ways from that of females (Walter 1983b). In all these species, females develop as endoparasitoids of Homoptera in four families (coccids, whiteflies, and related taxa). Males develop as ectoparasitoids on the same homopteran host species as the female, as hyperparasitoids on other parasitic wasps (sometimes, but not always, on the same homopteran host species used by the females), or as endoparasitoids of lepidopteran eggs. There is some evidence that developing males and females are differentially adapted to the defense responses of their respective hosts (Walter 1983a). These divergent ontogenies of males and females are another, albeit extreme, example illustrating that through ontogenetic specialization one species may simultaneously specialize on—and potentially coevolve with—more than one species.

For both complete metamorphosis and alternation of hosts, natural selection on specialization and the process of coevolution at one stage of these complex life histories could affect selection on specialization and coevolution at the other stages. Unfortunately, very little is known about how selection and constraints on alternating hosts, or constraints imposed during complete metamorphosis, influence evolving interactions. One possibility is that completely different blocks of genes turn on at these different stages, such that selection on the interactions is at least semi-independent. That is, a developmental switch channels individuals at different stages of different generations into different morphologies specialized for different kinds of habitats and interactions (Istock 1967; Wilbur 1980; West-Eberhard 1983). Feeding on one host may trigger activation of one block of genes rather than another. Or the genes controlling the choice of nectar hosts in butterflies may have no effect at all on the genes controlling use of larval hosts.

Moran's (1991) results, however, on *Pemphigus betae* aphids suggest that in some species genetics of adaptation and specialization at one stage of life may influence other stages. Clones of *P. betae* that alternate between *Populus* leaves and herbaceous roots grow more slowly, have higher mortality, and mature at a smaller size (an index of potential fecundity) in the laboratory than clones that have deleted the *Populus* phase. Similar results were obtained in field populations (Moran and Whitham 1988), indicating that the results are not an artifact of the laboratory environment favoring clones restricted to herbaceous roots. This difference between clones suggests that the genes controlling these different phases are not completely independent. Instead, adaptation to *Populus* has a negative effect on adaptation to herbaceous roots; that is, the genes for adaptation to these different hosts show antagonistic pleiotropy.

Alternation of hosts and complete metamorphosis are both solutions to the problem of having the most appropriate morphology and physiology for each stage in a life history. Both lifestyles offer the opportunity, through ontogenetic compartmentalization, of specialization to more than one other species during a lifetime. As a result, some species have at least the potential to coevolve with two or more other species, partitioning the genetics of their interactions into discrete stages within a life cycle.

Ontogeny and Specialization without Complex Life Cycles

Even without complete metamorphosis or alternation of hosts, individuals of some species interact with different species as they get older and larger. Werner and Gilliam (1984) have called this sequence of changes in interspecific interactions the ontogenetic niche and have cataloged a long list of examples, including pinfish (*Lagodon rhomboides*) and some turtles that change during development from carnivory to herbivory, and zooplankters that switch from herbivory to carnivory. Some of these changes are gradual and small relative to the extreme shifts found in host-alternating parasites or species with complete metamorphosis. In areas of high water salinity, for instance, juvenile common eiders *Somateria molissima* eat mostly taxa with relatively low saltwater content such as crustaceans and gastropods and gradually include more-saline bivalves only as they get older (Nyström, Pehrsson, and Broman 1991). During the first several days after hatching, the chicks of grey partridge *Perdix perdix* and red-legged partridge *Alectoris rufa* in Britain eat a combination of arthropods, seeds, and leaves, with arthropods making up half or more of the diet. During subsequent weeks the proportion of plant material in the diet gradually increases, and by three weeks of age the diets of both species average more than 80% plant material (Green, Rands, and Moreby 1987). These kinds of ontogenetic changes in diet occur in many species. It is possible that natural selection on diet specialization during the remainder of life is sometimes constrained by the requirements of a different diet in the early weeks of life.

In other species, different individuals within a population may follow different ontogenetic pathways in specialization. Individuals may be similar at birth but later, either rapidly or gradually, come to differ in morphology and specialization during development. In recent decades, one of the most striking results for the ontogeny of specialization in predators has been the demonstration that some species are either genetically polymorphic or ontogenetically plastic in jaw morphology. Moreover, the number of morphs may vary geographically.

Cichlid fishes have been used traditionally as one of the best examples of trophic specialization in predatory vertebrates, on the assumption that species have fixed differences in jaw morphology. That assumption is now known to be false for at least several species (Greenwood 1965; Kornfield et al. 1982; Meyer 1987, 1990a,b). In the Neotropical species *Cichlasoma citrinellum,* for instance, all individuals begin development with a papilli-form morph, in which the pharyngeal jaws are gracile, the teeth are slender and pointed, and the overall jaw and tooth structure is effective if the fish are feeding on soft diets (Meyer 1989). During ontogeny, however, some individuals develop into a molariform morph in which the jaws are more robust and the teeth are strong and rounded (fig. 6.2). This molariform morph is more effective at crushing hard prey such as snails (Meyer 1990b). Analysis of stomach contents shows that almost all molariform individuals eat snails, whereas few papilliform individuals include them in their diet (Meyer 1990a). In a related species, *Cichlasoma managuense,* different morphs can be induced by different diet early in development. Fish fed from birth on hard food rather than soft food have different jaw morphs by 8.5 months, which they retain if kept on those diets. If at 8.5 months they are given the same diet, they converge in jaw morphology by 16.5 months (Meyer 1987).

Trophic polymorphisms are increasingly being found in other taxa as well. The goodeid fish genus *Ilyodon* includes morphs once considered possibly to be separate species (Grudzien and Turner 1984). Some of the increasing

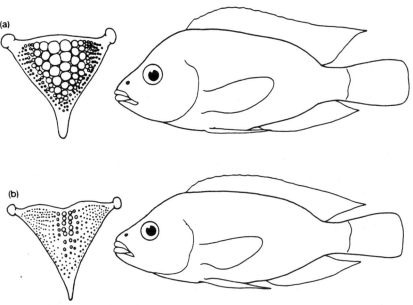

Fig. 6.2. Differences in body shape and pharyngeal-jaw morphology in *Cichlasoma citrinellum:* *a,* benthic form with molariform jaw morphology; *b,* limnetic form with papilliform jaw morphology, smaller jaw size, larger snout, narrower head, smaller eye, and thinner body. Pharyngeal jaws are shown from above. Reprinted, by permission, from Meyer 1990b.

number of known polymorphisms appear to result mostly from the effects of diet or growth early in ontogeny, as in the development of snail-crushing morphs in pumpkinseed sunfish, *Lepomis gibbosus* (Wainwright, Osenberg, and Mittelbach 1991), possibly the development of different foraging tactics in bluegill sunfish, *Lepomis macrochirus* (Ehlinger 1990), and the development of carnivorous, rather than omnivorous, morphs in tadpoles of spadefoot toads, *Scaphiopus multiplicatus* (Pfennig 1992). Others appear to have a combined genetic and ontogenetic basis. Arctic char (*Salvelinus alpinus*) develop recognizable trophic morphs, and in Iceland up to four morphs occur sympatrically (Skúlason, Noakes, and Snorrason 1989). The differences among morphs are discontinuous, so the designation of morphs is not simply the result of arbitrarily creating groups from a continuous distribution of phenotypes. Four morphs coexist in Thingvallavatn, Iceland's largest lake (fig. 6.3). Two of these morphs feed extensively in open water and differ in size: one feeds primarily on fish, especially three-spined sticklebacks (*Gasterosteus aculeatus*), and the other on zooplankton. The other two morphs are benthic feeders, preying mostly on molluscs (Malmquist et al. 1992). The two limnetic morphs may constitute one breeding population that separates trophically due to diet differences early in ontogeny. The two benthic morphs, however, differ in both growth rate and time of spawning and may be partially or completely reproductively isolated at least in some lakes (Sandlund et al. 1992).

In Norway's Salangen River system, where there are three morphs, matings within a morph can produce offspring of all three morphs (Nordeng 1983). The morphs are very closely related genetically, implying incomplete or very recent reproductive isolation even in those morphs that spawn at different times of the year. Different morphs within the same lake in Norway are more genetically similar to each other than the same morphs collected from different lakes (Hindar, Ryman, and Ståhl 1986), and the four morphs in Thingvallavatn, Iceland, have Nei's genetic distances of only 0.00004 to 0.00126 (Sandlund et al. 1992). Hence, the development of trophic morphs in Arctic char appears to have a complex mix of causes that range from phenotypic plasticity to genetically based polymorphisms to at least partial reproduction among some morphs in some populations.

Even more extreme polymorphisms resulting in specialized interactions occur in social insects. One level of specialization occurs in the caste system of insects such as ants and termites, which produce individuals specialized for foraging, others specialized for defense against enemies, and still others specialized to perform duties within the colony (Wilson 1975; Hölldobler and Wilson 1990). An additional level occurs among foragers of some species. Some of that specialization occurs through learning, as in the "majoring" of individual bumblebees from the same nest on flowers of different

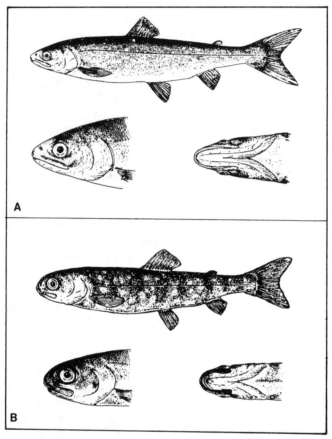

Fig. 6.3. Two of the four trophic morphs of Arctic char (*Salvelinus alpinus*) found in Thingvalla-vatn, Iceland's largest lake: *A,* the small limnetic form; *B,* the small benthic form. Reprinted from H. J. Malmquist, 1992, "Phenotype-specific Feeding Behaviour of Two Arctic Charr *Salvelinus alpinus* Morphs," *Oecologia* 92:354–61, by permission of Springer-Verlag.

species (Heinrich 1979), but some may also come from either genetic or on-togenetic differences among individuals within the same colony. Honeybee (*Apis mellifera*) colonies are commonly made up of a dozen or more subfami-lies of sisters, all with the same mother but with different fathers (Laidlaw and Page 1984). Oldroyd, Rinderer, and Buco (1992) established colonies with two identifiable subfamilies and recorded the plant genera collected by workers of these different subfamilies. They found that in four of the six experimental colonies, bees from different subfamilies majored on different plant genera. The differences in specialization that they found could be due to actual genetic differences in the floral preferences of bees from different fathers, or the differences could be due to genetic differences in the growth of bees from different fathers. As one possibility, bees from some fathers may grow larger than bees from other fathers under the environmental conditions within a single hive and, as a consequence, come to prefer flowers with different corolla depths or sizes. Intraspecific differences in floral preference due to bee size are known to occur in bumblebees. Given a choice, bumblebees with larger bodies and longer tongue lengths choose different flowers than smaller bees with shorter tongues (Barrow and Pickard 1984; Johnson 1986). Hence, at least in some eusocial insects, the spectrum of flower species chosen may reflect the range of differences in growth during development or genetically based differences in preference.

Conclusions

Alternation of hosts during ontogeny, complete metamorphosis, genetically based trophic polymorphisms, and ontogenetically induced trophic polymorphisms are all ways by which organisms compartmentalize and specialize in their interactions with other species during development. They illustrate several points about the evolution of specialization. First, as the different stages of these life histories are expanded, contracted, or even eliminated, the number and kinds of species that an individual is likely to encounter change, and so do the opportunities for specialization. Hence, small developmental shifts in life histories could have potentially large effects on the opportunities for specialization. Second, the complex life histories found in some of these species emphasize that even in organisms that are often recorded as interacting with many or widely divergent taxa, the interactions are often much more specialized than they appear at first glance. Third, these different life histories show that even a single population can evolve to specialize on—and potentially coevolve with—more than one other species simultaneously. Finally, the sequence of host species used during development and the number of morphs expressed can vary geographically within a species.

PART III

NATURAL SELECTION AND THE GEOGRAPHIC STRUCTURE OF SPECIALIZATION

CHAPTER SEVEN

Why Parasitism Is Special

The previous three chapters used studies of phylogeny, genetics, and ontogeny to suggest that specialization is an evolutionarily dynamic feature of populations and species rather than a static dead end. These studies suggest that a range of specialization exists within most phylogenetic lineages, and specialists are not always the terminal branches. Rather, specialization can increase or decrease, become compartmentalized during ontogeny, or become polymorphic among individuals within populations. Moreover, the patterns of specialization can differ among populations over evolutionary time as they encounter new species. These changes occur as the combination of genetic variation within and among populations and differing ecological conditions creates a spectrum of opportunities for specialization within most phylogenetic lineages. The resultant range of specialization that we observe within lineages, however, differs among interactions, and the following five chapters examine why extreme specialization is more common in some kinds of interaction than in others. These differences among interactions in the pattern of specialization influence the process of coevolution.

Most species are specialized to interact with only a few other species. That certainly is not the impression one gets from thinking only about birds and mammals and trees, but these are the showy exceptions in the evolution of diversity—and even some of these groups include extreme specialists. Most of the diversity of life is made up of pathogenic and mutualistic fungi, bacteria, viruses, rickettsiae, spiroplasmas, mycoplasmas, parasitic plants, parasitoids, acanthocephalans, nematodes, trematodes, phytophagous insects, and mites. Among these taxa, extreme specialization is not only common, it is the norm. Many of the extreme specialists within these taxa share a common lifestyle in which individuals complete development, or an entire stage of a complex life history, on a single host individual and often cause some detrimental effect to host fitness. They are parasites, and the single broadest generalization that seems possible about the evolution of extreme specialization is that it appears to be most common in parasitic species.

In a landmark book on the evolutionary biology of parasites, Price (1980) showed that the parasitic lifestyle produces evolutionary patterns that transcend taxonomic boundaries. Not just the traditional parasites of parasitology and pathology—trematodes, protozoa, rust fungi, and the like—but also species such as insects that complete development on a single host plant or animal share some common aspects of life history, population biology, and speciation. Extreme specialization extends to commensals and mutualistic symbionts that live on a single host individual (Thompson 1982), but it is in parasites that the pattern is most evident.

In this chapter I explore two aspects of specialization in parasites that are important foundations for the study of coevolution: the selection pressures that make extreme specialization especially common in parasitic taxa, and the geographic structure of specialization that is being discovered increasingly within parasitic taxa. I begin with parasites for three reasons: they exhibit the most common lifestyle on earth, they exhibit the greatest degrees of extreme specialization, and they affect—one way or another—nearly all other living species. In considering specialization in parasites, I have cast my net broadly to include any species that completes an entire stage of development on a single host individual and is likely to cause some decrease in fitness of the host, at least under some ecological conditions. Most of my examples are insects, because they are the most diverse form of life and allow the greatest number of comparisons in evaluating the effects of the parasitic lifestyle on the evolution of specialization.

The Problems of the Parasitic Lifestyle

For all parasites, the overall problem is the same: an individual must either meet all its nutritional demands and avoid all its enemies on that one host or it dies. Unlike free-living organisms, a parasite does not have the option of leaving its initial host and picking and choosing among other hosts. It must become securely attached to a host, it must face physiological, sometimes induced, changes in its host, and it must cope with potential predators, competitors, and its own parasites without leaving the host. Consider how this sequence of challenges faced by a parasite differs from those faced by free-living species. First of all, the seemingly simple problem of attaching safely to a host for a long period of time can be a major selection pressure for specialization (Southwood 1973). An internal parasite must be able to infiltrate host tissues, and an external parasite must be able to hold onto its host securely. For instance, two British aphids (*Tuberculoides annulatus* and *Myzocallis schreiberi*) that specialize on different oak species have trouble securely attaching themselves to each other's hosts because the leaves differ in texture (Kennedy 1986). Once attached, internally or externally, a parasite must then confront changes over time in its host's physiology. Over the period of time that a rust fungus or a caterpillar can take to complete development, a plant may change from a vegetative state, to flowering, and then to fruit production. During those changes, both the nitrogen content of plant parts and the concentrations of allelochemicals may vary tremendously. The umbelliferous herb *Lomatium dissectum,* for example, has 5% nitrogen in its floral buds, 4% nitrogen in the flowers once the buds have opened, and 3.5% nitrogen in the immature seeds (Thompson 1983d).

While feeding, parasites themselves induce other physiological responses in their hosts. Nearly all plants and animals damaged in any way respond somehow to the wound (Karban and Myers 1989), and parasites must cope with those responses regardless of whether the plant or animal is producing them specifically as a defense against that parasite. The vertebrate immune system, the invertebrate defenses of encapsulation and phagocytosis, and the induced defenses of plants are examples of the kinds of problems faced by parasites but not by free-living organisms (Harvell 1990b; Mitchison 1990). A free-living organism can nibble a few bites from a host and move on and is therefore less subject to host responses than a parasite. An indication of the importance of induced responses in the evolution of specialization in parasites is found in insect species that attack dying hosts or attack hosts in a way that prevents host physiological responses. Feeding on the dying or dead seems to allow for greater polyphagy than is often found in insects that complete development on a single plant individual. Ambrosia beetles that attack stressed or dying trees are generally polyphagous, whereas the few ambrosia beetles that attack living trees are highly host specific (Beaver 1989).

A similar pattern of specialization occurs in insect parasitoids, which lay their eggs in other arthropods. Some parasitoids paralyze or kill their hosts before ovipositing on them; others allow their hosts to continue development. Askew and Shaw (1986) found that these different tactics result in large differences in the number of families of leaf miners that a chalcidoid wasp can attack. Wasp species that paralyze or kill hosts at the time of oviposition attack 2.8 times the number of leaf-miner families as those that do not. An independent test within another family of parasitic wasps, the Ichneumonidae, has shown the same pattern. Wasps in the subfamily Pimplinae paralyze or kill hosts during oviposition, whereas Metopiinae do not. Pimplinae species in Canada average 10 host species distributed over an average of 7.6 genera, whereas Metopiinae average only 3 hosts distributed over an average of 2.3 genera (Sheehan and Hawkins 1991).

This same result occurs in phytophagous insects. Some species are able to prevent induced responses in their hosts by literally cutting off a plant's chemical defenses and feeding on dying leaf tissue and, just as in parasitoids, this tactic appears to allow higher levels of polyphagy than in insects that complete development on a single host. The best evidence comes from insects that interrupt the flow of secretory canals containing latex, gums, and resins. Such canals occur in the leaves of some thirty-five thousand plant species distributed over many plant families (Farrell, Dussourd, and Mitter 1991). By cutting quickly through the secretory canals and vascular bundles of a leaf or flower, and feeding only distal to the cuts, an insect releases the pressure in the secretory canals (Dussourd and Denno 1991) and prevents the flow into that section of leaf of any chemical defenses induced by feeding (Tallamy 1985). This trick of cutting off a plant's chemical defenses has evolved repeatedly in insects (Alexander 1961; Young 1978; Carroll and Hoffman 1980; Tallamy 1985; Dussourd and Eisner 1987; Dussourd and Denno 1991). The consequence for many of these species is the ability to feed on a wide range of unrelated plant species. Larvae of the noctuid moth species *Trichoplusia ni,* for instance, are able to feed on plants as different as carrots (Umbelliferae), dandelions (Compositae), and cucumbers (Cucurbitaceae) by trenching leaves and cutting off the induced defenses.

Even if all these problems of attaching and feeding on a single host are solved, a parasite attempting to complete an entire stage of development on a single host may be more susceptible to being killed by its own enemies— parasitoids, predators, competitors—on some hosts than on others. Hence, even if several hosts are equally good for attachment and growth, one host may be safest from enemies. This 'enemy-free space' can therefore narrow even further the range of hosts that results in highest fitness (Lawton and McNeill 1979; Price et al. 1980). Gilbert (1979) has used the phrase 'ecological monophagy' to describe the restriction of species to a single host for reasons other than host defenses, and a number of examples are now known from phytophagous insects (Smiley 1978; Atsatt 1981; Pierce and Elgar 1985; Gilbert 1991).

Extreme specialization has evolved so commonly in species that feed parasitically because the parasitic lifestyle produces a sequence of hurdles to interacting with multiple species not found in free-living species. For any one species, the major factor driving specialization toward one host may be long-term attachment to the host, induced host defenses, unbalanced nutrition, intake of large amounts of one or several toxic compounds, or avoidance of enemies. The past few decades have produced a large number of elegant and detailed studies showing how each of these factors can influence the choice of hosts. Nevertheless, directly or indirectly, all these studies show that for most extreme specialists, these specific problems are part of the general problem of completing development, or an entire stage of a complex life history, on a single host. At any moment in evolutionary time, or on any one host, one specific problem may be more important than others. The primary problem, however, is the parasitic lifestyle, which is expressed as different

Table 7.1. Relationship between continuous contact with hosts and extreme specialization in insects that feed as external parasites on birds and mammals

Parasite family	No. of species	Percentage of species restricted to:		
		1 host	2 or 3 hosts	> 3 hosts
Philopteridae	122	87	11	2
Streblidae	135	56	35	9
Oestridae	53	49	26	25
Hystrichopsyllidae	172	37	29	34
Hippoboscidae	46	17	24	59

Source: Adapted from Price 1980.
Note: Families are arranged from most to least continuous contact with hosts.

specific problems in different taxa and even in different populations of the same species.

Intimacy and Specialization: A Comparison of Insects with Marine Invertebrates

Completing development on a single host is one end of the continuum in the duration of contact between interacting species. In an analysis of the insect parasites of birds and mammals, Price (1980) found an overall relationship between continuous contact and specialization (table 7.1). At the one extreme, bird lice in the family Philopteridae spend their entire lives on their hosts, and the vast majority of these parasites are recorded from only one or two host species (Price 1980; Clayton, Gregory, and Price 1992). Bloodsucking streblid flies, which parasitize bats, are only slightly greater in mobility, living within bat colonies and exhibiting high degrees of host specificity, although not quite as high as philopterids. Yet lower in the proportion of monophagous species are botflies (Oestridae) in which females fly between hosts, and Hystrichopsyllid fleas in which females jump between hosts. Finally, the highly mobile hippoboscids, which are louse flies of birds and some mammals, have a low proportion of monophagous species.

Although the relationship between the parasitic lifestyle and specialization has been established repeatedly over the past decade, many analyses of patterns in specialization in insects and other taxa ignore this important influence. The result can be confusing conclusions about patterns of specialization within a lineage. When Gaston and Reavey (1989) analyzed the life history correlates of host specialization for larvae of the 927 species of British macrolepidoptera (26 families of butterflies and larger moths), they did not include completion of development on a single host as a variable. The variables they included were the effects of insect body size, overwintering stage, number of broods per year, the month in which larvae began development, the number of months each year that larvae are found in the field, and plant woodiness. Among these variables only body size was related to specialization: smaller species tend to be more host specific than larger species. This relationship between body size and specificity has also been observed in Lepidoptera elsewhere at least to the level of host genus (Wasserman and Mitter 1978). Size itself, however, may not be the cause of the relationship. Size is sometimes related to completion of development on a single host, and these two correlates of specialization need to be separated. Small size can be a consequence of the adoption of a parasitic lifestyle (Thompson 1983c).

More recently Gaston, Reavey, and Valladares (1992) analyzed specialization in 1137 species of British microlepidoptera, which are, on the whole, more host specific than British macrolepidoptera. This time, however, they included intimacy with individual host plants as a variable. They divided the microlepidoptera into internal and external feeders. Internal feeders included leaf miners, gall formers, case bearers, and borers into various plant parts such as stems, flower heads, and seeds. External feeders included species that feed exposed on the outside of the plant, either above ground or below ground, together with leaftiers, leaf rollers, and spinners. They found a significantly higher degree of host specialization in internal feeders than in external feeders (table 7.2). These results are not exactly an analysis of the effect on specialization of completing development on a single host: the great majority of the internal feeders probably feed strictly as parasites, but the external feeders undoubtedly exhibit a range of feeding styles from those that use only one host individual during development to those that graze among at least several plants. Nevertheless, the results of this broad survey reinforce the view that long-term intimate association with a single host individual favors specialization to a single host species.

There are indications from recent studies that the relationship between intimacy and specialization also occurs among invertebrates in marine environments. Extreme specialization in marine limpets seems to occur only in species in which individuals maintain a long-term association with a single host individual. Among the most host-specific are *Notoacmea paleacea,* which is restricted to surfgrass *Phyllospadix* spp. (Fishlyn and Phillips 1980), and the recently extinct *Lottia alveus,* which was apparently restricted to eelgrass *Zostera marina* (Carlton et al. 1991). (*Lottia alveus* also has the sad distinction of being the first marine invertebrate to become extinct in an ocean basin in historical time [Carlton et al. 1991].) *Notoacmea paleacea* is a tiny limpet, averaging only 6 mm, and exhibits a variety of striking morphological and behavioral traits that seem to be fine-tuned adaptations for living on this one host. Its parallel-sided shell precisely fits the narrow blades of surfgrass, and unlike many other gastropods, its response when physically disturbed is to

Table 7.2. The effect of feeding habit on specialization in British microlepidoptera

Insect stage and feeding habit	No. of species	Percentage of species feeding on:		
		1 species	1 genus	> 1 genus
Early-instar larvae				
Internal feeders	658	37	30	33
External feeders	448	25	23	52
Late-instar larvae				
Internal feeders	613	37	30	33
External feeders	519	27	23	50

Source: Data from Gaston, Reavey, and Valladares 1992.
Note: The values are the percentages of internal feeders (e.g., leaf miners) and external feeders (i.e., those feeding on the outside of plants) that are specific to one plant species, one plant genus, or more than one plant genus. Separate values are given for larval feeding habits early in development and late in development.

clamp down tightly onto the leaf surface rather than abandon the plant in an attempt to escape. Moreover, it harbors chemical compounds that it apparently gets from the host and that seem to provide it with some degree of chemical camouflage from searching predators (Fishlyn and Phillips 1980).

The link between intimacy and specialization appears again in sea slugs. *Elysia halimedae* from Guam (Paul and Van Alstyne 1988), *Elysia* sp. and *Cyerce nigricans* from Australia (Hay et al. 1989), and *Costasiella ocellifera* from Belize and the Bahamas (Hay et al. 1990) have all been reported as specific to particular seaweed species, as have two crabs, *Caphyra rotundifrons* in Australia and *Therandrus compressus* in Belize and the Bahamas (Hay et al. 1989, 1990). Although the extent to which individuals move among hosts is unknown, the details that are known of the life histories of these species suggest that they remain intimately associated with a single host for long periods of time. For example, unlike many marine invertebrates, *C. ocellifera* lays its eggs directly on host plants, so there is no larval dispersal stage (Hay et al. 1990). The crab *T. compressus* is so flattened, feltlike, and similarly colored to its host, that Hay et al. (1990) reported that they often could not detect them until the plants were returned to the laboratory and examined closely. In addition to these extreme specialists is an amphipod *Pseudamphitoides incurvaria* that eats and constructs domiciles from only a few species of the brown seaweed in the genus *Dictyota,* especially *D. bartayresii* (Lewis and Kensley 1982, Hay et al. 1990). It is one of only three herbivorous amphipod species known to construct domiciles from pieces of the host frond (Lewis and Kensley 1982) and is the most host specific of all amphipods.

In discussing specialization in *P. incurvaria,* Hay, Duffy, and Fenical

(1990) noted a continuum in specificity among amphipods from species with broad diets that feed mostly on detritus, microalgae, and small filamentous algae, through free-living species, to domicile-building species with very narrow diets. This suggests that, as for other taxa, the parasitic lifestyle in these animals is the primary determinant of the degree of specialization they exhibit to particular hosts. The specific reason that Hay and colleagues put forward, however, for evolution of extreme specialization in *P. incurvaria* is defense against predation: *D. bartayresii,* unlike other *Dictyota* species, contains compounds that deter reef fishes. Similarly, they have attributed specialization in the few known host-specific marine gastropods and crabs to the protection from predation afforded by feeding on seaweeds rich in compounds that deter their predators (Hay et al. 1989, 1990; Hay, Duffy, and Fenical 1990). Vermeij (1992), too, has argued that protection from predation may be a major cause of specialization in marine herbivorous molluscs that feed on kelp, in this case through the physical protection from predation afforded by excavating depressions in kelp stipes.

Nevertheless, as for phytophagous insects, decreased predation on chemically rich hosts and decreased predation on hosts that provide better physical protection seem inadequate as overall explanations for the origins of extreme specialization (Thompson 1988a). Predation may be the specific selection pressure at this time driving specialization in these marine species with a parasitic lifestyle, but the one common problem among small extreme specialists, whether marine or terrestrial, is not predation but rather completion of development on a single host. It is the process of adapting to an intimate association with an individual host throughout development, with all its attendant problems, including predation, that appears to be the reason why extreme specialization is common across all parasitic taxa.

The Geographic Structure of Specialization

The potential counterargument to the view that parasitism generally favors extreme specialization is the observation that some parasites have long lists of recorded hosts. The past decade, however, has produced a growing number of studies indicating that parasitic species with long, or even moderately long, host lists are often populations specialized to different hosts. This is one of the oldest observations in evolutionary biology, extending back to the publications of Benjamin Walsh on phytophagous insects in the middle 1860s (summarized in Brues 1924), but the genetic and molecular studies of the past decade have both reinforced and expanded that observation.

The emerging pattern is that there is a very strong geographic structure to specialization in the interactions between many parasites and hosts. In some cases, genetically differentiated populations of parasites are capable of inter-breeding. In other cases, the populations are turning out to be complexes of very closely related, sexually reproducing sibling species that are reproductively isolated from one another. In still other cases, the parasites are asexual demes, in which the distinction between genetically differentiated populations and sibling species is somewhat arbitrary. Not all parasites are extreme specialists, but together these studies are indicating that geographic structure in specialization is even more common than we once thought. This geographic structure of specialization is the raw material for a geographic view of coevolution. The following paragraphs catalog examples from a wide range of parasitic taxa to illustrate just how much hidden specialization has been discovered within the past decade within taxa previously considered to be polyphagous.

The fungal pathogen *Entomophaga aulicae,* which has been recorded attacking several moth families, was thought to be a single species until studies of infectivity of hosts, enzyme polymorphisms, restriction fragment length polymorphisms, and the size and number of primary conidia together demonstrated it to be a complex of species specialized to different hosts (Soper et al. 1988; Hajek et al. 1991). The variety of techniques needed to show that this species is actually a collection of specialist groups shows that the differences between species of specialists can be quite subtle. Similar subtle differences occur in the digenean *Echinoparyphium recurvatum.* This species was thought to be a relatively nonspecific parasite that alternates between hosts, using freshwater snails as the first intermediate host and waterfowl such as mallards and tufted ducks as the final hosts. It has now been shown that British populations are at least two morphologically indistinguishable sibling species (McCarthy 1990). One species uses a lymnaeid snail as the first intermediate host and lives as an adult in the anterior small intestine of mallards (*Anas platyrhynchos*). The other uses a prosobranch snail as the first intermediate host and lives as an adult in the posterior small intestine and rectum of mallards.

In other parasites, the populations are still considered to be one species, but they attack different hosts. The bacterium *Pasteuria penetrans* attacks a number of nematode genera, but individual populations have a more restricted host range. Moreover, some of these populations differ in the amount and nature of proteins on the surface of their spores, which affects their attachment to different hosts (Davies, Robinson, and Laird 1992). Similarly, the nematode *Heligmosomoides polygyrus,* which attacks about eighteen species of rodents, has been divided into four subspecies, three of which are host specific, although one of them has occasionally been recorded from two other rodents. The fourth subspecies has been recorded in Europe on fifteen rodent hosts, but recent studies have shown that at least some strains have a very low ability to infect hosts other than their usual host (Quinnell, Behnke, and Keymer 1991).

Perhaps more than for any other taxon, studies of phytophagous insects have produced over the past decade a continuous stream of results splitting species previously thought to be polyphagous into specialist populations or species. In 1981 Fox and Morrow reviewed the studies then available for insects and concluded that specialization is a populational phenomenon rather than a species property. Since then the evidence has become even stronger. For instance, the eastern tiger swallowtail *Papilio glaucus* has the longest list of recorded hosts of any of the 530 species of swallowtail butterfly, spanning seventeen plant families and thirty plant genera (Scriber 1984, 1988). Local populations, however, have a much smaller number of hosts, and some are restricted to a single host. Populations in southern Florida are closely adapted to feed on sweetbay (*Magnolia virginiana*) and survive poorly on most of the other food plants of the more northerly populations; similarly, the more northerly populations survive poorly on sweetbay (Scriber 1986b). In addition, Canadian and northern United States populations adapted to different host species are now considered to be a separate species from southern populations (Hagen et al. 1991). The two species differ in the spectrum of plants on which the larvae can survive, partly because they differ in their ability to detoxify the phenolic glycoside tremulacin, which occurs in quaking aspen and other Salicaceae (Lindroth, Scriber, and Hsia 1988; Scriber, Lindroth, and Nitao 1989). These two tiger swallowtail species still exhibit an intriguingly broad spectrum of hosts in some populations, but the range of hosts for individual populations is not nearly as broad as once thought and it differs geographically.

The same pattern has occurred in a wide variety of other insects as indicated by the following catalog of examples: one species considered to be polyphagous has often turned out to be a geographic mix of specialized populations including, in some cases, some sibling species. Hence, a polyphagous species of small ermine moth (*Yponomeuta padellus*) in Europe is now considered to be a collection of populations specific to different hosts (Water 1983; Menken, Herrebout, and Wiebes 1992). Colorado potato beetles *Leptinotarsa decemlineata* in North America, which have been recorded from twenty species in the Solanaceae, are now known to be subdivided geographically into populations adapted to different hosts (Hsiao 1978, 1988; Hare and Kennedy 1986; Hare 1990). The whitefly *Bemisia tabaci,* which is responsible for major damage to a number of plant species in North America, is now considered to be two species based upon recent studies of amplified DNA fragments, allozymic frequencies, and crossing experiments (Perring et al. 1993). The birch-feeding leafhopper *Oncopis flavicollis* in Britain is now considered to be three sibling species that differ in their preferences for birch species (Claridge and Nixon 1986). Similarly, the rice brown planthopper has turned out to be two species, each with more-specialized diets—one feeding on wild and cultivated rice (*Oryza* spp.) and the other on another grass (*Leersia hexandra*) (Claridge, Den Hollander, and Morgan 1988). The planthopper genus *Ribautodelphax* in western Europe was split from a group of six species into a complex of eleven species, most of which are monophagous or narrowly oligophagous (Bieman 1987). Finally, the tephritid fly *Tephritis conura*, which feeds in the flower heads of at least seven species within the thistle genus *Cirsium* from the European Alps to Fennoscandia, is now known to be a geographically differentiated complex of populations, possibly including sibling species, adapted to different hosts (Zwölfer 1988; Zwölfer and Romstöck-Völkl 1991).

In addition to broad geographic differences in specialization among populations, some studies have shown significant local differences in gene frequencies among individuals feeding upon different hosts, implying that they are sibling species or at least that the population is not a panmictic unit made up of polyphagous individuals. The treehopper *Enchenopa binotata* was previously thought capable of attacking deciduous trees in six genera in eastern North America. Now it is considered to be a complex of six species, each with a different host. These sibling species differ in phenology, host selection, and allozyme frequencies (Wood and Guttman 1983; Wood and Keese 1990; Wood, Olmstead, and Guttman 1990). Similar differences in host selection and allozymes have been found in populations of the galling sawfly *Euura atra*, which has been recorded in Finland on a variety of willow species (Roininen et al. 1993), and in populations of the apple maggot fly (*Rhagoletis pomonella*), some populations of which attack apple whereas others attack hawthorn in the midwestern United States (Bush 1969). The host-specific populations of the apple maggot are at least partially reproductively isolated (Feder, Chilcote, and Bush 1988, 1990a,b; McPheron, Courtney Smith, and Berlocher 1988; Prokopy, Diehl, and Cooley 1988).

This long list of examples showing specialization within taxa once thought to be polyphagous parasites makes suspect the conclusion of true polyphagy in any parasitic species that has not been thoroughly studied. Some studies are finding differences in specialization in areas less than 1 km apart. Via's (1990, 1991a,b) detailed experimental field studies of pea aphids (*Acyrthosiphon pisum*), a polyphagous species that attacks a range of legumes, have shown differences in specialization among clones over such short distances. Clones of these aphids are specialized to particular legume species and have lower fitness, as measured by longevity and fecundity, when raised on a legume other than their normal host. Other studies are beginning to show differences in preference hierarchies among individuals within populations (e.g., Wiklund 1981; Ng 1988), although it is not yet certain whether these differences among individuals are genetically based. If they are, they reflect yet another level of specialization contributing to differences in the range of specialization found among populations.

As increasing numbers of studies continue to partition once polyphagous species into more host-specific complexes of sibling species, differentiated populations, and even families within populations, the problem will be to understand how different life histories of both parasites and hosts favor these different patterns of geographic structure in specialization. Some hosts, for example, may elicit less specialization than others. Temperate forests often have several hundred species of gilled mushrooms (Agaricales) and tens of species of pored mushrooms (Polyporaceae) (Hanski 1989). Gilled mushrooms are small, moist, relatively soft, and ephemeral, lasting from a few days to a few weeks. Pored mushrooms, in contrast, are typically dry, hard, and last from months to years. In comparing *Drosophila* attacking gilled mushrooms, Jaenike (1978) argued that the ephemerality and unpredictable availability of gilled mushrooms (something experienced by every mushroom collector) together with low chemical diversity among many mushroom species may allow for fairly high levels of polyphagy. In a preliminary study of broader patterns in specificity of fungivorous insects, Hanski (1989) concluded that polyphagy is common, but that species feeding on pored mushrooms, which are much more predictably available as hosts than gilled mushrooms, tend to use fewer hosts.

Even if some mushrooms elicit low levels of local and geographic specialization from their parasites, they seem to be the exceptions rather than the rule. Price (1977, 1980) has argued that the parasitic lifestyle is more likely than other kinds of life histories to lead to population subdivision, geographic differentiation, and formation of sibling-species complexes. Small demes resulting from the parasitic lifestyle have been found in some, although not all, studies (e.g., McCauley 1991; Mulvey et al. 1991). We as yet have no general theory that predicts the geographic structure of specialization in parasites, whether insects or other taxa, based upon differences in the life histories of the parasites or the hosts. What is clear, however, is that there is extensive geographic structure in the patterns of specialization among parasite taxa, which must be part of our theory of how interactions evolve between species.

Conclusions

The parasitic lifestyle favors extreme specialization and, commonly, geographic differences in specialization. The specific reasons that parasites are specialized to one host or a small number of hosts vary among taxa and are worth study in themselves, but all the specific reasons have their common origin in the problem of completing development, or one stage of a complex life history, on a single host. The past decade has produced a constant stream of studies showing that species once thought to be polyphagous are genetically differentiated groups of specialists. Some of these groups are complexes of sibling species, others are differentiated populations that are still capable of interbreeding, and still others are sympatric families specializing on different hosts. To be sure, there are generalists among parasites, but there is more hidden specialization within many parasitic species—especially among populations—than we once thought. This hierarchy in the geographic structure of specialization provides the basis for a geographic view of evolving interactions.

CHAPTER EIGHT

Choosing among Multiple Victims

The moment when a parasite must leave its initial host and travel to a second host to complete development, the opportunities for the evolution of extreme specialization change. In this chapter I examine the continuum between parasitism and the free-living lifestyle of grazers and predators to develop three points about the evolution of specialization and the geographic structure of interactions. First, natural selection usually favors less specialization in species that require two or more host individuals to complete a particular stage of development as compared with related parasitic species. Second, extreme specialization within local populations for one prey or victim species evolves in free-living species only under a set of four uncommon ecological conditions. Third, where extreme specialization in free-living species does occur, it may lead to pronounced geographic structure, producing populations highly adapted to only one victim species. If these points are generally true, then some aspects of the coevolutionary process will differ between parasites and free-living species, but where coevolution does occur, it will sometimes be as much a geographic as a local process.

The chapter begins with the evolutionary transition from parasitism to grazing and then expands to include grazers and predators that attack many victims during their lifetimes. Early in the chapter I draw examples from small grazers such as marine limpets and some phytophagous and predatory insects that move among victims rather than from more-obvious examples such as antelope and lions, because I wish to emphasize that the lifestyles of parasitism, grazing, and predation create general patterns in the evolution of specialization that apply across taxa.

Moving among Victims: The Transition from Parasitism to Grazing

Transitions between parasitism and grazing and predation have undoubtedly occurred many times during evolutionary history. I use 'grazing' here in the general sense of moving between and feeding on two or more individual victims during a particular stage of development without necessarily killing each victim; and I use 'predation' here to mean the rapid killing of individual prey. Hence, an insect that completes all its larval development, from egg hatch to pupation, on a single host individual and causes some damage to its host is a parasite in an evolutionary sense; a larva of a related species that must crawl between two or more host individuals to get enough food to complete larval development is a grazer; and an adult beetle that picks up seeds one after another, eating each within a few minutes, is a predator.

Moving between two or more victims may seem to be a small difference from the parasitic lifestyle, but it changes at least two important aspects of natural selection. First, if an individual must feed on two or more individuals to complete a particular stage of development, then natural selection can favor diet mixing. By including two or more species within a diet, an individual can achieve a whole range of benefits unavailable to an extreme specialist that attacks only one species: reduced searching time, balanced nutrient intake, minimized ingestion of any single toxin, continued sampling of foods that are changing in availability and quality, and increased digestive efficiency (Freeland and Janzen 1974; Westoby 1978; Robbins 1983; Bjorndal 1991). Through these benefits, the opportunity of diet mixing can become a requirement for species, if the fitness of individuals that actively mix their diets is higher than the fitness of individuals that do not.

There is now ample evidence that some species grow better on a mixed diet than on a single-food diet, and that some species with mixed diets actively mix their foods. The plate limpet *Acmaea scutum* in California maintains a mixed diet of two algal species, even when one or the other of these two species is much more abundant (Kitting 1980). In this case, the cause of diet mixing is unknown, but it seems clear from Kitting's studies that the limpets actively mix their diets. Similarly, the sea hare *Dolabella auricularia,* a large (up to 1 kg wet mass) gastropod that grazes on algae, actively mixes its diet. When offered pairs of algal species in different ratios, individuals choose to consume certain proportions of each. Hence, the mixtures of algae they consume (i.e., the ratios) are more similar than the mixtures that are offered (Pennings, Nadeau, and Paul 1993).

The grasshopper *Schistocerca americana* also appears to be able to mix its diet, in this case to overcome inadequacies in any one food. When nymphs are offered inadequate but complementary artificial diets with unique flavors, they switch between these foods more often than nymphs offered adequate diets (Bernays and Bright 1991). The grasshopper *Taeniopoda eques* also switches frequently between plants and survives much better on a mixed diet than on a single-plant diet (Bernays et al. 1992). Similar benefits of diet mixing appear in other taxa. The chicks of European bee-eaters (*Merops apiaster*) put on more weight per gram of food eaten when fed a mixed diet than when fed a diet of either bees or dragonflies (Krebs and Avery 1984).

The second change from the parasitic lifestyle that moving between two or more victims allows is the evolution of learning through experience. Learning what species to eat is commonly an important part of the ontogeny of specialization in grazers and predators; there is no reason for natural selection on parasites to favor the evolution of this type of learning. This difference between lifestyles means that grazing species may often be less genetically specialized to one host, because they can confront the challenge of finding additional hosts by learning which hosts are edible and which should be avoided.

For example, there is a clear difference in at least one type of learning ability between phytophagous insects that feed as grazers and those that feed as parasites in the few species that have been studied for this ability so far. Various forms of learning from experience have been found in six orders of insects that include phytophagous species (Papaj and Prokopy 1989). Two of those types of learning are known to occur in insects searching for food: induction and aversion. Some insects can be induced, by feeding on a plant, to increase their relative preference for that plant, either making it more acceptable than other plants or making it as acceptable as plants that are higher in the preference ranking of naive individuals (Jermy 1987; Dethier 1988). Similarly, some insects acquire an aversion for plants that have caused temporary illness (Dethier 1980, 1988). Yellow woolly bear caterpillars (*Diacrisia virginica*), which graze among herbs, show aversion learning (Dethier 1980), whereas tobacco hornworms (*Manduca sexta*), which are more sedentary, show much less ability to learn (Dethier and Yost 1979). When *D. virginica* caterpillars, and those of another arctiid moth, *Estigmene congrua,* are given a choice of three plants, including *Petunia,* which makes them temporarily ill, they avoid *Petunia* once they have recovered. In contrast, recovered tobacco hornworms show little avoidance of plants that made them ill. Like grazing caterpillars, grazing grasshoppers such as *Locusta migratoria* and *Schistocerca americana* learn to avoid distasteful or toxic foods after an initial encounter (Blaney and Simmonds 1985; Bernays and Lee 1988). Similar avoidance learning occurs in other taxa that feed as grazers on plants (e.g., mammals) (Bryant et al. 1991).

Diet mixing and learning, then, are two of the evolutionary opportunities available to grazers but not to parasites, and these affect patterns of specialization within and among populations. The effects of these opportunities on specialization can be seen in species such as yellow woolly bear caterpillars. These caterpillars feed as true grazers, moving among low herbs and taking bites from a number of plant individuals and species. The caterpillars move freely through the vegetation, spending very little time on any one plant. When a caterpillar encounters a plant, it generally eats a part of a leaf, takes a postprandial rest, then moves on to another plant (Dethier 1988, 1989). Yellow woolly bears epitomize how parasitism and grazing differ in their effects on the evolution of specialization. These caterpillars are among the most polyphagous of phytophagous insects. More than one hundred plant species have been recorded as food for this species, and individual larvae commonly use multiple plant species during development.

Although many insects that feed as grazers feed on plant species that are too small to permit a larva to complete development on a single plant in, some other insect grazers feed on trees, and these grazers, too, commonly show high levels of polyphagy. Trees would seem to be the kind of large hosts that would allow the evolution of a parasitic lifestyle, and many insect species do in fact have life cycles in which larvae always complete development on a single host and are highly host specific. But some do not, and it is in these species that move between hosts that polyphagy is most common. This relationship between grazing and polyphagy is especially evident in four families of moths that include some of the most-polyphagous tree-feeding insects in the north temperate forests: Geometridae (inchworms), Lymantriidae (gypsy moths and allies), Psychidae (bagworm moths), and Tortricidae (including many leaf rollers, leaf tiers, and budworms). Many of these moths share a syndrome of life history traits. Wingless adult females eclose in fall or winter and lay their eggs; these eggs or the first-instar larvae overwinter; and in the spring the larvae disperse on silken threads to neighboring trees if their emergence on the host tree has not coincided with bud break (Barbosa, Krischik, and Lance 1989). Dispersal of larvae on silken threads (often called ballooning) has evolved among insects only in the Lepidoptera and is almost exclusively found in these forest moths, which specialize on the spring flush of new leaf tissue that is high in water content and nitrogen and low in secondary compounds (Roff 1990).

After they have ballooned onto a host, larvae of some tree-feeding moths remain there throughout development unless forced to move because they have stripped the tree of leaves. Gypsy moths (*Lymantria dispar*) in the northeastern United States, however, sometimes change hosts during development even when population levels are low (Lance and Barbosa 1982; Rossiter 1987). Oaks (*Quercus*) are the preferred hosts for early development, but in mixed forests of oaks and pitch pine (*Pinus rigida*) the moths oviposit on both hosts, even though early instars cannot survive on the pine (Rossiter 1987). Fourth-instar and later larvae, however, are able to survive on pitch pine and actually suffer lower parasitism on this host than on oaks. These later-instar larvae commonly (but not invariably) disperse from oak to pine to complete development. Similar shifts in later instars between oaks and other conifers have been observed in various parts of the northeastern United States (Rossiter 1987). In some cases, this switching of hosts results in faster development, larger size, or higher fecundity than diets of one plant species provide (Barbosa, Martinat, and Waldvogel 1986).

The association of grazing with polyphagy is also evident among species within particular families of moths. In his study of saturniid and sphingid caterpillars in Santa Rosa National Park in Costa Rica, Janzen (1988) found that most are highly host specific. *Hylesia lineata,* however, is unusual in being extremely polyphagous. Janzen (1984a) has found the caterpillars of this species feeding on forty-six plant species in seventeen families, although most egg batches are laid on only a small subset of these potential hosts. An ovipositing female lays a batch of about 300 eggs on a woody host, and the larvae live gregariously for the first several instars, spending their days in a group on the trunk or a large stem. At night they disperse throughout the canopy to feed. On a large plant the larvae are able to complete development without searching for another host. But on smaller plants, they sometimes defoliate the host, then march off in search of a new one. In addition, when a caterpillar is physically disturbed, it will often release its hold and fall off the plant, after which it must find its way back onto the original host or onto a new one. Unlike some other caterpillar species, *H. lineata* does not drop on a silk thread when escaping. So, it cannot use this lifeline for getting back onto its original host. The defoliation of trees, subsequent movement to a new host, the escape response to disturbance, and the local richness of plant species, which makes neighboring plants unlikely to be the same species as the original host, all would seem to act in concert to favor polyphagy in this species at the parasite/grazer interface. Similar movement between hosts also occurs in the small number of other polyphagous saturniid moths in Santa Rosa: *Automeris rubrescens, A. io, A. zugana,* and *Periphoba arcaei* (Janzen 1984b).

There is an association, then, between the lifestyle of moving among two or more victim individuals and selection against extreme specialization to one species in tree-feeding moths and other insects. This association between lifestyle and polyphagy undoubtedly evolves through selection acting both ways: the grazing lifestyle favors polyphagy, and the ability to attack a variety of different victim species allows the possibility of a grazing lifestyle. So long as a grazing lifestyle is favored by natural selection over parasitism, however, extreme specialization should be rare except for species living in virtual monocultures of a victim species.

Polyphagy, however, in insect taxa that move between hosts does not imply that there is no geographic structure in specialization in these species. Just as for parasitic species, research in recent years has shown more genetic differentiation in specialization within these species than was once suspected. The fall webworm (*Hyphantria cunea*), which has been recorded from thirty plant families in North America, may encompass genetically differentiated populations of at least two species, each with a broad host range but with different preferences (Jaenike and Selander 1980). Similarly, the fall armyworm (*Spodoptera frugiperda*), which feeds on a large number of grasses, is now known to include genetically differentiated strains that differ in the grasses they eat (Pashley 1986; Pashley and Martin 1987). Some species even differ locally in host use. The fall cankerworm (*Alsophila pometaria*), which has been recorded from a large number of deciduous trees in eastern North America, is polymorphic for host preference (Futuyma, Cort, and Noordwijk 1984; Futuyma and Philippi 1987; Futuyma 1991). None of these tree-feeding grazers or other phytophagous insects approaches the degree of extreme specialization commonly found in species that always complete development on a single host individual, but there is both local and geographic structure to specialization in these polyphagous taxa.

The same links between moving among victims and polyphagy occur in insects and other arthropods that are predators of other insects. In some cases, however, there is an evolutionarily middle ground in which predatory insects feed in succession on the offspring of one female or on the immature insects found in one nest. These species feed as predators, but they are in more long-term, intimate contact with a group of related prey individuals than more free-living predators. Species that feed on ants, for instance, include both free-living species and others that live in ant nests. Those predators that live in intimate, long-term contact with a single ant colony tend to show more extreme specialization than free-living species (Hölldobler and Wilson 1990; Jackson and Olphen 1991). Similarly, some other invertebrate predators that spend all of their larval development feeding within a clonal colony of prey show high levels of specialization. These include some green lacewings, ladybird beetles, and several other insect taxa that are predators on aphids or related homopterans that form large families of asexually produced individuals. These colonies of aphids are often only one genotype and are essentially one evolutionary individual. A larval lacewing can spend all its development killing one genetically identical aphid after another within a single aphid clone. This unusual habit can favor extreme specialization, and some predators on aphid clones are among the most prey-specific insect predators. For example, larvae of the green lacewing *Chrysopa slossonae* have been recorded feeding only on woolly alder aphids (*Prociphilus tessela-tus*). Both the adults and the larvae feed on these aphids, and females lay their eggs on branches near aphid clones (Tauber and Tauber 1987; Milbrath, Tauber, and Tauber 1993).

Regardless of whether the host or prey is plant or animal, then, the transition from parasitism to a free-living lifestyle of grazing or predation seems generally to result in decreased specialization to one species. The problem of finding more than one victim of the same species, the absence of many of the selection pressures that directly favor specialization in parasites, and the opportunities for learning and diet mixing all act together to favor decreased specialization in species that must leave their original host to complete a major stage of their development.

Specialization in Free-Living Grazers and Predators

Although a free-living lifestyle often seems to preempt specialization to one victim species, specialization at least to a very small number of victims has evolved in some free-living taxa. Nearly all free-living grazers and predators that approach extreme specialization appear to feed upon victim species that share a set of four uncommon ecological conditions (modified from Thompson 1982):

1. The victim species is abundant throughout the year (or the part of the year that the forager is searching for food) or during periods of time when other potential foods are scarce.
2. The victim species is predictably available, locally or regionally, year after year.
3. Successful capture, handling, or digestion of the victim as a major component of the diet requires specialized foraging techniques, morphology, or physiology.
4. The victim moves slowly or is completely sessile, and it is easy to find.

These conditions do not guarantee extreme specialization in grazers and predators; rather they seem to be the minimum set of conditions allowing habits approaching extreme specialization to occur. Examples of victim species that meet these conditions and have at least one fairly specialized free-living species that exploits them are one-seeded juniper, ponderosa pine and some other conifers, bamboo, eucalypt leaves, ants and termites, and *Pomacea* snails. (In addition, some mutualistic resources, including certain kinds of flowers and fruits, meet these criteria and are discussed in later chapters.) Even with these taxa, however, specialization is rarely to one victim species. Usually, the grazers and predators specialized for feeding on these species feed on two or more species within one genus or one family, and they may vary geographically in the particular species that they feed upon. The following sections use mammals, birds, and reptiles to examine where and when these ecological conditions allowing specialization to one or a few species have been met.

Extreme Specialization in Predators of Animals

Ants and termites are perhaps the prey species that most commonly meet the set of four conditions allowing extreme specialization in free-living predators of animals. These social insects are the most abundant animals in tropical environments, are among the most predictably available prey, and are relatively slow moving. In addition, many species are chemically or morphologically highly defended against predators. The bizarre morphologies of free-living species that specialize on ants and termites and the abundance of these prey suggest that efficient exploitation of these prey as a sole or primary food requires a high degree of specialization. The formic acid found in ants and the terpene-based secretions produced by the nasute soldiers of some termites suggest the need for physiological as well as morphological specialization. The African aardvark and aardwolf, the South American anteater, the Asian and African pangolin, and the Australian short-beaked echidna and numbat all show specialization to a diet made up almost exclusively of termites or ants. Moreover, all these specialists—whether monotremes, marsupials, or eutherians—have converged on similar physiological mechanisms for coping with this specialized diet (McNab 1984). In contrast to the ant and termite specialists among mammals, most other insectivorous mammals are much less taxon specific in their choice of prey (Hanski 1992).

None of the myrmecophagous mammals appears to be restricted throughout its geographic range to a single prey species. Rather there is a geographic structure to specialization in these interactions, with different populations eating different species or groups of species. The aardwolf (*Proteles cristatus*) is perhaps the most specialized in that it is virtually restricted to one termite genus, *Trinervitermes*. Only during winter months do aardwolves sometimes supplement their diet with a small proportion of another harvester termite genus, *Hodotermes*. Aardwolves use different *Trinervitermes* species in different parts of their geographic range (Richardson 1987), but it is unknown whether different populations are adapted for different *Trinervitermes* species. Aardwolves differ from other ant and termite specialists by capturing termites on the soil surface, and this difference may be one of the reasons why they are more prey specific than other species that feed on termites or ants. Their *Trinervitermes* prey are common, surface-foraging harvester termites and are the only African termites that regularly move in large numbers on the soil surface (Kruuk and Sands 1972). Although these termites are highly chemically defended and avoided by most other myrmecophagous species, aardwolves have broken through those defenses and are able to tolerate the terpenes produced by the nasute caste of these termites, allowing individuals to consume up to 300,000 termites during a single night of foraging (Anderson, Richardson, and Woodall 1992).

Like aardwolves, numbats (*Myrmecobius fasciatus*) are also termite specialists, but few studies of their degree of specialization to particular termite species have been conducted (Friend and Kinnear [1983] 1991). They regularly eat two termite species, but more than a dozen termite species have been found in the scats of this rare marsupial. Calaby (1960) found that the species composition of termites in scats approximately matched the relative abundance of termite species in the numbat's habitats. The most abundant two termite species appear in most of the scats, with other species appearing more or less according to their commonness. It is possible that the numbat actually prefers these most common species, or avoids some of the other species, but understanding the preferences for prey will require detailed studies of individuals within single populations.

The other ant and termite specialists all show considerable geographic variation in the prey species they eat. Giant anteaters (*Myrmecophaga tridactyla*), which range from Guatemala and Belize south to Argentina, eat both termites and ants. The relative proportion of these two taxa in the diet varies geographically, but this may reflect differences in availability rather than genetic differences in specialization (Redford 1985). In Emas National Park in southwestern Brazil, giant anteaters feed on at least eight species of termites and six species of ants, spending almost 90% of their time on termites. They spend more time feeding on some species than on others, but determining preferences is complicated by the fact that the mounds of *Conitermes* termites may harbor as many as five termite species and four ant species (Redford 1985). Individuals move rapidly among colonies, generally spending only a few minutes at most at any one mound, then leaving before the colony can fully mount its defenses. They wander over home ranges estimated in different regions as three to twenty-five square kilometers (Shaw, Machado-Neto, and Carter 1987). Similarly, the two medium-sized Neotropical anteaters (*Tamandua mexicana* and *Tamandua dactyla*) average less than one minute per feeding bout, and they adopt different foraging techniques for different prey. When feeding on termites such as *Nasutitermes,* which have specialized nasute soldiers, they often forage in logs and other sites away from the termite nests and the soldiers (Lubin and Montgomery 1981).

As in mammals, avian taxa with highly specialized diets feed on predictably abundant prey that require specialized morphology, physiology, or behavior to harvest as a major component of a diet. Few are restricted to one prey species, but some show highly developed adaptations to a small number of prey taxa. The most extreme specialist predator among avian species may be snail kites (*Rostrhamus sociabilis*), which in Florida are restricted almost exclusively to the snail *Pomacea paludosa* and in Central and South America to several *Pomacea* species (Snyder and Snyder 1969; Beissinger 1983, 1990; Bourne 1985). Just as in aardwolves, then, there is a geographic structure to specialization in this predator, which feeds on different species within a single prey genus in different parts of its geographic range. *Pomacea* snails are predictably abundant year-round, and the highly modified, slender, hooked bills of snail kites make these birds highly efficient at extracting the snails from their shells. During rare periods, such as droughts, when their normal prey are less available, these kites do take other prey, including small turtles, freshwater crabs, and another snail (*Viviparus georgianus*), but there is little question that they are normally restricted to *Pomacea* snails. All the alternative prey that they sometimes use are similar to *Pomacea* in having shells or carapaces and moving slowly, but the kites are less efficient at handling these alternative prey (Beissinger 1990).

Specialization to one prey genus also occurs in the Hispaniolan colubrid snake *Darlingtonia haetiana*. This snake is thought to be restricted almost entirely to the frog genus *Eleutherodactylus,* at least as indicated by the stomach contents of preserved specimens (Henderson and Schwartz 1986). This diet, however, is not quite as narrow as it sounds: *Eleutherodactylus* is the most species-rich vertebrate genus on Hispaniola and includes forty-nine named species. Of the eleven colubrid snakes on Hispaniola (distributed over six genera), *D. haetiana* is the only species that does not feed mostly on lizards, but it is also the only snake to occur commonly at higher elevations where *Eleutherodactylus* is most common (Henderson 1984; Henderson and Schwartz 1986). Similar restrictions to a single prey genus, based upon stomach analyses, have been suggested for several other snake species, such as the Australian burrowing snake *Vermicella annulata* (Shine 1980) and the red-naped snake *Furina diadema* (Shine 1981).

All these examples of specialization to one prey genus provide one more argument for analyzing the evolution of interactions in general, and coevolution in particular, from a geographic perspective. Different populations of a predator have the potential to evolve with different but closely related prey species, creating a geographic mosaic in relationships between a predator species and a prey genus.

Most other predatory vertebrates that are highly specialized to one type of prey remain flexible in their choice of food, but they too have the potential to differ geographically in their interactions with a small group of taxonomically related or ecologically related species. Bee-eaters, whose very name suggests a high degree of prey specialization, have a diet that spans a number of bee and wasp families and includes other insects as well. In summarizing the results of studies of prey use by populations of sixteen species, Fry (1984) reported that Hymenoptera make up 20–96% of the prey individuals. In most bee-eater species Hymenoptera, especially bees and social wasps, make up most of the diet, and honeybees are important prey for many of the species. The diet of nestlings has been studied in detail for only two species. Both studies showed that bees and wasps are common in the diets of even young birds. Fecal and pellet samples taken from six nesting burrows of rainbow bee-eaters *Merops ornatus* on Rottnest Island off Perth indicated that 95% of the prey individuals are Hymenoptera, and these include species from at least five families of bees, wasps, and ants (Calver, Saunders, and Porter

1987). In contrast, Hymenoptera make up only 36% of the diet of nestling European bee-eaters *Merops apiaster* on the island of La Camargue, France, with dragonflies composing most of the rest (46%) of the diet (Krebs and Avery 1984).

Extreme Specialization in Grazers and Predators of Plants

Plants more commonly than animals fit the criteria for species that allow for specialization to one species or genus in grazers and predators. Several bird and mammal species live in a virtual sea of the one plant species they eat and have become highly adapted to grazing on that one species. Among these are sage grouse (*Centracercus urophasianus*), which is one of the most specialized birds in the world. These birds have evolved in the large expanses of big sagebrush (*Artemisia tridentata*) found in western North America. They have a winter diet that is almost exclusively big sagebrush and a summer diet that is still mostly this one species (Klebenow and Gray 1968; Peterson 1970; Roberson 1986). Not only is big sagebrush the dominant plant in this bird's habitat and predictably available throughout the year but careful selection of plants is required to make it a major part of a diet. The birds preferentially choose plants high in crude protein and low in monoterpenes (Remington and Braun 1985).

The same unusual set of ecological conditions has allowed the evolution of extreme specialization in Stephens' woodrat (*Neotoma stephensi*), which is probably the most species-specific mammalian grazer. This species of the southwestern United States feeds almost exclusively on one-seeded juniper (*Juniperus monosperma*), which is by far the dominant plant in the woodrat's habitat and is one of only a few species in those communities that retain green leaves throughout the winter (Vaughan 1982). During a two-year study at one site, Vaughan found that 90% of this woodrat's diet was one-seeded juniper, and shorter term samples from other sites yielded similar percentages. These wood rats pay a price for this extreme specialization. Compared with most other rodents, they grow slowly and have small litters (Vaughan and Czaplewski 1985).

Abert's squirrel (*Sciurus aberti*) is similar to Stephens' woodrat in its extreme specialization to a single, abundant conifer species that dominates the habitats in which it lives. This squirrel of southwestern North America feeds most heavily on ponderosa pine (*Pinus ponderosa*) throughout the year, although it also feeds on underground truffles, seeds of other conifers, and sometimes lupines (Keith 1965; Hall 1981). Like one-seeded juniper, ponderosa pine often grows in large stands where it is the only major tree species. From late autumn through early spring, Abert's squirrels eat mostly the inner bark of these trees and, to a lesser extent, the terminal buds, switching to mainly ovulate cones for the rest of the year. As in some other extreme specialists, specialization for these squirrels appears to require individuals to be highly selective in choosing plants. The trees they use have significantly lower xylem oleoresin flow rates, significantly lower levels of ß-pinene and ß-phellandrene in the xylem oleoresin, and higher concentrations of structural carbohydrates in the phloem than uneaten trees (Snyder 1992).

The places to search for other extreme specialists to one plant species among vertebrate herbivores are in environments similar to those inhabited by sage grouse, Stephens' woodrat, and Abert's squirrel: communities dominated by a single plant species that is available throughout the year (or throughout the parts of the year that the herbivore is actively searching for food) and that requires specialized morphology, physiology, and behavior for successful harvest as a major component of a diet. Similar, although less extreme, specialization is also likely to be common in environments dominated by groups of plants within a single genus or group of related genera. The other commonly cited specialists among mammals of which I am aware—such as koalas, greater gliders, and pandas—are all of this sort. They use at least several dominant, related plant species in their diets, often differing geographically in the particular species or mix of species they eat.

Koalas are restricted to a diet of a small number of species in the plant genus *Eucalyptus*. These plants are abundant, but their foliage—high in fiber, tannins, and essential oils but low in nitrogen—appears to be a poor food for mammals. About twenty eucalypt species from the five hundred in the genus make up the bulk of the koala's diet, with particular species differing in importance in different parts of the geographic range and at different times of year. Individual koalas probably feed on more than one species during each year, varying seasonally in the species they eat (Hindell, Handasyde, and Lee 1985), but there have been no long-term studies of the feeding behavior of individual koalas in natural populations (I. Hume, personal communication).

Even though eucalypts completely dominate much of the forested parts of Australia, only three other marsupials include eucalypt leaves as a high proportion of their diet (Pahl and Hume 1990), indicating how difficult it may be to become a specialist on some genera even when they are predictably abundant. The greater glider (*Petauroides volans*) seems to come the closest to the koala as a eucalypt specialist, but the extent of its dietary restriction within *Eucalyptus* is incompletely known. In a two-year study of one population, greater gliders were observed to feed on the leaves and buds of six eucalypt species, with more observations recorded for three of the species than would be expected from their relative frequencies in these habitats (Kavanagh and Lambert 1990). Neither of the other two eucalypt-feeding species are as exclusive year-round to eucalypts. The common ringtail possum (*Pseu-*

docheirus peregrinus) also feeds largely on eucalypts (61–98% of its diet by volume) but also includes other plant genera (Pahl 1987). The common brushtail possum (*Trichosurus vulpecula*) spends, at least in one population studied in detail, up to 34% of its time feeding on noneucalypt species (Freeland and Winter 1975). All these eucalypt specialists have elaborate morphological or physiological adaptations for coping with their phenol- and fiber-rich diets (Hume 1982, Foley and Hume 1987a,b). Even so, they routinely eat plants other than eucalypts and differ among populations in the combinations of plants they eat.

Giant pandas (*Ailuropoda melanoleuca*) and red pandas (*Ailurus fulgens*) are the other extreme specialists among free-living herbivores, feeding primarily on bamboo in two genera. Bamboo fits all the criteria as a resource for extreme specialization: it is abundant, predictably available year-round, requires a specialized morphology and physiology, and is easy to find. The panda's thumb, which in both species is an elongated radial sesamoid bone rather than a true thumb, has now become the mainstay example in evolutionary biology of the jury-rigged nature of adaptation to a specialized function (Gould 1980b). Unlike many grasses, the nutritional quality of bamboo varies little throughout the year, thereby providing a food that is both abundant and predictable in quality (Schaller et al. 1985; Johnson, Schaller, and Hu 1988). In the Wolong Reserve in China, the bamboo species *Bashania fangiana* is found in 94% of the fecal droppings of red pandas (Reid, Jinchu, and Yan 1991). In most months this species of bamboo is found in 100% of the droppings, and only in the fall, when fruits ripen, does it decrease to as low as 79%. The remainder of the diet includes shoots of a mostly lower-elevation bamboo species (*Fargesia robusta*) and, in August and September, fruits of seven plant genera.

Giant pandas feed within the Wolong Reserve primarily on *B. fangiana* stems and leaves but also use *F. robusta* shoots (Schaller et al. 1985). Prior to 1983 the pandas used *F. robusta* mostly at times of high snowfall, when they moved to lower elevations where this species is more common. During 1983 *B. fangiana* mass-flowered and died, decreasing its standing crop by 82%. Mass flowering is characteristic of bamboo species (Janzen 1976), and in *B. fangiana* it apparently occurs about every forty-two to forty-eight years (Schaller et al. 1985). In the two years following 1983, the giant pandas did not change their diet (Johnson, Schaller, and Hu 1988). By the winters of 1985–1986 and 1986–1987, however, they had decreased their use of *B. fangiana* stems and increased their use of *F. robusta* leaves (Reid et al. 1989). This sequence of events highlights the difficulty of extreme specialization in a large mammal. In this case, it indicates that pandas need at least two bamboo species to support populations through a flowering event (Schaller et al. 1985; Reid et al. 1989).

None of the other grass-feeding mammals has evolved the degree of extreme specialization found in pandas. Unquestionably, grasses have driven the evolution of specialization in many mammals, and mammals have influenced the evolution of many grass species. Nevertheless, virtually all these grazers have mixed diets; they eat not one but rather a number of grass species and most include dicots in their diet as well. Their diets also vary geographically. Grass species compose up to 77–98% of the diets of some populations of eastern grey kangaroos (*Macropus giganteus*) and wallaroos (*Macropus robustus*) during all seasons of the year (Taylor 1984), although in other populations of eastern grey kangaroos, as well as in western grey kangaroos (*Macropus fulginosus*), dicots can compose a higher percentage of the diet (Halford, Bell, and Loneragan 1984; Caughley et al. 1988). Hence, the geographic structure of interactions between mammals and grasses involves more species than some other interactions between plants and specialized vertebrates.

These examples of specialization in vertebrate herbivores indicate that specialization to one plant family, one plant genus, and even one plant species has evolved repeatedly among these free-living species. The conditions for extreme specialization to a single species are rare, and they seem to result in grazers with limited geographic ranges. Specialization to one genus is probably more common, and the examples cited here suggest that such interactions have a geographic structure, with populations differing in the combinations of species they eat.

Geographic Specialization in Specialist Grazers and Predators

As more studies of interactions have been completed at broader geographic scales in recent years, more examples of geographic structure in specialization have become known. Studies of two groups of birds—black-cockatoos and crossbills—show a pattern of geographic structure in specialization that may apply to some other taxa of grazers and predators once they are studied in more detail.

Black-cockatoos are a complex group of species and subspecies in which populations differ considerably in bill morphology and diet. Until recently there were four recognized Australian black-cockatoo species (fig. 8.1), but current taxonomy has now increased that number to five species, and there are major geographic differences among populations within some of these species. Glossy black-cockatoos (*Calyptorhynchus lathami*), which appear to be the most specialized among the black-cockatoo species, have very broad bills, which they use to dehusk the cones of *Allocasuarina*. Casuarinas are abundant trees along rivers and in some other habitats in Australia, and glossy black-cockatoos use *Allocasuarina* throughout their geographic

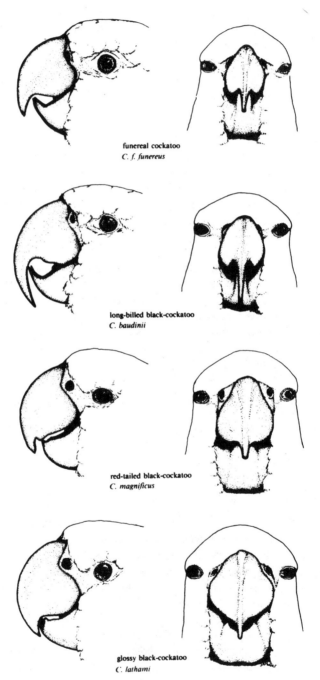

funereal cockatoo
C. f. funereus

long-billed black-cockatoo
C. baudinii

red-tailed black-cockatoo
C. magnificus

glossy black-cockatoo
C. lathami

Fig. 8.1. Differences in bill shape among black-cockatoo (*Calyptorhynchus*) species in Australia. *Calyptorhynchus funereus* has now been divided into two species, *C. funereus* in southeastern Australia and *C. latirostris* in southwestern Australia. Reprinted, by permission, from Joseph 1986.

Fig. 8.2. Profiles of a red crossbill, a red crossbill with the cross experimentally removed, and a pine siskin. Reprinted, by permission, from Benkman and Lindholm 1991.

range. Locally, they may feed exclusively on one species. On Kangaroo Island these cockatoos are recorded feeding solely on *A. verticillata* (Joseph 1982a). The other black-cockatoo species have bills adapted to different plant taxa. Red-tailed black-cockatoos (*C. magnificus*) from southwestern Australia are recorded feeding only on the large fruits of *Eucalyptus calophylla* (Ford 1980), and other populations of this species have bills adapted for feeding on yet other plant taxa (Ford 1980; Joseph 1982b, 1986).

As with black-cockatoos, recent detailed studies of crossbills show very high degrees of extreme specialization to the seeds of particular plant species. Moreover, the study of specialization in crossbills has been supplemented with elegant experimental analyses showing how specialization in bill morphology influences efficiency in harvesting seeds. Crossbills are highly adapted for extracting seeds from partially closed conifer cones. The narrowness of their bills allows them to slide between the scales of cones, and the cross in their bills makes them very effective at separating the scales and extracting the seeds (Benkman 1987a, 1988a). If the cross in the bill is removed experimentally from these birds (fig. 8.2), they become less efficient at extracting seeds from closed cones than control birds with crossed bills (Benkman and Lindholm 1991). As the bill regrows on the experimental birds, they become increasingly efficient at harvesting seeds from closed cones. This gradual increase in efficiency as the bill regrows mimics to some extent the increase in efficiency that crossbills must have experienced over evolutionary time as natural selection favored those birds with increasingly crossed bills.

Bill size is correlated with cone size in mainland populations of both Eurasian and North American crossbills, and that relationship varies geographically (Lack 1944; Benkman 1989, 1993). In Newfoundland, the relationship between bill size and cone size in island populations of red crossbills (*Loxia curvirostra*) differs from that in mainland populations, perhaps resulting from the lack of competition from red squirrels on the islands (Benkman

1989). Some populations of red crossbills have been suggested to be sympatric without interbreeding (Knox 1992), and recent studies of red crossbills throughout North America have suggested that this species may actually be a complex of as many as eight species or subspecies (a boon or headache to 'life listers', depending upon perspective), each specialized to a different conifer (Groth 1991, unpublished doctoral dissertation, in Benkman 1993). These crossbill types differ in call, morphology (skeletal and bill measurements), and the conifers in which they are commonly found. If crossbills are similar to other birds, the differences in morphology among types are likely to be strongly heritable (Grant and Grant 1989) and not simply the result of different diets during development. Crossbill offspring produced in captivity resemble their parents in bill and body measurements (C. W. Benkman, personal communication), suggesting that these measurements are highly heritable in crossbills.

Benkman (1993) has tested four red crossbill types from northwestern North America to evaluate how efficiently they handle seeds from different conifer species. These four crossbills types were collected from western hemlock (*Tsuga heterophylla*), Douglas fir (*Pseudotsuga menziesii*), ponderosa pine (*Pinus ponderosa*), and lodgepole pine (*Pinus contorta*). The optimal combination of bill depth and husking groove width differs for each of these four conifer species and produces an adaptive landscape with four discrete peaks. No one combination would allow efficient harvest of two or more of these conifers (fig. 8.3). Each of the four tested crossbill types has either the optimal bill size or the optimal husking groove width, or both, for the conifer species on which it feeds during winter and spring, which is the time of greatest food shortage.

The specialized bill morphologies of these crossbills adapt them to seed extraction from different kinds of partially closed cones. The bill does not increase their efficiency at extracting seeds from open cones: crossbills lacking the cross and pine siskins, which have less specialized bills (fig. 8.2), are as efficient as normal crossbills in extracting seeds from open cones (Benkman and Lindholm 1991). Moreover, it makes crossbills inefficient at harvesting other kinds of seeds, such as sunflowers and thistles, when they are forced to do so experimentally (Benkman 1988b). The advantage of the crossed bill is that it allows these birds to harvest seeds from conifer species that sometimes hold their seeds in closed cones, including spruce, pine, and hemlock (Benkman 1987b). Conifers such as western hemlock may open their cones during autumn, but some of the cones may reclose later due to high-moisture conditions. The seeds in these reclosed cones can become the key resource for crossbills during the lean periods of late winter and spring, and harvesting them efficiently requires their specialized bills.

Benkman has argued that, among the many conifer species in North

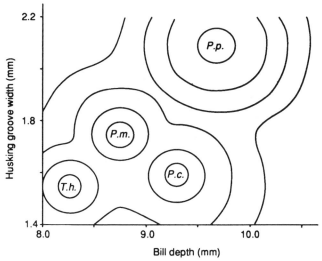

Fig. 8.3. An adaptive landscape for bill depths and widths in red crossbills. The contours repre-sent different fitnesses, and each circle encompassing an abbreviation for a genus and species corresponds to an adaptive peak based upon predicted optimal bill depths and husking groove widths needed to harvest each of four conifer species: *T.h.* = western hemlock (*Tsuga hetero-phylla*), *P.m.* = Douglas fir (*Pseudotsuga menziesii*), *P.p.* = ponderosa pine (*Pinus ponderosa*), and *P.c.* = lodgepole pine (*Pinus contorta*). Optima are based upon regressions resulting from experimental tests of harvesting efficiency as a function of bill morphology. Redrawn from Benkman 1993.

America, only a few have the potential to support the evolution of specialized crossbill types. These "key conifers" must have three characteristics. They must produce seeds yearly or hold them from one year to the next; they must hold their seeds in cones throughout late winter and early spring, when seeds are most limited in availability; and their seeds must be encased in cones that are difficult for competitors to manipulate. This list of specific conifer characteristics fits the list of general conditions for extreme specialization I gave earlier, with one exception. It includes competition as a necessary pre-requisite for the evolution of extreme specialization. Benkman argues that competition for a limited resource is a necessary prerequisite for selection for specialization because it favors specialists that are able to harvest cones more efficiently than generalists.

Although competition for a limited resource provides one route to the evo-lution of specialization and may be an important driving force in the evolu-tion of crossbills, I do not think that it is a necessary general requirement for the evolution of extreme specialization. Even in the absence of competition, individuals better adapted at exploiting resources will survive better, produce more offspring, and increase in frequency in the population. Hence, extreme specialization is common in phytophagous insects, but there is little evidence that competition has been a major driving force in specialization in these taxa. Some kinds of interactions, resources, and lifestyles seem intrinsically to favor extreme specialization. Unquestionably, as decades of experimental research and mathematical models have shown, competition has the potential to favor increased specialization. It has been, for example, an important evolutionary force shaping patterns of diet specialization in Galápagos finches, some species of which are extreme specialists on *Opuntia* cactus (Grant 1986; Grant and Grant 1989). Nevertheless, competition is only one of the causes of specialization, and it does not appear to be a general prerequisite.

Conclusions

Extreme specialization in free-living animals is undoubtedly less common than in parasites. Nevertheless, it does occur under some ecological conditions. Moreover, recent studies suggest that when more species are studied in detail throughout their geographic ranges, additional examples of extreme specialization within local populations will become known. What is increasingly clear from studies of free-living predators and herbivores is that, just as in parasites, a number of species show differences among populations in the species they attack, thereby creating a geographic structure that should be part of the theory of evolving interactions.

CHAPTER NINE

Coping with Multiple Enemies:
The Geography of Defense

Although parasites, grazers, and predators can all become specialized to one species under some ecological conditions, their victims must often contend simultaneously with multiple enemies, whether parasites, grazers, predators, or competitors. That asymmetry has commonly been used to argue that coevolution between pairs of species must be very rare. In this chapter, I examine how individual populations can evolve defenses simultaneously against multiple enemies, and how the waxing and waning of defenses in populations faced with different enemies can lead to a geographic mosaic in the pattern of defense within species.

Specialization in Defense against Multiple Enemies

There are now thousands of papers that show or suggest how specific forms of defense thwart specific parasites, grazers, or predators. The largely unexplored ground in the evolution of defense is how populations simultaneously maintain specific defenses against several different enemies. With extensive recombination, panmixis, and consistent directional selection by each enemy species, a population should be able to maintain specific defenses against multiple enemies, but there are almost no studies actually demonstrating that. Dodson (1988), however, has made a significant inroad in how defense evolves against multiple enemies through his work on cladocerans and their predators.

Daphnia species show species-specific and population-specific shifts in their positions within a water column in response to particular predators (Dodson 1988; De Meester 1993). Dodson subjected clones of seven *Daphnia* species taken from different lakes to three types of predator: small (2.5 cm) bluegill sunfish (*Lepomis macrochirus*), adult backswimmers (*Notonecta undulata*), and fourth-instar phantom midges (*Chaoborus americanus*). The different lakes harbor different combinations of these predators. The clones were divided and some individuals subjected to one predator, whereas others were faced with two or all three predators. *Daphnia* detect each of these predators through water-soluble compounds. Dodson found that the number of predator types to which a clone responded depends upon two things: the average body size of individuals within the clone and a history of presence of predators in the lake from which the clone was taken. Body size influenced defense in that smaller *Daphnia* species and clones tend to rise in the water column when predators (or water that had contained predators) are present during daylight hours, whereas larger *Daphnia* sink. Dodson argued that these differences in response make biological sense in that the smaller *Daphnia,* which are particularly prone to predation by tactile-foraging predators like *Chaoborus* midges, can best escape these predators during daylight by rising in the water column. During daylight hours the midges search for prey mostly in the deeper water, rising in the water column only at night. In contrast, the larger *Daphnia* species are more prone to predation by visually searching sunfish and backswimmers, which search mostly higher in the water column. Therefore, escape from these predators is in deeper water.

If small *Daphnia* always rose and large *Daphnia* always sank, these interpretations would have a hollow ring. But two additional results suggest that the responses of individual clones have been shaped by the past history of predation by multiple predators in the different lakes from which the clones were taken. In a test of eight clones against all combinations of all three predators (twenty-four combinations of clones and predators), in all but two cases clones responded to all the usual predators in their lakes. Most clones (five of eight) did not respond to predators not found in their lakes. Moreover, several clones that are subject to all three predators differed in their response depending upon the predator, moving up in the presence of midges and down in the presence of the sunfish or backswimmers. Additionally, two of the clones also differed in their responses to sunfish and backswimmers. Together, these results suggest that these *Daphnia* species are able to evolve specific defenses against up to three predators and harbor all these defenses simultaneously or in various combinations appropriate to the range of predators in their lake.

Maintaining many defenses simultaneously can be costly, and some cladocerans have evolved the ability to produce some defenses only when particular predators become abundant. These induced defenses, in cladocerans and in other taxa, have large benefits in increased survival in the presence of predators, but they come at a cost to growth or reproduction, at least under some environmental conditions (Kerfoot 1977; O'Brien and Vinyard 1978; Harvell 1986, 1992; Lively 1986; Walls and Ketola 1989; Riessen 1992). These large benefits and large costs, together with fluctuation in densities of predators in some environments and the ability of cladocerans to detect when their predators are reaching high densities, must be the driving force favoring induction of these defenses only in the presence of specific enemies (Dodson

1989; Harvell 1990a). For *D. pulex* spine production is an induced defense that develops only when the numbers of its predator *Chaoborus americanus* exceed a threshold of 0.5 larvae per liter (Havel 1985). The juvenile *D. pulex* that produce these spines do so at a significant cost: their development is slowed by 8–15% and their intrinsic rate of natural increase is reduced by 8–9% compared with individuals lacking the spines. Riessen and Sprules (1990) estimated that, in the absence of predation, individuals lacking spines could increase from 0.1% to 99% of a *Daphnia* population in only 3.4 years. These results, then, suggest that *Daphnia* populations differing in predator densities could quickly evolve differences in their response to predators, thereby creating a geographic structure in the levels of defense found among populations.

Genetics of Defense against Multiple Enemies

Studies of the molecular genetics of defense genes are adding another level to our understanding of how defense against multiple enemies evolves within populations. Recent cytogenetic and molecular mapping studies of plant resistance have made tremendous progress in identifying regions of chromosomes conferring resistance against parasites (e.g., Debener et al. 1991; Gebhardt and Salamini 1992; Dickinson, Jones, and Jones 1993). These studies have indicated that resistance genes against different pathogens are frequently clustered onto small regions of plant chromosomes. In some cases clusters of genes confer resistance against different genotypes (isolates) of a particular pathogen species (Bennetzen and Hulbert 1992). For example, many of the twenty-five genes that confer resistance to the common rust in maize map to one small region of chromosome 10 (Hulbert, Sudupak, and Hong 1993). Some of these resistance genes were previously mapped to one locus (e.g., *Rp1*) and were thought to be allelic series at the locus. Further work on the *Rp1* locus has shown it to be a very closely linked cluster of genes (Pryor 1987; Hulbert and Bennetzen 1991; Hulbert, Sudupak, and Hong 1993).

In other cases, these clusters of genes confer resistance against two or more species. The French wheat VPM1 carries resistance against three wheat rusts: stripe rust (*Puccinia striiformis* f. sp. *tritici*), stem rust (*P. graminis* f. sp. *tritici*), and leaf rust (*P. recondita* f. sp. *tritici*). Resistance to each of these rusts is controlled by single genes, and they are very closely linked on the short arm of chromosome 2A (Bariana and McIntosh 1993). Lines artificially selected for resistance to one of these rusts are also resistant to the other rusts (Bariana and McIntosh 1993). In yet other cases these clusters of resistance genes confer resistance not only against related parasites but also against very different parasite taxa. For example, the genes *Cf-2* and *Cf-*

5 code for tomato resistance to the fungus *Cladosporium fulvum* and map to the same small region of chromosome 6 as the gene *Mi,* which confers resistance to root-knot nematodes, *Meloidogyne* spp. (Dickinson, Jones, and Jones 1993).

Clustering of resistance genes onto small areas of chromosomes is therefore now known for genes conferring resistance against different pathogen isolates, different but related pathogen species, and even very different parasite taxa. Gene clustering therefore seems to be one of the ways in which plants contend with the problem of the evolution of defense against multiple enemies, but just why resistance genes are clustered is presently unknown. One possibility is that these clustered resistance loci are actually complex gene regions within which rearrangements or recombinational events, such as unequal crossing over, can lead to novel combinations of resistance specificities (i.e., combinations of genes specific to particular parasite genotypes) (Pryor 1987). These regions may allow plants to respond to the evolution of new parasite genotypes by creating new, but functionally similar, combinations of resistance genes (Pryor 1987). Alternatively, these regions may provide a localized area for creating genetic variation, producing resistance against a number of parasite taxa (Dickinson, Jones, and Jones 1993). Some of these regions (e.g., the *Rp1* complex resistance locus in maize) are known to be meiotically unstable (Bennetzen et al. 1988; Bennetzen and Hulbert 1992).

Whatever the exact causes for clustering of plant resistance genes, the molecular results are revealing a sophisticated system of response to the twin problems of rapid evolution of parasites and simultaneous attack by multiple parasites. Some researchers have even begun to draw analogies between certain aspects of disease resistance in plants and the mammalian immune system, which is remarkably adept at handling these twin problems (Pryor 1987; Dangl 1992). What seems clear at present is that organisms as different as angiosperms and mammals have developed genetic mechanisms that allow populations to fine-tune to some degree their defenses against the local suite of enemies with which they must contend.

Geographic Distributions of Species and the Accumulation of Enemies

Biogeographic analyses of how populations accumulate multiple enemies add yet another level to our understanding of the problem of the evolution of defense. Beginning with Southwood's 1961 analysis of the number of insects associated with British trees, dozens of studies have attempted to partition the causes of variation in the number of parasites attacking plant and animal species. For plant species, geographic range has consistently been found to be one of the best overall predictors of the number of associated insect or

Table 9.1. Summary of the percentage of variation in number of insects associated with plant species explained by the geographic range of plants

	No. of studies	Proportion of variation in number of associated insect species explained by plant geographic range (r^2)	
		Mean	Range
Trees and shrubs			
All insect orders combined	7	0.67	0.49–0.91
Particular insect taxa	11	0.37	0.13–0.90
Herbs			
All insect orders combined	6	0.48	0.18–0.71
Particular insect taxa	5	0.45	0.24–0.79

Sources: Compiled from Tahvanainen and Niemelä 1987 and Leather 1991.

pathogen species (table 9.1), although other factors, such as plant size and number of related, coexisting plant taxa, also influence the total (e.g., Lawton and Shröder 1977; Lawton and Price 1979; Kennedy and Southwood 1984; Strong, Lawton, and Southwood 1984; Tahvanainen and Niemelä 1987; Leather 1991). Similarly, geographic range has also been found to be a major determinant (or correlate) of the number of helminth parasites associated with British freshwater fishes (Price and Clancy 1983), although for the monogenean gill ectoparasites of cyprinid fish in West Africa the size of individual fish rather than host geographic range accounts for most of the variation (Guégan et al. 1992). Some of the differences among studies on the effects of geographic range may result from differences in the geographic scale over which some taxa range or the geographic scale at which studies are conducted. Some geographic areas have larger pools of potential parasites than others, and some of the differences in the size of the parasite pool come from differences in the number of closely related species of hosts within a geographic area.

Generalizing from this island biogeographic approach to the number of parasite species on hosts, Price (1987, 1990) has hypothesized that the composition of the parasite assemblage attacking host species undergoes continual change over evolutionary time, with new parasite species being recruited as others are lost. Parasites may evolve to specialize on a host species, and hosts may evolve defenses against these and other, less host-specific parasites, but meanwhile new parasite species may colonize a host and some long-established parasites may become extinct either locally or globally over evolutionary time. A consequence of this scenario should be that the turnover of species is likely to result in different assemblages of parasites in different geographic areas. The *Prunus* hosts used by macrolepidoptera species found

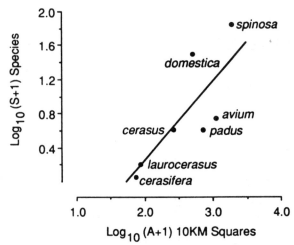

Fig. 9.1. Relationship between plant geographic distribution (A) and the number of species (S) of macrolepidoptera feeding on *Prunus* in Britain. The regression accounts for 62% of the variance. A similar regression for the five *Prunus* species that occur in Finland accounts for 79% of the variance. Redrawn from Leather 1991.

in Britain and Finland support this view. Forty-six macrolepidopteran species have been recorded as feeding on the genus *Prunus* in Britain and Finland. More of them feed on widely distributed *Prunus* species than on more locally distributed species (fig. 9.1), and the species composition of the herbivores varies geographically: nineteen species feed on *Prunus* only in Britain and another eighteen feed on it only in Finland (Leather 1991).

If the parasite assemblages on hosts are evolutionarily and geographically dynamic, as indicated by the results for *Prunus,* then current defenses against parasites should reflect both the current selection pressures imposed by the mix of enemies now attacking and the past selection pressures imposed by enemies now gone, all swirled into a geographic mosaic that reflects the fact that different host populations will be attacked by different suites of enemies. For defenses that may evolve quickly, there will be few evolutionary lags in how fast populations respond to their current suite of enemies, but for other defenses the time lags may be longer.

There are surely also evolutionary time lags in the ability of parasites, grazers, or predators to respond to new defenses, which will contribute to the geographic structure of interactions. In almost all analyses of the number of parasites associated with plant or animal host species, outlier taxa have been found that seem to harbor fewer enemies than other species similar in geographic distribution, life form, and the kinds of environments they inhabit. Jones and Lawton (1991) analyzed one aspect of the cause of these outliers, which is also one aspect of the general problem of how hosts evolve to defend themselves against multiple enemies. They asked whether plants with unusual or diverse allelochemicals interact with fewer insect species than would otherwise be expected. Unusual and diverse allelochemicals have both been commonly invoked as ways in which plants may defend themselves against all but the most specialized herbivores. A single unusual allelochemical may act as a strong barrier to most herbivores. Alternatively, or additionally, evolution of an increasingly diverse array of allelochemicals may strip away one herbivore species after another, leaving only a few specialists and preventing colonization by new herbivores.

Against these views Jones and Lawton also considered two alternatives, suggesting how the presence of unusual or diverse allelochemicals may actually increase, rather than decrease, the number of insect species associated with a plant species. One is that plants with diverse arrays of allelochemicals will tend to accumulate disproportionately more insect species over time, because such plants will likely share some of the classes of compounds found in other plant species, which will facilitate host shifts by preadapted insects. The other is that phytophagous insects will disproportionately colonize plants with unusual allelochemicals, because such plants offer an escape from parasites or predators that use plant cues to find their victims. In addition to these four hypotheses, they considered the null hypothesis that unusual or diverse allelochemistry has no consistent effect on the number of phytophagous insects that plants accumulate over evolutionary time. They chose for their comparison the insect fauna associated with British Umbelliferae, and their analyses suggested that neither unusual nor diverse allelochemistry predictably influenced the number of insects associated with a plant species. The only hypothesis that received any support at all, and then only very weak support, was that chemically diverse plants harbor more insect species than less chemically diverse species. The correlation, although significant, accounted for only 5% of the variance.

It may be that all four of the factors considered by Jones and Lawton influence the number of insects attacking a plant species and that they cancel out one another. But another reason that an unusual defense or a high diversity of defenses will not necessarily lower the overall number of attacking species may be that the expression or efficacy of defenses against enemies can vary among environments. For instance, the pinyon pines (*Pinus edulis*) that have colonized the cinder fields of Arizona's Sunset Crater since it stopped erupting about seven hundred years ago are a genetic mix of individuals differing in their degrees of resistance to phytophagous insects, but, in addition, the cinder-field population generally suffers a much more severe overall loss to insect herbivores than do nearby plants growing on sandy-loam soils (Whitham and Mopper 1985; Mopper, Witham, and Price 1990).

The stresses imposed by the physical environment apparently make many of the cinder-field plants more susceptible to the herbivores, which in turn is favoring plants with higher degrees of heterozygosity and resistance (Mopper et al. 1991). In the meantime, herbivore populations remain chronically high in the cinder-field population.

In the midst of such variation in expression of defenses among populations, it is easy to envision how a parasite population could colonize and initially adapt to a new host in an environment in which the expression of the host's defenses is unusually weak and alternative hosts are rare. Once adapted to the host, the parasite may then be able to spread to other environments, in which the host's defenses had previously been too strong for colonization.

This view of initial attack on relatively poorly defended victim populations requires that defenses commonly vary in their phenotypic expression or efficacy among environments. The varying patterns of resistance and virulence found in crop plants and pathogens in which the genetics of the interaction are well known support this view. For example, resistance in barley and virulence in the fungus *Ustilago hordei* are governed by a gene-for-gene interaction in which specific alleles for virulence in the fungus are countered in some barley genotypes by specific alleles for resistance. Plants with the genotype R_2R_2 are usually susceptible to pathogens with the genotype V_2V_2. Nevertheless, tests in California and British Columbia showed that this genotypic combination resulted in no pathogenicity in California and low levels of pathogenicity in British Columbia (Ebba and Person 1975).

Resistance to herbivory in natural communities can show similar variation among environments. Snowshoe hares (*Lepus americanus*) browse heavily in Alaska on the winter-dormant twigs of Alaska paper birch (*Betula resinifera*), green alder (*Alnus crispa*), and feltleaf willow (*Salix alaxensis*). All these plant species produce chemical compounds that affect their palatability to hares. The production of these compounds, however, can vary with the level of soil fertility and the amount of shade. If paper birch is fertilized with nitrogen or a combination of nitrogen and phosphorus, the plants reduce the amount of papyriferic acid in their internodes, and their twigs become more palatable to hares (Bryant et al. 1987). Similarly, when feltleaf willow is fertilized with a combination of nitrogen, phosphorus, and potassium, the plants reduce the concentrations of soluble carbohydrates and phenolics in their winter-dormant twigs, which makes them more palatable to hares than twigs on control plants (Bryant 1987). Shading also increases palatability, but a combination of fertilization and shading reduces palatability compared with that found in unfertilized plants in the sun (fig. 9.2).

In contrast, neither soil fertility nor shade affects the palatability of winter-dormant green alder twigs (Bryant et al. 1987). But this lack of effect of environment on green alder is, in fact, uncommon in studies of interactions

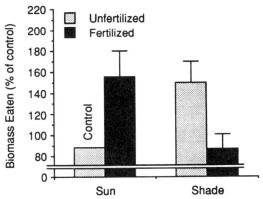

Fig. 9.2. Effects of nitrogen/phosphorus/potassium fertilization and shade on the palatability of winter twigs of feltleaf willow (*Salix alaxensis*) to snowshoe hares in Alaska. Values are the mean and 95% confidence intervals (*n* = 10 hares). Redrawn from Bryant 1987.

between plants and herbivores. Changes in soil fertility, temperature, plant density, water availability, and even atmospheric CO_2 and ozone have all been shown to influence these interactions by influencing the physiology and growth of the plants or feeding by herbivores (e.g., Maddox and Cappuccino 1986; Lincoln, Couvet, and Sionit 1986; Karban, Brody, and Schnathorst 1989; Fajer, Bowers, and Bazzaz 1989; Jones and Coleman 1988; Bazzaz 1990b). Hence, in some interactions vigorous plants grown under high levels of nitrogen availability may suffer higher levels of herbivory than plants grown under low nitrogen levels (White 1984; Room, Julien, and Forno 1989). In other interactions, environmental change may increase one plant's susceptibility to herbivory while decreasing that of another. For example, following desiccation the palatable red alga *Gracilaria tikvahiae* becomes less susceptible to grazing by the sea urchin *Arbacia punctulata,* whereas the commonly unpalatable brown alga *Padina gymnospora* becomes more susceptible (Renaud, Hay, and Schmitt 1990). Together, all these results from a number of different taxa show that the efficacy of defenses can vary among environments.

Divergence in Defenses: Setting the Stage for a Geographic View of Coevolution

In addition to geographic and local variation in the expression or efficacy of defenses, there is yet another reason why species with broad geographic ranges may continue to be subject to attack by large numbers of enemies despite unusual or diverse arsenals of defenses. Many species with broad ranges are subdivided into populations with only small levels of gene flow among them. Some of these populations may happen to be free of some or all of their usual parasites, grazers, or predators. In the absence of these antagonistic interactions, hosts should be expected to lose their defenses either through genetic drift or natural selection. Some defenses, such as production of spines in *Daphnia* or some secondary compounds in plants, will be disfavored by natural selection in the absence of antagonistic species, because they result in lower growth or reproduction than in individuals lacking these locally unnecessary defenses. Even sexual reproduction, to the extent that it is maintained as defense against parasites, may vary geographically depending upon the presence of parasites. Lively (1987, 1989) found that populations of the snail *Potamopyrgus antipodarum* differ in the frequency with which they produce males, and the frequency is positively correlated with the incidence of infection by a digenean trematode (*Microphallus* sp.). Because this trematode has adapted to local populations of its snail host, Lively argued that sexual reproduction is favored in females within populations with high infection rates as a defense against these local adaptations. The genetic diversity among the offspring of sexual females is more likely to include novel defenses against the parasite than the occasional mutation in asexual lineages.

Even if different local populations are faced with the same enemies, and similar spectra of genotypes in their enemies, host populations may still diverge in defense through three causes: the harboring of resistance genes may place those individuals at an intraspecific competitive disadvantage under some environmental conditions; defenses may be correlated with other life history traits; and the outcome of the interaction may vary among environments. The experiments of Lenski and Levin (1985) on coevolution between bacteria and bacteriophages suggest how the first of these causes—intraspecific competitive disadvantage—may lead to geographic structure in defense. In a chemostat study of coevolution between *Escherichia coli* and virulent phage T4, Lenski and Levin measured the rates of evolution of resistance and virulence. They found that *E. coli* rapidly evolved resistance to T4 phage, but that no T4 mutants appeared to counter that resistance during the long runs of their experiments. These experiments indicate that the bacteria can acquire resistance more easily than the phage can extend its host range to include more host genotypes. Lenski and Levin suggested that the cause of this difference is that bacterial resistance could evolve simply through changes in single bases, whereas attack of a new bacterial mutant requires more-specific changes in the phage's configuration or in the mechanisms triggering the injection of the genetic material.

Under such conditions, it would seem that all *E. coli* populations should rapidly evolve toward complete resistance against T4 phage, but the results of Lenski and Levin's experiments suggest that this will not always be the case. They argued that there are selective constraints that can prevent the complete establishment of resistant genotypes. They found that when glucose is not limiting, resistant bacteria show the same intrinsic rate of increase as susceptible bacteria. When glucose is limiting, however, the resistant bacteria are at a competitive disadvantage to their susceptible competitors. Consequently, resistance should be expected to vary among populations, depending upon the degree to which glucose limits population growth.

Further experiments have indicated that the reduced competitive fitness due to resistance to T4 is caused by pleiotropic effects caused by the resistance genes (Lenski 1988a). When resistant populations were allowed to evolve for four hundred generations in the absence of T4, their competitive fitness approached that of nonresistant populations evolving under identical conditions (Lenski 1988b). Surprisingly, this change in fitness occurred not from the loss of the resistance genes but from other genetic changes that compensated for the presence of the resistance genes. Hence, in some populations, the 'ghost of parasites past' could add another layer to the geographic structure of defense by maintaining some geographic variation in levels of defense among populations even in the absence of parasites.

Linkage of resistance genes with other life history traits can further increase differences in defense among populations and is the second reason why defenses could evolve differently among populations despite attack by similar enemies. For example, extensive selfing in populations could produce nonadaptive differences in defenses resulting from the restricted opportunities for recombination. This appears to be what has happened between local populations of the annual legume *Amphicarpaea bracteata* in Illinois that contain genetically divergent lineages for disease resistance to the specialist pathogen *Synchytrium decipiens*. Over 90% of the flowers produced on the plants are cleistogamous flowers, which result in self-pollination. Parker (1991) found that, as a consequence of this mating system, disease resistance was strongly correlated with other ecologically important characters such as plant morphology, phenology, and patterns of resource allocation. In one population disease resistance decreased over two years despite the continued presence of the pathogen. He attributed these nonadaptive changes in disease resistance to the strong correlation with other characters influencing fitness in these plants.

The third reason for divergence in defense despite similar patterns of attack by enemies is that the effects of attack on fitness may vary broadly among environments. There is now considerable evidence that the outcomes of interactions between species vary among environments. The 'interaction norm' of an association between two species (analogous to the term 'reaction norm' as used in population genetics) can vary from antagonism to commensalism to mutualism among environments even given the same genotypes

Table 9.2. Effect of grazing by deer and elk on growth and reproduction in scarlet gilia (*Ipomopsis aggregata*) in Arizona

	Mean (±SE)				
	Grazed		Control		$P <$
Flowers per plant	262	(50.6)	114	(14.6)	.0003
Fruits per plant	83	(11.9)	55	(7.7)	.050
Seeds per fruit	8	(0.8)	10	(1.2)	.174
Seed dry mass (mg)	1.4	(0.2)	1.3	(0.2)	.898
Stem and leaf dry biomass (g)	16.9	(3.8)	11.9	(2.2)	.134
Root dry biomass (g)	4.0	(0.5)	2.6	(0.4)	.007
Stem diameter (mm)	4.4	(0.3)	4.5	(0.4)	.888
Plant height (cm)	53.7	(2.5)	58.4	(3.3)	.456
Number of stems	5	(0.6)	1.0	(0.0)	.0001

Source: Data from Paige 1992.
Note: P = significant differences as indicated by least significant difference multiple range test. (The test also included another treatment—simulated herbivory by clipping—which closely matched naturally grazed plants for most plant characters.)

(Thompson 1986c, 1988d). The studies of Paige and Whitham (1987; Whitham et al. 1991; Paige 1992), which suggest that herbivores can actually have direct, mutualistic effects on the plants they attack under certain environmental conditions, provide one example of why attack may favor the evolution of strong defense in one population and reduced or different defenses in another environment. In some populations, grazing by deer and elk on *Ipomopsis aggregata* causes increased branching, greater flower production, and higher seed output than in control plants, with little or no loss in plant growth (table 9.2). Indirect, positive effects of herbivory in some environments, such as reduction of competitors or nutrient enhancement (McNaughton 1986), are another reason why selection for the level and type of defense may vary among host populations. The timing of the interaction during a life history, which may vary among environments, also has major effects on the outcome. Monthly grazing by the vole *Microtus montanus* on seedlings and young plants of the winter annual *Bromus tectorum* results in the death of most plants but causes only reduced growth in plants that are thirty days old and almost no effect on survival and seed output in plants that are at least ninety days old when first grazed (Pyke 1987).

Some populations may occur in environments free from enemies and therefore lose their defenses through natural selection, genetic drift, or a combination of the two. Hawaiian plants, which have been protected for much of their evolutionary history from most herbivores, are an extreme case: they are extraordinarily nonaromatic, nonprickly, and nontoxic. Carlquist (1970) has noted that although plants in the mint family are often strongly fragrant, the Hawaiian species are either odorless or nearly so. Almost all the plants cataloged in *Poisonous Plants of Hawaii* (Arnold 1968) are plants introduced to the islands over the past several hundred years. The same applies to morphological defenses. Anyone walking along the trails in the Hawaiian Islands notices the strong thorns on raspberries, but these are the introduced, rather than the native, raspberry species.

Similar geographic variation in defenses occurs in animals. Unlike ground squirrels in temperate North America, Arctic ground squirrels (*Spermophilus parryii*) have probably been free of snake predation for about three million years. When naive Arctic ground squirrels are experimentally confronted with snakes, they do not show the antisnake defenses that are found in California ground squirrels (*Spermophilus beecheyi*), which must defend themselves against gopher snakes and rattlesnakes (Goldthwaite, Coss, and Owings 1990). The extraordinary tameness of many vertebrate species on the Galápagos Islands in comparison with related species on mainlands is another example of how defenses may be lost in populations in some geographic areas.

Although the geographic differences in defense in these cases are interspecific, similar kinds of geographic variation in defense also occur among populations within species. Populations of the newt *Taricha granulosa* along the Pacific coast of Canada and the United States generally have high concentrations of a neurotoxin, tetrodotoxin, in their skin. This toxin is lethal to a wide range of potential predators (Brodie 1968; Brodie and Brodie 1990). The newts on Vancouver Island, British Columbia, however, are much less toxic than those on the mainland (Brodie and Brodie 1991). In this particular case, the lower defense in the island population is not due to the absence of their major predator, garter snakes (*Thamnophis sirtalis*). Instead, the defense and counterdefense in the newts and their predators seem to have evolved differently on the island than on the mainland. The snakes attack newts on the island just as they do on the mainland. But, for reasons that are not yet completely understood, both newt toxicity and garter snake resistance to the toxin are lower on the island than on the mainland.

The evolutionary loss of defenses in the absence of enemies, correlations between defense characters and life history characters, geographic variation in expression of defenses caused by differences in environment, differences in selection pressures caused by different suites of enemy species and competing intraspecific genotypes, and genetic drift must often act together to create a genetic mosaic in the evolutionary arms races between antagonists. The same mosaic of variation in counterdefenses must also commonly occur in parasites, grazers, and predators, creating geographic variation in specialization and in the kinds of counterdefenses deployed against hosts. The combined variation in defense and counterdefense must therefore give the evolution of many antagonistic interactions a geographic aspect that depends upon a mix of natural selection, gene flow, genetic drift, and the degree of overlap in the geographic distributions of the species.

Conclusions

Two conclusions are possible about the evolution of specialization in defense. First, at least some populations can evolve specific defenses against two or more enemies simultaneously. Hence, the observation that a host or prey species is attacked by multiple enemies is not, by itself, evidence against the possibility of pair-wise coevolution between a species and its enemies. Second, there is a great potential for geographic variation in defense within species. A major consequence of this variation is that much of the coevolution occurring between species may occur through the geographic interplay of populations differing in defense, counterdefense, and specialization rather than through reciprocal change within local populations.

CHAPTER TEN

Extreme Specialization in Mutualists

Although mutualisms appear superficially to be different from antagonistic interactions, they are built upon the same genetically selfish principles. Just as in antagonistic relationships, mutualisms involve the use and manipulation of the morphology, physiology, and behavior of other species in ways that increase an individual's fitness. Nevertheless, the evolution of mutualism sets in train unique selection pressures that make some aspects of specialization and coevolution in these associations different from antagonistic interactions. In this chapter I examine the relationship between antagonism and mutualism, the evolution of specialization in intimate mutualisms between species, and the different selection pressures shaping mutualisms between free-living species and their hosts. All so-called mutualistic interactions have the potential to vary broadly in outcome among populations, ranging from mutualism to commensalism and even antagonism in different parts of the geographic ranges of interacting species.

The terminology used to describe mutualisms remains unsettled, with no general agreement over even the most basic words such as 'symbiosis'. I think that it is useful to use 'symbiosis' to refer to any long-term, intimate relationship between two species whether it is parasitic, commensalistic, or mutualistic. This was the original usage of the word before some authors began to make it synonymous with 'mutualism'. This usage makes it possible to distinguish between symbiotic mutualisms (i.e., intimate mutualisms) and nonsymbiotic mutualisms (i.e., those between two free-living species). This distinction is helpful for the same reasons that the distinctions between parasite, grazer, and predator were helpful in the previous chapters for understanding how natural selection shapes specialization in antagonistic interactions.

Antagonism, Mutualism, and the Distribution of Outcomes

Few interactions described as mutualistic are likely to always be mutualistic for all individuals. The outcomes of interactions depend upon the age, size, social, and demic structures of populations and upon the community contexts in which they occur. Even an interaction between two local populations can vary in outcome in different years, ranging sometimes from antagonism or commensalism to mutualism (Thompson 1988d; Cushman and Addicott 1991; Cushman and Whitham 1991). Almost half the studies of interactions between ants and plants with extrafloral nectaries show no apparent benefits to the plants (Becerra and Venable 1989). There is therefore a distribution of outcomes in interspecific interactions both within and among populations. The distribution will be skewed more toward mutualism in some populations than in others, and the differences in the distributions among populations are the raw material for a geographic view of specialization and coevolution in these relationships.

In some cases the evolution of mutualism may spread from a local population to other populations, but there is no intrinsic direction in the evolution of mutualism. Many mutualisms appear to evolve directly from preexisting antagonistic interactions (Thompson 1982), but this is quite different from saying that antagonistic interactions generally evolve toward commensalism or mutualism, as once held by some parasitologists and ecologists. Reduced antagonism can evolve in some populations either through increased defenses or through the evolution of traits that have less of a harmful effect on the fitness of the other species. Both mechanisms appear to have been involved, for example, in the evolution of reduced antagonism between the European rabbit (*Oryctolagus cuniculus*) introduced into Australia and myxoma virus, which was introduced subsequently to control the rabbit (Fenner and Myers 1978, Dwyer, Levin, and Buttel 1990).

There is now a well-developed theory of the conditions that favor the evolution of reduced antagonism, at least for internal parasites, and the conceptual theory of mutualism for many kinds of interaction is also advancing. One of the conditions favoring reduced antagonism in long-term interactions is high partner fidelity (the pairing of individuals over a long series of exchanges) and low availability of alternative partners (Axelrod and Hamilton 1981; Bull and Rice 1991). For internal parasites the availability of alternative partners depends upon both the modes and the rates of transmission between hosts. Low transmission rates and strictly vertical transmission (parent to offspring) generally appear to favor parasite genotypes that are less detrimental in their effects on host survival; high transmission rates and horizontal transmission (between unrelated individuals in a population) favor parasite genotypes that maximize their growth rate in the host and thereby remain highly antagonistic to their host (Anderson and May 1982; Ewald 1983, 1987; Gill and Mock 1985; May and Anderson 1990). Hence, evolution of these interactions is likely to vary among populations, depending upon local transmission rates.

Two recent studies of the relationship between virulence and the opportunity for horizontal transfer have provided strong support for current theory. Bull and colleagues (Bull, Molineux, and Rice 1991; Bull and Molineux 1992) have directly tested the theory by experimentally manipulating three strains of *Escherichia coli* and several variants of filamentous bacteriophages. When the interaction was maintained in culture in a way that minimized horizontal spread (infectious transfer of the phage), natural selection favored less-virulent forms of the phage. When the cultures, however, allowed horizontal spread, then the advantage of the less-virulent forms was lost. This interaction is therefore likely to evolve in different ways in different populations.

The same pattern occurs in the interactions between fig wasps and their nematode parasites (Herre 1993). In this case, the opportunity for horizontal transmission of the parasite, and hence the evolution of virulence, depends upon the number of fig wasp females that normally enter a fig inflorescence (synconium). Each fig species is pollinated by its own species of fig wasp. Depending upon the fig species, one or more gravid female fig wasps enter each enclosed inflorescence, which eventually becomes the fig. The females pollinate and lay eggs within the flowers and then die within that fig. The offspring develop within the seeds, eclose as adults, mate, and then the females fly off to lay their eggs in another fig inflorescence. In some fig species, only a single fig wasp enters each fig, whereas in some other species, at least two fig wasps enter most figs.

In the vicinity of the Panama Canal, each of eleven species of fig wasp is parasitized by a distinct species of nematode (Herre 1993). The nematodes lay their eggs within the figs, and the eggs hatch synchronously with the emergence of adult female fig wasps. The young nematodes crawl onto a fig wasp, enter the body cavity, and are transported with the female as she flies off in search of a fig in which to lay her eggs. As the nematodes grow, they eventually consume the adult female, emerge from her body, and renew the cycle by laying their eggs in the fig. If only one fig wasp lays her eggs within a fig, the nematodes must rely solely upon the offspring of that female as hosts for the next generation. That is, transmission of the nematodes is strictly vertical, from a fig wasp female to her offspring. If, however, several fig wasps oviposit into the same fig, the nematodes have the opportunity to attack the offspring of several females (horizontal transmission).

Herre (1993) found that vertical transmission has favored nematodes that have no effect on the number of offspring produced by their host. By comparison, horizontal transmission has favored nematodes that are more destructive to their hosts. In fact, among the eleven species he studied, there was an almost linear relationship between virulence and the proportion of figs colonized by two or more fig wasp females (fig. 10.1).

The theory and experiments on reduced antagonism therefore suggest that

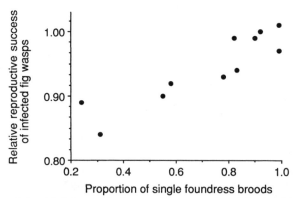

Fig. 10.1. The relationship between virulence and the proportion of figs in which all offspring are from one fig wasp female. Virulence was measured as the relative number of offspring produced by nematode-infected fig wasp females compared with the number produced by uninfected females, in figs colonized by only one female. Redrawn from E. A. Herre, "Population Structure and the Evolution of Virulence in Nematode Parasites of Fig Wasps," *Science* 259:1442–45, copyright 1993 by the AAAS.

the direction of evolution depends partially upon the structure of populations, which will affect factors such as partner fidelity and transmission. We should therefore expect a continuum between antagonism and mutualism among different populations of two or more interacting species. This continuum should be true not only of the interactions between hosts and their symbionts but also of interactions between free-living organisms. That expectation is met in the relationships between flower visitors and flowers, ants and extrafloral nectary-bearing plants, birds and fruits, and similar mutualisms in which the local outcome of a pair-wise interaction depends upon population structure or the presence of other species within the community (e.g., Cushman and Whitham 1989; Koptur and Lawton 1988; Schemske and Horvitz 1988; Breton and Addicott 1992; Rashbrook, Compton, and Lawton 1992; Thompson and Pellmyr 1992; Howe 1993).

In addition to variation in outcome among populations within species, we should expect that both antagonism and mutualism should be sprinkled throughout phylogenetic lineages—appearing here and disappearing there—as the structure of populations and communities changes over evolutionary time. It is fairly easy to find lineages of related species whose interactions with other taxa range from antagonism to mutualism.

For example, the fungal family Clavicipitaceae parasitizes a diverse group of hosts ranging from grasses to insects to other fungi. Those attacking grasses vary from causing localized infections (e.g., *Claviceps*) to causing systemic infections (e.g., *Balansia* and *Epichloë*), which often sterilize their

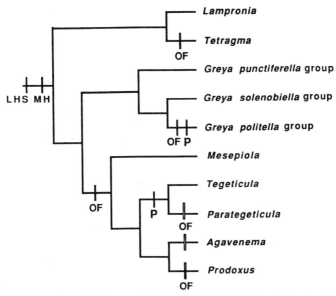

Fig. 10.2. Partially collapsed cladogram of the moth family Prodoxidae, indicating the evolution of life history traits critical to the evolution of pollination mutualism. LHS = local host specificity; MH = mating on host; OF = oviposition in flower; P = pollination by adult moths during visits for oviposition. All genera from *Mesepiola* through *Prodoxus* feed on Agavaceae. The filled bars indicate gain, open bars indicate loss, and the mixed bar indicates both states are present within the genus. Reprinted from Pellmyr and Thompson 1992.

hosts. There are many species both of clavicipitaceous fungi and of grasses, and an almost complete spectrum from parasitism to mutualism and associated adaptations is found among the species (Clay 1988, 1990). The basis of the mutualism appears to be the alkaloids produced by the fungi, which render the plants toxic to grazing mammals and some insects. Clay's comparisons among grass and fungal taxa suggest that through the selective advantage of defense against herbivores, some of these interactions have become highly coevolved and increasingly intimate. In some cases, the grass species have overcome sterilization, and the fungal species have lost the ability to sexually reproduce themselves, relying now upon transmission through host seeds.

Similarly, prodoxid moths include some species that are antagonistic to their hosts and some that are mutualistic. Prodoxids include two lineages—the mutualistic yucca moth genera *Tegeticula* and *Parategeticula* and the *Greya politella* species group—that are pollinators of their host plants. Both these lineages have evolved from related taxa whose larvae bore into plant tissues, including stems and seedpods (fig. 10.2). Ovipositing females of the two pollinating prodoxid lineages lay their eggs within flowers, and the larvae feed by boring through the floral tissues and eating developing seeds. The major difference between these moths and their antagonistic relatives is that, as adults, they either passively or actively pollinate the same flowers in which they oviposit (Pellmyr and Thompson 1992; Thompson and Pellmyr 1992).

In the pollinating yucca moths, some individuals 'cheat' by ovipositing into at least some flowers without actively pollinating them (Tyre and Addicott 1993). The extent to which this cheating behavior varies among populations is unknown, but these observations of cheating strengthen the view that these mutualisms are evolutionarily dynamic, allowing for the possibility of the evolution of cheating lineages within some yucca moth populations.

Specialization in Intimate Mutualisms

The extreme specialization found in taxa such as the mutualistic prodoxid moths appears sometimes to be built upon the pattern already developed in their antagonistic relatives. In the case of the yucca moths and their relatives, both the antagonistic and the mutualistic species include populations that are specific to one plant species or a very small group of closely related plant species (Davis, Pellmyr, and Thompson 1992; Thompson and Pellmyr 1992). The pollinating species are parasites that happen to have become mutualists, and just as parasites more commonly show extreme specialization than grazers and predators, mutualists that remain in intimate contact with a single host over long periods of time more commonly show extreme specialization than mutualists that are in contact only briefly (e.g., most mutualisms between birds and fruits, pollinators and plants, cleaner fish and host fish)—that is, nonsymbiotic mutualisms. The legacy of parasitism, then, appears to be the primary reason why, among pollinator species, the most extreme specialists are generally pollinating floral parasites that are symbiotic mutualists (such as the yucca moths and *Greya* moths, fig wasps, and globeflower flies) rather than the pollinators that are nonsymbiotic mutualists and whose only contact with hosts is through their short-term visits to collect pollen or nectar (Thompson 1982).

As mutualisms become even more long term and intimate than those between prodoxid moths and their host plants, selection on specialization may come to differ from that found in antagonistic interactions. Long-term intimate mutualisms are among the most pervasive interactions in nature. Mycorrhizal associations in plants have been found for at least some species—and usually the majority of species—in most gymnosperm and angiosperm families that have been studied so far. In some major plant families (e.g., Pinaceae, Betulaceae, Salicaceae, Dipterocarpaceae) 100% of the species for which data are available are mycorrhizal (Newman and Reddell 1987). Almost the entire higher-plant flora of some geographic areas is capable of forming mycorrhizal associations: at least 80% of British angiosperms are potentially mycorrhizal (Harley 1989). An even more diverse array of long-term symbiotic associations occurs in animals. At least thirteen invertebrate phyla include taxa harboring chronic endosymbionts (i.e., endosymbionts inhabiting 100% of the host population) (Saffo 1992). Many of these endosymbionts contribute to host nutrition, detoxification of foods, bioluminescence, or other functions that improve host survival, growth, or reproduction. They are so intimately part of the life of their hosts that, in some cases, they account for as much as 50% of the biomass or volume of the host (Saffo 1992).

Many of these intimate mutualisms show extreme specialization. Nevertheless, Law (1985) has suggested that the process of natural selection in some of these interactions may produce genotypes of symbionts so accommodating to hosts that they could transfer to other, unrelated hosts (Law 1985). There are few data to test Law's hypothesis directly, and there is no indication of whether such mutualists could maintain, over the long term, an association with two or more unrelated hosts. In support of the idea, Law noted that some single strains of mutualistic symbionts of plants (e.g., root-nodule-forming rhizobia) seem more capable of being readily inoculated into a broad range of plants than antagonistic symbionts such as fungal pathogens. Inoculation is only one aspect of the establishment of a successful association, but it is the first step.

The distribution of zooanthellae provides some additional support for the hypothesis. Zooanthellae are unicellular algae (yellow-brown dinoflagellates) that live as endosymbionts within hundreds of marine invertebrate species. Most zooanthellae have been classified into a small number of species in the genus *Symbiodinium,* but the taxonomy remains uncertain. Rowan and Powers (1991) confronted the problem of specialization in zooanthellae by isolating 131 individuals from twenty-two host taxa and characterizing them by the restriction fragment length polymorphisms (RFLP) on their ribosomal RNA. These molecular analyses indicated that individual hosts contained only one algal RFLP type, and the samples included six RFLP types. These six types, however, were distributed across phylogenetically distinct hosts. That is, closely related algae occurred in phylogenetically dissimilar hosts, which is consistent with the part of Law's hypothesis suggesting that mutualistic symbionts should be capable of shifts among phylogenetically unrelated hosts.

Law's hypothesis of reduced host specialization in symbionts is also tied to another, related hypothesis. Law (1985; Law and Lewis 1983) has argued that endosymbiotic mutualisms are more likely to favor not only reduced specificity to a single host but also loss of sexual reproduction and less speciation than in their hosts. He reasons that parasites are continually faced with new defenses produced by their hosts, which favors extreme specificity to a single host and sexual reproduction to confront the defenses, whereas endosymbiotic mutualists live in an environment that gets better rather than worse over evolutionary time as hosts adapt to accommodate the symbiont.

Law's overall argument therefore appears to cut both ways in its implications for the evolution of extreme specialization—favoring first increased, then decreased, specialization to one host. Extreme specialization would evolve early in the evolution of a mutualism, which would in turn lead to abandonment of sexual reproduction in the endosymbiont, because it is living generation after generation in the predictable environment of a single host species. Decreased specificity to that host, however, would develop later, once the symbiont has been molded over evolutionary time into an organism so innocuous that it can now even reside as a commensal or a mutualist in other hosts. If so, at any point in time different populations could be at different points in this evolutionary process.

The problem in evaluating the evolution of extreme specialization in this sequence of events is that, with the loss of sexual reproduction, the limits of the symbiont 'species' become difficult to define. Speciation events in the host will not produce parallel events of reproductive isolation in the symbiont. Consequently, the symbiont could appear to become less of a specialist as it resides in all the descendent species in a host lineage. Whether the symbionts residing in each of the hosts should be considered part of the same species, or different species, becomes somewhat of a judgment call. What matters is whether the symbiont genotypes residing in different hosts are capable of associating with more than one host species.

Evaluating further these ideas on specialization will therefore require a better understanding of the taxonomy of symbionts through the use of new molecular techniques, and evaluations of 'compatibility' that go beyond successful inoculation. The problem of extreme specialization in intimate mutualisms will demand sorting out the specificity of individual clones of symbionts and their evolutionary dynamics in natural populations. This complication of comparing sexual hosts and asexual symbionts lessened the power of Martin's (1992) attempt to evaluate Law's prediction of decreased specialization in long-term intimate interactions. Martin examined ectosymbiotic interactions between insects and their associated fungi. A number of insect species have associated fungi that they tend and eat (e.g., attine ants, macrotermitine termites) or use to digest cellulose in the food they eat (e.g., some cerambycid beetles). Martin noted that, in each of these associations, the number of fungal species involved is less than the number of insect species. The implication is that the fungal species are commonly able to reside in two or more hosts. But in evaluating specialization within the fungal taxa,

Martin also noted that the taxonomy of some of these lineages has been difficult: some do not produce fruiting bodies, which have formed the basis for much of fungal classification. Hence, although Martin concludes that these associations broadly conform to Law's predictions, the conclusions rest for now, as he realized, on the uncertainties of the taxonomy of asexual fungal species. Even if specialization in these mutualistic symbionts is lower than that in parasitic species, the vast majority still appear to have a very narrow range of hosts.

Specialization in Short-term (Nonsymbiotic) Mutualisms: Visitors

In addition to the highly adapted, intimate mutualisms found in the vast majority of multicellular organisms, there are shorter-term mutualisms that have major effects on patterns of specialization within biological communities. The mutualisms between pollinators and plants, frugivores and fruits, and cleaner fish and their hosts all have produced specialized lifestyles that play major roles in shaping the organization of communities. In most of these short-term mutualisms, one species acts as a visitor and the other as a host. This distinction between the participants is useful, because natural selection on visitors and hosts can differ markedly (Thompson 1982; Cushman and Beattie 1991), creating asymmetries in specialization and geographic differences in these interactions.

Extreme specialization is rare in mutualistic visitors for two reasons. One is that the resources exploited by mutualistic visitors are almost always ephemeral and do not permit a lifestyle of extreme specialization to a single host. The other reason is that mutualisms diversify in communities over evolutionary time in ways that select against extreme specialization. The limitations imposed by resources are present right from the start of an interaction, whereas the opportunities created by diversification of interactions appear gradually as more species are drawn into the association. Together these two influences on specialization prevent extreme specialization in all but a very few short-term mutualisms. Fruit-eating vertebrates, flower visitors that are not also floral parasites, ants that visit extrafloral nectaries (but derive nothing else directly from the plants), and cleaner fish—to name four of the most common kinds of short-term mutualistic visitors—almost all have more than one host and they commonly vary geographically in the species they visit.

Just as among free-living predators and grazers, some free-living mutualistic animals approach extreme specialization. But unlike free-living predators and grazers, mutualistic visitors rarely, if ever, appear to be able to evolve long-term specialization to one host species. The following paragraphs summarize the degrees of specialization found in the most specialized vertebrate frugivores and nectarivores to illustrate the kinds of limitations to specialization found in visitors within short-term mutualisms. Some of these species are highly dependent upon one plant family, or even one plant genus, although only one has been suggested to feed on only one plant species. (In considering these species, I lump them together as mutualists as a convenience for describing organisms that are at least potentially mutualistic, while realizing that under some ecological conditions particular species, or populations within species, are not always mutualistic.)

The most extreme specialist reported among vertebrate frugivores is Mac-Gregor's bird of paradise (*Macgregoria pulchr*), which has been called a specialist on the fruits of one tree species, *Dacrycarpus compactus* (Beehler 1983b). Because of the inaccessibility of its habitat in the central highlands of New Guinea, however, very little is known about this species and no long-term study of its diet has been undertaken (Mayr and Rand 1937; Rand 1940; Beehler 1983b).

Other than MacGregor's bird of paradise, the frugivores most specific to single plant taxa seem to be those that feed on mistletoe berries or figs. There are about 1,500 mistletoe species distributed over four plant families world-wide (Reid 1986). Some euphonias (*Euphonia* spp.) in Central and South America, mistletoebirds (*Dicaeum hirundinaceum*) and spiny-cheeked honeyeaters (*Acanthagenys rufogularis*) in Australia, other flowerpeckers (*Dicaeum* spp.) in the tropical Pacific, and some populations of phainopeplas (*Phainopepla nitens*) in North America, all rely heavily upon mistletoe species in their diets (Walsberg 1975, 1977; Anderson and Ohmart 1978; Davidar 1983; Reid 1990; Pérez-Rivera 1991). These birds often feed upon two or more mistletoe species, varying geographically in the species they have available to eat. Some species take both nectar and fruit from mistletoes, and in addition, most species include arthropods in their diets. The diets of plain-colored flowerpeckers (*Dicaeum concolor*) in western India include various combinations of the fruits and flowers of six mistletoe species as well as a variety of arthropod taxa (Davidar 1983). Some populations of mistletoebirds in South Australia feed primarily on the fruits and nectar of one mistletoe species, *Amyema quandang,* throughout the year, but other populations include the fruits of another mistletoe, *Lysiana exocarpi* (Reid 1990).

Figs are the other fruits that have favored the evolution of specialists. As with mistletoe-feeding species, fig specialists often eat more than one plant species, and they vary geographically in the fruits or the proportions of types of fruit they eat. Figs are eaten by perhaps more bird and mammal species than any other kind of fruit in tropical forests (Janzen 1980). Because they are often available year-round, these plants have been called keystone mutualists in some Central and South American rain forests (Gilbert 1980; Terborgh 1986). Similarly, they are often crucial resources for frugivores throughout the tropical Pacific (Leighton and Leighton 1983; Lambert and Marshall

1991). Among the 231 avian species within a 2.1 km^2 study area in a Malaysian forest, Lambert (1989a) observed 60 species from eighteen families feeding upon figs at least occasionally. Figs are available year round in many of these tropical forests in the tropical Pacific and in Central and South America because many of them show asynchronous fruiting patterns (Innis 1989; Bronstein et al. 1990). Moreover, they are relatively easy to find and often abundant. Consequently, they have many of the features that would allow extreme specialization, at least to the level of specialization on a group of fig species rather than one fig species.

At least some fig species are of poor nutritional quality relative to some other tropical fruits (Herrera 1981) and it may therefore be difficult to specialize on them as the sole item in a diet. None of the known fig specialists eats only one fig species, and all probably eat some fruits besides figs. Among the most extreme fig specialists are green pigeons (*Treron* spp.) in Malaysia, which feed on an array of fig species and digest both the fruits and the seeds (Lambert 1989a,b), and two congeneric birds of paradise, the trumpet manucode (*Manucodia keraudrenii*) and the crinkle-collared manucode (*M. chalybatus*), which include figs as more than 80% of their diets (Beehler 1983a). In contrast to the tropical Pacific and tropical America, figs in the rain forests of Gabon appear to be too rare to favor specialization among frugivorous birds and mammals (Gautier-Hion and Michaloud 1989). In these communities two species of Myristicaceae and one species of Annonaceae utilize figs to make year-round frugivory, although not extreme specialization, possible, because the fruits are available even during the lean periods of fruit abundance.

Except for the fig and mistletoe specialists, obligate frugivores generally eat fruits from a variety, often dozens, of plant species. The short-tailed fruit bat (*Carollia perspecillata*) includes at least eighteen fruit species in its diet (Fleming 1988). The same breadth of diet is commonly found in avian frugivores, although plants in the Lauraceae are particularly important for some obligately frugivorous birds in both the tropical Pacific and tropical America. Resplendent quetzals (*Pharomachrus mocinno*) (Wheelwright 1983), oilbirds (*Steatornis caripensis*) (Snow 1962), and bearded bellbirds (*Procnias averno*) (Snow 1970) in tropical America all rely heavily on lauraceous plants, as do some species of fruit pigeons in tropical and subtropical Australia (Crome 1975). All these frugivores take fruits from other plant families as well.

Specialization on nectar and pollen is similar to specialization on fruits. Some thrips that live in flowers and feed on pollen (Kirk 1984, 1985) and some solitary bee species that provision their nests with pollen (Linsley, MacSwain, and Raven 1963) are restricted to a very narrow range of hosts. But most colonial and social bees that provision nests over a longer period of time and other free-living flower visitors that move from flower to flower seem to include in their diets plants from at least two or more genera. The Galápagos carpenter bee, *Xylocopa darwini,* which is the only bee species in the Galápagos Islands, has an especially broad diet and has been recorded visiting flowers of at least sixty plant species in twenty-eight families (Linsley, Rick, and Stephens 1966). Similarly, almost all hummingbirds, sunbirds, honeyeaters, honeycreepers, pteropodid bats, and nectar-feeding marsupials that have been studied so far use more than one plant species during each year, although much of the evolutionary fitness of some of these flower visitors may depend heavily on one or two plant species.

With one known exception, all vertebrates specialized to a diet of nectar and pollen are able to fly, moving readily between flowering plants that are sometimes widely dispersed. The exception is the honey possum, *Tarsipes rostratus,* a small (7–12 g) marsupial with a slow developmental rate and an average life span of about one year (Richardson, Wooller, and Collins 1986; Russell 1986). Although some other marsupials include pollen and nectar as a regular part of their diets (Rourke and Wiens 1977; Sussman and Raven 1978; V. Turner 1984), only the honey possum is restricted to this diet. It is so different from other marsupials that it has been placed in its own superfamily. Honey possums occur only in the coastal heath sand plains of southwestern Australia, which are rich in year-round availability of flowers in the Proteaceae and Myrtaceae. They feed primarily on the pollen and nectar of proteaceous plants, especially *Banksia* spp., which they gather with highly specialized brush-tipped tongues (Richardson, Wooller, and Collins 1986), but during most months of the year individuals use at least two or more flowering species (Wooller et al. 1983).

The dietary flexibility retained by almost all these vertebrate frugivores and nectarivores suggests that natural selection for extreme specialization occurs even less commonly in free-living mutualistic visitors than in predators and grazers. The kinds of resources they exploit almost never appear to allow complete specialization to only one host species and rarely to one host genus. Instead, species rely locally upon combinations of hosts, and those combinations vary geographically.

The Spread of Mutualism and Selection on Visitors

The limits to specialization on mutualistic visitors are reinforced by the opportunities for despecialization that arise as new hosts evolve to exploit preexisting mutualisms. The interactions between hawkmoths (Sphingidae) and flowers within local communities show clearly how convergence of host species can produce a complex local network of mutualistic interactions within which there is little selection on visitors to become specialized to one host species. In the tropical dry forest at Cañas in northwestern Costa Rica, Haber and Frankie (1989) recorded sixty-five hawkmoth species and thirty-one native plant species adapted primarily for hawkmoth pollination. The plants visited by the hawkmoths include about 10% of the trees, together with some shrubs, herbs, lianas, and epiphytes. Most, but not all, the plants pollinated by hawkmoths at Cañas have the combination of floral characters generally associated with hawkmoth pollination: white tubular corollas and flowers that open or mature their stamens in the evening and that produce a sucrose-rich nectar and a sweet-smelling scent also in the evening (Faegri and van der Pijl 1971; Baker and Baker 1983; Grant 1983).

There is, however, some variation in floral characters among the plant species. Twenty-one of the thirty-one species have tubular flowers with either narrow or flared corollas (fig. 10.3a,b), eight have brush flowers with reduced corollas (fig. 10.3c), and two have free petals (fig. 10.3d). Most flowers open in the evening: two-thirds of the plant species do not completely open their flowers until after dark, and most of those that do open during the day do not dehisce their anthers until dusk or later. All but three of the species have white flowers, and all but one of the species tested for nectar composition have nectar dominated by sucrose. (The one exception is a species with hexose-dominated nectar, which is common in bat-pollinated plants.) What is noteworthy for understanding the spread of mutualisms is that these thirty-one plant species are distributed over fourteen families. Despite their differences in floral characters, they are all convergent solutions to exploiting the large local assemblage of hawkmoths as pollinators.

Unlike the plants, the hawkmoths are all in one family and are all variations on the same basic hawkmoth design. The local assemblage at Cañas spans twenty-seven genera, and tongue lengths among the species range from 1 to 20 cm. The short-tongued species are restricted to flowers with short corollas, but the long-tongued species visit flowers with short corollas as well as flowers with long corollas. Haber and Frankie found little evidence of floral specificity among the hawkmoth species. The moths move among the various plant species that have evolved adaptations to attract these moths. Additionally, they also visit a large number of other plant species. Haber and Frankie found pollen from over one hundred other plant species on the moths they collected in the field during the study, and some of these species were visited more commonly than 'hawkmoth flowers'.

The view that this study, and other similar studies, creates of mutualisms between hosts and visitors is one of a shifting, dynamic relationship in which species are collected and dropped out of the interaction over evolutionary time as new species evolve or arrive locally. Any one species may interact mutualistically with different groups of other species in different parts of its geographic range. Once evolution has produced a group of species with simi-

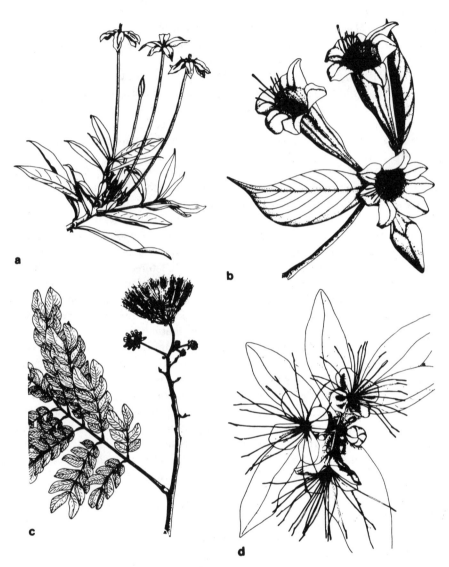

Fig. 10.3. Shapes of hawkmoth-pollinated flowers in the tropical dry forest at Cañas in north-western Costa Rica: *a,* tube flower of *Lindenia rivalis; b,* expanded tube flower of *Coutarea hexandra; c,* brush flower of *Pithecellobium samán;* and *d,* the free-petaled, tubeless flower of *Capparis indica.* Reprinted, by permission, from Haber and Frankie 1989.

lar traits, either through radiation from a common ancestor or convergence of unrelated taxa on a suite of characters, an additional impetus is created for yet other species to exploit the mutualism. Hence, bat pollination, bat dispersal, avian pollination, avian dispersal, and most of the other animal-assisted pollination and dispersal syndromes have all arisen multiple times throughout the angiosperms.

The Spread of Mutualism and the Origin of New Lifestyles

A major consequence of this spread of mutualism is the evolution of new visitor lifestyles that are specialized to that type of interaction without being specific to a single visitor or host. Long-tongued hawkmoths and eusocial bees that rely upon a succession of plant species throughout the year are one result of the convergence of hosts attracting particular groups of mutualistic visitors; obligate frugivores are another. Visitors that rely upon a succession of fruiting or flowering species throughout the year are possible only once a number of similar mutualisms have evolved and now co-occur within single communities or throughout the migration range of a species (Thompson 1982).

It is no surprise then that most obligate frugivorous and nectarivorous vertebrates are birds and bats that can fly in search of these scattered resources. Some species, such as resplendent quetzals and wompoo pigeons (*Ptilinopus magnificus*) move relatively short distances among communities throughout the year in search of fruit. Resplendent quetzals in Costa Rica move seasonally among plant communities at different elevations (Wheelwright 1983) and wompoo pigeons in subtropical Australia move erratically over short distances, searching for fruiting figs, which make up the bulk of their diet (Innis 1989).

As the number of species involved in these interactions increases, the obligate mutualists sometimes develop specialized lifestyles to exploit the diversity of their hosts. This process, however, does not appear to lead to specialization to any one plant species. For example, frugivorous bats have evolved two different lifestyles, both of which allow the use of a diversity of plant species (Spencer and Fleming 1989). Three separate evolutionary lineages of bat have converged on a lifestyle that includes three common features: consumption of figs (along with other fruits and sometimes nectar), in which individual figs are harvested one at a time and carried away to be eaten elsewhere, foliage-roosting as solitary individuals or as small groups, and frequent changes in the roosting sites over small areas of forest. This suite of behaviors is found in the tropical American phyllostomatid subfamily Stenoderminae, in African epomorphorines, and in at least one species of Australian pteropodid, the tube-nosed bat *Nyctimene robinsoni*. In contrast, the larger Australian flying foxes *Pteropus,* which feed on figs as well as other fruits and nectar—and in at least one species sometimes even the juices from green leaves chewed into a bolus that is expelled as a pellet (Lowry 1989)— roost in large colonies often kilometers away from available food, move roosts only seasonally, and sometimes defend parts of fruiting trees.

Hummingbirds, too, have specialized into different lifestyles, and in the process they have evolved to exploit subsets of the wide range of flowering plants available in many communities—without becoming specialized to any one plant species (Feinsinger and Colwell 1978; Stiles 1985). In the pre-montane primary forest of La Montura in Costa Rica, Stiles (1985) found four different kinds of association between seventy plant species and twenty-two hummingbird species. Some of these interactions are 'generalized' associations between short-billed hummingbirds and flowers with short corollas. But there are, in addition, three associations involving more-specialized lifestyles. The green-fronted lancebill (*Doryfera ludovicae*), which has a long, straight bill, relies upon five species of large shrubby epiphytes (four ericads and a mistletoe) with deep, straight corollas. Two hermit species with long, curved bills—the green hermit (*Phaethoris guy*) and the white-tipped sickle-bill (*Eutoxeres aquila*)—rely upon plants with long, curved corollas. Green hermits include nineteen such plants in their diet, and white-tipped sickle-bills only two. Sicklebills, however, are short-term visitors to the community at times of the year when only two plant species it can exploit with its exceptionally curved bill are in flower. The fourth lifestyle is an association between the green-fronted brilliant (*Heliodoxa jacula*) and three species of moth- or bat-pollinated *Marcgravia.* This hummingbird relies upon these three species, but it is a nectar thief rather than a pollinator.

In general, the evolution of visitors that rely upon a succession of hosts appears to be a common result of the spread of short-term mutualisms among species. It is part of the process by which mutualisms diversify within biological communities. It produces groups of mutualistic visitors that are specialized for a particular lifestyle without being specialized to any one host species.

Specialization in Hosts: Evolving a Selective Sieve

The cheater lifestyle of the green-fronted brilliant at La Montura illustrates that selection on hosts in mutualisms differs from selection on visitors. For mutualisms involving short-term visits, in which a host exchanges a reward for some service done by a visitor, the problem for the host is to create a selective sieve. It must allow visits by mutualistic visitors but simultaneously minimize visits by thieves and cheaters that steal the rewards without providing any of the intended services in return. Hence, plants must contend with both true pollinators and nectar robbers, and they must attract frugivores that disperse seeds while minimizing visits by other species that digest the seeds or drop them all below the parent tree. Similarly, a host fish must distinguish between a cleaner fish and the cheating species, which look and act like cleaners but actually take bites out of the side of a host once it has adopted the cleaning position.

A selective sieve is much more difficult to maintain than an absolute barrier, especially in an interaction in which visitors must come and go repeatedly. Consequently, even if selection favors extreme specialization to particular visitors, it may be difficult to maintain over long periods of evolutionary time. The evolutionary moment a host offers a new resource specifically to attract one visitor species, that host becomes a potential target for other species that may begin to use that resource opportunistically and even evolve to specialize on it. Somewhere in the geographic range of a broadly distributed host species, a cheater is almost bound to evolve to exploit it. The cheater population may then spread to other host populations. By adopting a permissive behavior that lets cleaner fish move unmolested over its skin, a host fish opens up a world of possibilities for other kinds of fish that might prefer to attack or attach to the host. By presenting pollinators with readily harvested, high-energy rewards such as nectar and pollen, a plant lays itself open to nectar robbers and nonpollinating pollen collectors. This concentration of resources and the visitors it attracts even give rise to new, specialized predatory lifestyles: crab spiders, ambush bugs, and flower-mimicking praying mantids are all specialized exploiters of mutualisms that use flowers as a platform for capturing prey.

A host can use morphological defenses in an attempt to try to screen out the thieves and cheaters, but that simply sets up the conditions for an evolutionary arms race. Long floral spurs, which make it difficult for short-tongued insects to reach the nectar, favor the evolution not only of long-tongued pollinators but also of thieves that chew in through the side of the corolla to reach the rewards. The alternative, adopted by some plants that have become extreme specialists to particular pollinators, is to eliminate most of the usual rewards that attract cheaters. Hence, yucca species, which are pollinated only by the yucca moths that lay eggs in their flowers, have highly reduced or absent nectaries (Trelease 1902). Orchids that are pollinated by males of a single wasp or bee species use the trick of pseudocopulation and have abandoned rewards and rely upon mimicking the female wasps or bees to attract their specific pollinators. Other species use fragrances that are collected by particular pollinator species and may be of little utility as rewards for other species.

Even if the cheaters and thieves can be thwarted through defense or elimination of most rewards, hosts are still faced with the problem of just how

Table 10.1. Variation among years in the percentage of visits to *Lavandula latifolia* in southeastern Spain by pollinators within three insect orders

Insect order	Percentage of visits each year					
	1982	1983	1984	1985	1986	1987
Hymenoptera	66	52	36	50	51	58
Lepidoptera	26	35	58	41	40	29
Diptera	9	13	6	11	9	13

Source: Adapted from Herrera 1988.

selective to make the sieve among the remaining visitors. If the relative frequency of visits and effectiveness of pollinators were constant over time within plant populations, then selection would favor specialization to attract only the most frequent and effective pollinator(s)—what Stebbins (1970) called "the most effective pollinator principle." But studies of pollinator assemblages have now shown that the local composition of pollinator taxa can vary markedly within seasons, between years, and over short distances (e.g., Herrera 1987a; Schemske and Horvitz 1988; Thompson and Pellmyr 1992). The large variation possible in a local pollinator assemblage is evident in Herrera's (1987a, 1988) study of visitation frequency of pollinator species to an insect-pollinated Mediterranean shrub, *Lavandula latifolia* (Labiatae), during six consecutive years. In the relatively undisturbed montane woodlands of southeastern Spain, *L. latifolia* is pollinated by up to eighty-five species of Hymenoptera (especially relatively long-tongued bees), Diptera (especially syrphid flies), and Lepidoptera (especially nymphalid, satyrid, and lycaenid butterflies). About forty of these species have been shown to be effective pollinators, although they differ in their efficiency (Herrera 1987a), and another forty-five have been inferred as pollinators (Herrera 1988). Herrera found that only 36% of the pollinator species were recorded visiting the flowers in all six years of the study.

These results indicate how widely the composition of the pollinator assemblage can vary between years. Even within a year only 22% of the pollinator species were present throughout the entire flowering season. If the taxa are grouped into orders—Hymenoptera, Lepidoptera, Diptera—similar variation is evident. Although flies composed 6–13% of visits each year, bees and wasps varied from 36 to 66% and butterflies and moths varied from 26 to 58% (table 10.1). Consequently, even selection for specialization on bees as a group, or butterflies as a group, probably varies among years. Moreover, only 41% of the pollinator species were recorded at all four study populations, indicating that the relationship between this plant and its pollinators varies geographically as well.

The lack of predictability in the mutualist assemblage, the tendency for mutualisms to spread within communities, and the penetration of the selective sieve by thieves and cheaters are therefore all major counterbalances to natural selection favoring extreme specialization throughout the geographic range of a species in short-term mutualisms (Thompson 1982; Feinsinger 1983; Schemske 1983; Howe 1984; Herrera 1985). Natural selection on specialization in mutualisms probably shifts routinely over evolutionary time, sometimes favoring increased and sometimes decreased specialization, as new traits appear in individuals and new species evolve and immigrate to exploit an interaction.

Conclusions

There is now a substantial body of theoretical and empirical work on the ecological conditions favoring reduced antagonism and mutualism. Intimate interactions, which often derive from parasitic relationships, appear to evolve toward reduced antagonism, and sometimes perhaps mutualism, when transmission rates between hosts are low and are generally between a host and its offspring (vertical transmission). Hence, the evolution of these interactions is likely to vary among hosts and host populations that differ in spatial and social structure.

Natural selection on specialization is different in short-term (nonsymbiotic) mutualisms than in intimate (symbiotic) mutualisms. Selection is less likely to favor extreme specialization in short-term mutualisms than in long-term intimate interactions. Specialization in visitors is limited by the kinds of resources they exploit and by the opportunities for despecialization that develop as new hosts evolve to exploit preexisting mutualisms. As mutualisms spread within and among communities, selection can favor new visitor lifestyles that exploit the very diversity of hosts available rather than any one host. Specialization in hosts to one visitor species is similarly limited by the tendency for mutualistic resources to attract new visitor species, unpredictability in the visitor assemblage, and the inability of hosts to evolve a selective sieve that permanently allows visitation by specialist visitors but not by thieves and cheaters. As a result, sustained local opportunities for pair-wise coevolution appear to be uncommon in short-term mutualisms.

Instead, the spread of short-term mutualisms within and among communities creates the opportunity for a very dynamic local and geographic structure to these interactions. Visitors readily adjust their choice of hosts depending upon the local spectrum of hosts available, and hosts are confronted with an ever-changing spectrum of visitor species.

CHAPTER ELEVEN

Further Limitations on Specialization in Mutualisms

Other limitations, arising from the spread of mutualisms and, in some cases, quirks of the biology of particular taxa, can prevent extreme specialization in short-term mutualisms. In this chapter I use interactions between pollinators and plants to consider how those limitations can prevent hosts from becoming specialized to one mutualistic visitor species or restrict specialization to some populations but not to others.

Swamping of Mutualism and Prevention of Specialization in Hosts: A Case Study

Even if a visitor is a predictable and effective mutualist, natural selection may not be able to favor specialization in hosts once there is an assemblage of visitor species. The reason is that the mutualistic effects of any particular visitor may simply be swamped by the combined effects of all the other species. This must be a common situation in mutualisms as they spread within communities. The evolutionary problem is to understand how and when extreme specialization between local populations of a particular host and visitor can evolve in the midst of co-mutualists.

Studies of the interactions between the relatives of yucca moths and their host plants are beginning to show how selection for specialization depends not only on the pair-wise interaction between two mutualists but also on the community context in which the interaction occurs. The most reciprocally specific interactions between insects and plants are those involving pollinating floral parasites that oviposit into the plants they pollinate: yuccas and yucca moths, figs and fig wasps, and globeflowers (*Trollius*) and globeflower flies (e.g., Riley 1892; Davis 1967; Janzen 1979; Bronstein 1989; Herre 1989; Pellmyr 1989; Addicott, Bronstein, and Kjellberg 1990; Powell 1992). Each of these interactions involves an insect that lays its eggs in some of the same flowers that it pollinates.

These insects are commonly referred to as either seed predators or seed parasites. They are seed predators in the sense that the larvae quickly kill individual seeds. The focus of natural selection in these interactions, however, is on the relationship between the insects and the parent plant. Natural selection on specialization in the insects comes from their parasitic relationship with the parent plants, and natural selection on the plants comes from the symbiotic (both parasitic and mutualistic) relationship with the insects. Hence, a better term for these insects is perhaps floral parasites.

The reciprocal extreme specialization in these interactions sets them apart from most plant–pollinator mutualisms (Feinsinger 1983; Schemske 1983) and from most other mutualisms between plants and insects or vertebrates (Thompson 1982; Wheelwright and Orians 1982; Howe 1984). Most fig wasps and globeflower flies are specific, at least locally, to only one plant species and, in turn, local plant populations rely exclusively on that one pollinator or on a group of congeneric pollinators (Ramirez 1970; Bronstein 1987; Pellmyr 1989). Extreme specialization similar to that found in fig wasps and globeflower flies is also becoming apparent as studies reveal more sibling species within the yucca moths and related prodoxid moths (Davis 1967; Miles 1983; Addicott, Bronstein, and Kjellberg 1990; Davis, Pellmyr, and Thompson 1992).

All these well-known interactions have evolved to be so exclusive of other pollinators that it is difficult now to tease apart how extreme specialization to a pollinating seed parasite could have evolved in plant populations against a background of co-pollinators. But the recently discovered interactions between the moth genus *Greya* and their host plants provide the conditions necessary for understanding how extreme specialization can evolve, or, alternatively, be constrained by the community in which the interaction takes place. *Greya* includes species in which the moths are pollinating floral parasites but share the plants with co-pollinators (Pellmyr and Thompson 1992; Thompson and Pellmyr 1992). *Greya* appears to be the prodoxid genus closest to the Agavaceae feeders, which includes the yucca moths (Nielsen and Davis 1985; Wagner and Powell 1988; Davis, Pellmyr, and Thompson 1992). All pollinating *Greya* species develop as larvae on the plants that their parents pollinate. But the plants they use are in the Saxifragaceae, especially the genus *Lithophragma*.

In eastern Washington the flowers of *Lithophragma parviflorum* are pollinated by *Greya politella*, bombyliid flies, and bees, and it is this multispecific interaction that shows how selection for specialization can depend upon community context. *Lithophragma parviflorum* is the most widely distributed species in the genus, growing throughout western North America in steppe, grassland, and savanna. Each plant produces several small basal leaves and a floral stalk with tubular flowers that open in sequence from bottom to top. Pollinators probe the narrow tubular corolla, which includes ten anthers and long, fused styles, to reach the nectar at the base. Below the nectary, the inferior ovary contains 150–400 ovules. The life cycle of *Greya*

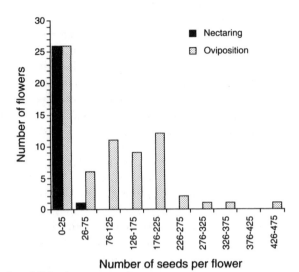

Fig. 11.1. Number of *Lithophragma parviflorum* seeds resulting from a single bout of nectaring or oviposition by *Greya politella*, a relative of yucca moths (*n* = 27 for nectaring; *n* = 70 for oviposition). Redrawn from Pellmyr and Thompson 1992.

politella is centered on *L. parviflorum* plants (Thompson and Pellmyr 1992). The adults mate on the flowers, the females oviposit into the ovary, the early-instar larvae feed on the developing seeds, and after a summer aestivation (or possibly feeding below ground), the later-instar larvae feed on the new basal leaves produced in late fall or winter.

Both male and female *G. politella* probe flowers in search of nectar, but these probes with their proboscises result in little or no pollination. Only when females probe the flowers with their ovipositors in an attempt to lay eggs in the ovary does significant pollination occur (fig. 11.1). During oviposition a female inserts her abdomen deep into the corolla, cuts through the nectary into the ovary with her ovipositor, and lays several eggs stacked one on top of the other within the ovary. Pollination occurs as the pollen on the female's abdomen, picked up during oviposition into others flowers, is deposited onto the stigmatic surface (fig. 11.2).

Each developing *Greya* larva eats about four to ten developing seeds, and together larvae eat on average 15–27% of the seeds within attacked flowers. These eaten seeds are one of the costs to the plants of pollination by this *Greya* species. The values are within the range found for yuccas, figs, and globeflowers, which are all pollinated exclusively by floral parasites (Janzen 1979; Addicott 1986; Bronstein 1988; Herre 1989; Pellmyr 1989). Hence, *G. politella* has the potential to be a mutualist with its host plant: females are

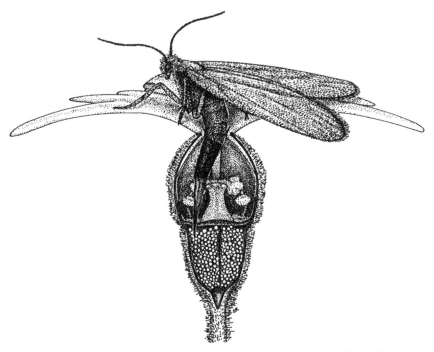

Fig. 11.2. *Greya politella* ovipositing into the ovary of a *Lithophragma parviflorum* flower. Reprinted, by permission, from Pellmyr and Thompson 1992.

efficient pollinators and they impose only a modest cost in consumed seeds relative to the number of ovules they fertilize during oviposition. Moreover, they commonly visit about half the flowers in a population (Thompson and Pellmyr 1992).

The mutualism, however, can be swamped locally by the other insects that visit and pollinate these flowers. During four years of study of this interaction at Granite Point in eastern Washington State, at least eighteen insect taxa (species and species complexes) visited the flowers of *L. parviflorum*. In systematic censuses of the pollinators during two of those years, the most frequent visitors were bombyliid flies, which accounted for 66–88% of the 5,522 recorded visits. Bees, mostly solitary bees, made up most of the remaining visits, whereas *Greya politella* accounted for only 1–2% (Thompson and Pellmyr 1992). This low percentage of visits by *Greya* results from two causes: the greater relative abundance of the co-pollinators and the faster rate at which these co-pollinators visit flowers. An ovipositing *Greya* can take

Table 11.1. Differences among pollinator species in the mean number of ovules fertilized during a single visit to *Lithophragma parviflorum*

| Flower visitor | Number of visits | Number of mature seeds | | | |
		Median	Mean	SD	Range
Diptera					
Bombylius albicapillus	50	69	93	101.0	0–371
Bombylius major	155	106	130	125.8	0–530
Bombylius washingtoniensis	37	71	110	117.3	0–443
Hymenoptera					
Andrena + *Anthophora*	4	162	176	203.8	0–378
Evylaeus cooleyi	10	0	25	54.9	0–154
Osmia spp.	8	61	64	42.6	0–125
Lepidoptera					
Greya politella:					
Probing for nectar + ovipositing	72	75	89	95.1	0–453
Nectaring only	28	0	4	7.0	0–27

Source: Data from Thompson and Pellmyr 1992.

several minutes to oviposit into a flower and then wait an hour or more before visiting another flower. Meanwhile, a bombyliid fly can visit ten or more flowers within a minute.

The 1–2% of visits to *Lithophragma* flowers by *Greya* stands in stark contrast to the near monopoly that the closely related yucca moths maintain in visitation to yucca flowers. If bombyliid flies and bees visited *Lithophragma* flowers as cheaters rather than as co-pollinators, then that 1–2% would be critically important for pollination. The bombyliids and bees, however, are at least as efficient in effecting seed set as *Greya*. A single visit by any of the bombyliid flies results on average in fertilization of 93–130 ovules, or about 25–50% of the total, as compared with an average of 89 ovules for *Greya* (table 11.1). Not only are these co-pollinators as efficient as *Greya,* they do not, unlike *Greya,* eat *Lithophragma* seeds as larvae. Hence, they are more purely mutualistic in their interaction with *Lithophragma* than is *Greya*.

The four years of study at Granite Point indicated that pollination in this *L. parviflorum* population does not rely upon visitation by *Greya*. At only one study plot in one year was there any discernible effect of *Greya* pollination over and above that of the other co-pollinators. In that one instance, flowers with *Greya* eggs were more likely to have some developing seeds than flowers without *Greya* eggs. When seed consumption by *Greya* larvae is included in the analysis, this positive effect of *Greya* disappears (Thompson and Pellmyr 1992). Specialization for pollination by *Greya* would therefore seem possible only in a *L. parvifloum* population in which co-pollinators were rare or unpredictable relative to *G. politella*. That is, selection for reciprocal specialization in this kind of mutualism relies at least as much on the community context of the interaction as it does on the costs and benefits of the pair-wise interaction. Both *L. parviflorum* and *G. politella* occur broadly over the northwestern United States and southern Canada, and co-pollinators may be rare or unpredictable in some of the populations. Hence, specialization and pair-wise coevolution may be possible in some parts of the geographic range of these two species but not in others.

Direct and Indirect Effects on Selection for Specialization

The community context of an interaction between visitors and hosts can pre-empt the evolution of specialization in yet another way. Not only must specialization for or against particular visitor species often occur in the midst of a host's interactions with other mutualistic visitors, but it must also develop amid yet other interactions (e.g., parasites, grazers, predators) that may directly or indirectly influence the mutualism. By studying each pair-wise interaction, we can achieve some understanding of how this network of species fits together, but this method gives only an incomplete picture of the rich variety of direct and indirect ways in which a group of interacting species can influence each other. The statistical procedure of path analysis provides one way to understand how an interaction between a pair of species is influenced, both directly and indirectly, by other species within a community. It is useful as a way of visualizing how the network of potential selection pressures may act against the evolution of extreme specialization for particular mutualists or permit it only in local populations where particular combinations of species are present.

Originated by Wright (1934), the method makes use of standardized partial regression coefficients (called path coefficients), each of which is the expected magnitude of the direct effect of an independent variable on a dependent variable when all other independent variables are held constant. The procedure then uses each of these direct effects to analyze all the direct and indirect pathways by which one variable affects another. Hence, an insect herbivore can directly influence the number of mature seeds that a plant produces by feeding on seeds. Alternatively, it may affect the number of mature seeds by chewing on leaves, which affects the amount of photosynthetic resources that the plant has available for flower production, which in turn affects the attractiveness of the floral display to pollinators and hence the number of floral visitors, which in turn affects the number of flowers that are eventually pollinated. The influence of the herbivore, however, may not end there. The decreased level of available photosynthate caused by herbivore damage could directly influence the number of pollinated ovules that the plants can sustain to maturity, thereby decreasing any direct relationship between pollinator behavior and seed set. These tortured pathways can be analyzed for the magnitude of their direct effects, indirect effects, and the total correlation (combined direct and indirect effects). (Li [1976] and Pedhazur [1982] discuss the procedures in detail, and Kingsolver and Schemske [1991] review a number of the uses of the procedures in evolutionary biology.)

For the study of interspecific interactions and the evolution of specialization, the method provides a way of grasping some aspects of the network of interactions between species and the constraints on natural selection for specialization in mutualisms. It is not a panacea but rather just one more useful tool for helping to understand the sometimes subtle ways in which the effects of one species on another can be influenced by the larger assemblage of interacting species. Since the pathways are unidirectional, the procedure cannot be used to produce a path diagram that is reciprocal and, hence, coevolutionary. Bidirectional effects can be included, but these cannot be decomposed to analyze how they alone affect the model.

Because path analysis can analyze both direct and indirect effects, it can be more useful in studying interspecific interactions than multiple regression, which measures only the direct effects of a number of independent variables (e.g., the independent direct effects on seed production of visits by pollinator species A, B, and C). In path analysis, one can begin with a group of observations of or experiments on an interaction and construct a flow diagram of how one species affects another and how yet other species might influence that interaction. The flow diagram is a hypothesis about the direct and indirect ways in which one species affects another. Multiple versions of the flow diagram can be tested to analyze how different views of the relationship affect the outcome. It can also be used to examine how small geographic differences in the structure of communities could potentially influence the pair-wise interactions between species.

No one has actually attempted to use path analysis to study the reciprocal effects of a pair of species on each other in a community context. There have, however, now been several attempts to analyze how an assemblage of species influences traits important for fitness in one particular species. These studies show the kinds of limitation on the evolution of specialization in mutualism that results when interactions take place in the context of other interactions. The most detailed attempt has been the studies of Schemske and Horvitz (1988), who were interested in how variation in the interactions between a Neotropical perennial herb, *Calathea ovandensis,* and its mutualists and herbivores affects the number of fruits that a plant matures. Although they studied the yearly variation in the outcome of interactions at one site as the structure of the community changed, their results can also be used to envision how the outcome of an interaction can vary geographically among communities.

Calathea ovandensis occurs in the understory of lowland forests. The inflorescences produce extrafloral nectaries that are harvested by ants. The larvae of a riodinid butterfly, *Eurybia elvina,* feed on the inflorescences but are tended rather than killed by the ants because they, like the flowers, produce secretions that the ants harvest (Horvitz, Turnbull, and Harvey 1987). The presence of ants on the plants is positively correlated with that of *Eurybia* larvae, with ants having a positive effect on flower and fruit production and *Eurybia* having a negative effect (Horvitz and Schemske 1984, 1988). These relationships can be built into the path model as a correlation between ant and caterpillar numbers and unidirectional effects of ants and caterpillars on the number of flowers and fruits (fig. 11.3).

Calathea flowers are visited mostly by hymenopterans and lepidopterans, but only some of the hymenopterans are effective pollinators (Schemske and Horvitz 1984). The number of flowers on a plant influences the numbers of visits by each pollinator, which in turn can influence the number of fruits that a plant initiates. Schemske and Horvitz knew from earlier studies that pollinators did not differ in their influence on the number of fruits that matured from initiated fruits. Consequently, their path model did not include any direct link between pollinators and the number of mature fruits. Both the ants and the *Eurybia* larvae, however, had the potential of influencing the number of mature fruits both directly and indirectly, so both kinds of links were included.

The interactions during 1984 and 1985 differed in the insect assemblage visiting the plants. Although ant presence was similar in the two years, the number of *Eurybia* larvae in 1984 was twice that in 1985, and the more effective pollinator species were less common in 1984 than in 1985. Consequently, the resulting path diagrams of how insects influence the production of mature fruits were also quite different between the two years (fig. 11.3). In both years, the presence of ants had a direct positive effect on the number of flowers produced and, through that influence, a positive indirect effect on the number of mature fruits. Hence, ants affected the number of flowers, which affected the number of visits by pollinators and the number of fruits initiated, which in turn affected the number of fruits matured.

The links between flowers, pollinators, and initiated fruits differed between years. No one pollinator species had a significant effect on fruit initiation in 1984, when the two most effective pollinators, *Rhathymus* sp. and *Euglossa* spp., were uncommon. These effective pollinators, however, were much more common in 1985 and had a significant positive effect on fruit production. This variation between years in the direct effects of particular pollinator species on plant reproduction indicates the kind of variation in selection for pollinators that a plant population might experience over time. Moreover, it is not just the assemblage of pollinators but also the mix of pollinators and herbivores that influences mature fruit production. *Eurybia*

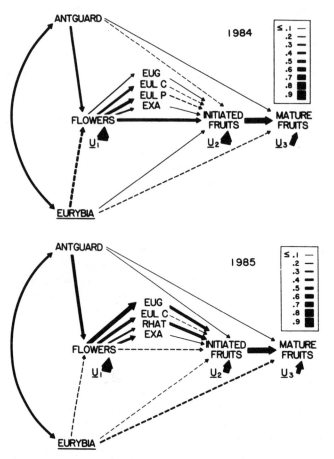

Fig. 11.3. Path diagrams for *Calathea ovandensis* in 1984 and 1985. Positive effects are indicated by solid lines, and negative effects by dashed lines. The keys give an indication of the magnitude of the path coefficients. EUG = *Euglossa* spp.; EUL C = *Eulaema cingulata;* EUL P = *Eulaema polychroma;* EXA = *Exaereta smaragdina;* RHAT = *Rhathymus* sp. Reprinted, by permission, from Schemske and Horvitz 1988.

larvae had the largest overall effect of animals on mature fruit production in 1984, with about 63% of the effect direct and the remainder indirect. In 1985, *Eurybia* larvae and the two most effective pollinators, *Rhathymus* sp. and *Euglossa* spp., had the largest effects on mature fruit production. The negative effect of *Eurybia* larvae was about equal to the positive effect of each of these two pollinators. If feeding by herbivores routinely disrupts the relationship between pollinator visits and seed set, then natural selection for specialization to particular pollinators would be slowed considerably or even prevented.

These path diagrams give an indication of how these insects interact directly and indirectly among themselves and with *Calathea* in ways that influence plant reproduction. It is not a complete model, but it is not intended to be. It provides a way of testing ideas about how these species interact, and it suggests how much of the variation in mature fruit production remains unexplained by our view of these interactions. In this case, the path model explained 53% of the variance in mature fruit production in 1984 and 59% in 1985. Differences in plant age, size, past history of flowering and herbivory, phenology of flowering, soil conditions, and competition from neighboring plants could all contribute to the remaining, unexplained variance. The path analysis shows how these interactions with both mutualists and antagonists influence mature fruit production amid all these additional sources of variation. It also shows just how small and variable the effects of individual mutualists—and hence natural selection for specialization—can be when all the other direct and indirect effects of other interactions are superimposed onto a pair-wise interaction. A pair-wise interaction may cover wide geographic areas, but natural selection favoring specialization and coevolution may occur only in a small fraction of the populations.

A path analysis of seed output in the herbaceous perennial *Lomatium salmoniflorum* produces a different picture of the direct and indirect effects of interspecific interactions between plants, pollinators, and herbivores (Thompson and Pellmyr 1989). This umbellifer is restricted to basaltic outcroppings in steppe along the Snake and Clearwater rivers, which border Washington, Idaho, and Oregon. Almost all species within this community flower in the spring, after the cold of winter and before the drought of summer. Most of the pollinator community, including solitary bees, flies, beetles, moths, and butterflies, are also most active as adults in spring. Consequently, there is a riot of flowering and insect activity constrained to the period from March to May, after which most plants turn brown and die back to the ground. Early-flowering plants encounter few pollinators, whereas late-flowering plants encounter more pollinators but also more competition for those pollinators from other plant species. In addition, insects that lay their eggs within immature seeds are common throughout the community and are more generally common among the plant species that flower in the middle to latter part of spring.

Within this community, *L. salmoniflorum* is among the earliest plants to begin flowering each year, the first flowers opening in early to mid-February. Flowering individuals produce sequentially one to nineteen umbels, with each umbel containing ten to three hundred flowers. On most plants, the first umbel has only male flowers. Most later umbels have a mix of male and hermaphroditic flowers. The flowers of *L. salmoniflorum* are visited by a variety of solitary bees, flies, and beetles, which differ in phenology and undoubtedly in efficiency as pollinators. After pollination, the immature seeds are eaten by the larvae of a weevil, *Smicroynx* sp. *cinereus* group (Curculionidae). Adult weevils insert their eggs through the soft ovary walls of recently pollinated flowers, and the larvae must finish development before the seeds mature (Ellison and Thompson 1987). Hence, the adult weevils must face the problem of eclosing too early, when only male flowers are available, and of eclosing too late, when seeds are past the stage suitable for oviposition and larval development.

The number of viable seeds that a plant disperses at the end of spring is therefore an interplay between the start date of flowering, the number of hermaphroditic flowers produced, the percentage of those flowers that set seed, and the percentage of seeds eaten by weevil larvae, and not mutualism with any particular pollinator. The activity of particular pollinator species appears to have no effect on overall seed output, and hence there seems to be little opportunity for the evolution of specialization to any one pollinator species in this population. The path diagram that best describes how these factors interact with one another includes both direct and indirect effects (fig. 11.4). Indirect effects among the four factors in the model are important. The over-

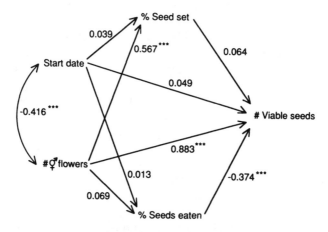

Fig. 11.4. Path analysis of the factors affecting the number of seeds produced by *Lomatium salmoniflorum* plants. Single-headed arrows indicate causal direction between variables in the path model. The double-headed arrow indicates correlation between two variables without implied direction of causation. Standardized path coefficients are shown between variables. *** = $P < .0001$. Reprinted from J. N. Thompson and O. Pellmyr, 1989, "Origins of Variance in Seed Number and Mass," *Oecologia* 79:395–402, by permission of Springer-Verlag.

all path model accounts for 93% of the variance in viable seed production, whereas a simple multiple regression of the direct effects accounts for 86%. If analyzed separately, the start date of flowering is significantly correlated with the number of viable seeds that a plant produces. Yet, when placed in a path analysis with the other factors influencing viable seed production, the effect of start date is seen to be almost entirely indirect. The number of hermaphroditic flowers and the percentage of seeds eaten by weevils have the strongest direct effects on the percentage of viable seeds that a plant produces. The negative effect of beetles on viable seed production is strong and significant and, unlike that found in *Calathea,* entirely direct.

Unlike the results for *Calathea* in the second year of that study, seed set in *Lomatium* does not depend directly on the mix of pollinators, and this factor was not included in the model. Experimental hand-pollinations had shown that the percentage of flowers setting seeds in this population is limited by available resources rather than by pollinator visitation, with large plants producing more flowers and setting a higher percentage of seeds than small plants (Thompson and Pellmyr 1989). Any selection for particular pollinator species would have to come from any differential effects that different species had in dispersing pollen to neighboring plants ('plant male function') rather than from seed set ('plant female function').

The studies of both *Calathea* and *Lomatium* give an indication of how the evolution of specialization in pair-wise interactions between potential mutualists may be limited within some local populations by the other sources of variation within natural communities. It is currently fashionable in evolutionary ecology to give more credibility to experimental studies, in which other sources of variation are carefully controlled or eliminated, than to observational studies. As Schemske and Horvitz (1988) point out, however, we need both. It is easy to show powerful effects of one species on another in a carefully manipulated environment, and there is an important place for such experiments in understanding the evolution of interactions. Both the *Calathea* and *Lomatium* studies were supplemented with hand-pollination experiments on seed set to test whether the percentage of flowers setting seed was limited by the number of pollinators visiting the plants or by the resources available to the plant for filling seeds. But these experiments alone were insufficient for understanding the ways in which the outcomes of interactions are twisted and turned in natural communities where plant reproduction depends on interactions with species other than the pollinators. Careful experiments on interactions between pairs of species must go hand-in-hand with studies that analyze interactions amid the full richness of variation produced by the entangled web of interactions among species. It is those kinds of studies that can show how the outcome of pair-wise interactions can vary tremendously among populations, depending upon the community mix of other species.

Limitations on Specialization from the Quirks of Biology: The Lesson from the Rarity of Ant Pollination

Natural selection on life histories and the community context of an interaction may have powerful influences on the evolution of specialization in mutualism, but there is one additional limitation on the evolution of specialization: that imposed by simple quirks of biology and by characters that may have evolved from other interactions. It is the kind of limitation that makes extreme specialization to some potentially mutualistic taxa more difficult than to other taxa except in some populations under some rare ecological conditions.

The rarity of pollination by ants may be one of the major examples of how the quirkiness of biology and the past history of a taxon can prevent the evolution of extreme specialization in most populations. Why pollination by ants is so rare has long been a puzzling question in the evolution of mutualism and specialization. It is especially puzzling because plants have co-opted ant behavior in so many other ways: to increase nutrition, repel enemies, and disperse seeds (Beattie 1985). Admittedly, worker ants—with their lack of wings, relatively smooth bodies, and limited local movements to and from nests—lack some of the obvious traits common to many other pollinators. But their sheer abundance and common presence on plants stand as a potential counterbalance to these deficiencies as pollinators. Nevertheless, very few plants have enlisted ants as pollinators.

Only about two dozen plant species in sixteen families have been suggested as potentially ant pollinated, but the ability of ants to effect pollination has actually been shown for only three plant species (review in Peakall, Handel, and Beattie 1991). Only two of these species, both Australian, are known to be exclusively pollinated by ants. In *Leporella fimbriata,* the flowers mimic reproductive females of *Myrmecia urens,* and pollination occurs as winged males move from flower to flower attempting to copulate (Peakall, Beattie, and James 1987; Peakall, Angus, and Beattie 1990; Peakall and James 1989). In *Microtis parviflora,* worker ants pollinate the minute flowers, which are borne in a tight cluster on a single stalk (Peakall, Beattie, and James 1987; Peakall and Beattie 1989, 1991).

Beattie, Peakall, and colleagues have offered a novel explanation for the lack of ant pollination. Their explanation falls outside the usual mainstays of abundance and efficiency in discussions on the evolution of pollination systems and relies instead on a quirk of the biology of ants. The thoracic metapleural glands of ants produce antibacterial and antifungal secretions that also happen to reduce pollen germinability and pollen tube growth (Beattie et al. 1984, 1985, 1986). Even contact with the integument of male ants lacking these glands can result in damage to the pollen grains, although the cause is unclear: secretions from another gland or transfer of metapleural gland secretions from other ant castes are possibilities (Peakall, Angus, and Beattie 1990).

If Beattie and his colleagues are right, then most plants have been prevented from using their most common visitors as pollinators by a chance consequence of the ants' defenses against microorganisms. The question then becomes one of how the Australian orchids *L. fimbriata* and *M. parviflora* overcame this problem. *L. fimbriata* attaches its pollinia to winged male ants with a stigmatic secretion, which protects the pollen from damage. Loose pollen adhering directly to an ant's integument quickly loses viability (Peakall, Angus, and Beattie 1990). Hence, a common mechanism used by orchids to attach pollinia to pollinators may happen to make pollination by ants possible. The pollen of *M. parviflora* is carried on a short stalk, but in addition, the major pollinator at the study site near Sydney, *Iridomyrmex gracilis,* has an integument that does not damage the pollen of this species. This lack of damage does not seem to result from some special coating on the pollen of this orchid to prevent damage. The integument of *I. gracilis* also does not affect *Brassica* pollen in experimental trials. Instead, unlike at least one other *Iridomyrmex* species (Beattie et al. 1984), *I. gracilis* seems to have reduced levels of these secretions or slightly different secretions that happen not to damage pollen.

For other plants to be pollinated by ants, the number of ant visits would need to be very large to overcome the loss of pollen grains caused by the ants' metapleural secretions. Such conditions may appear only in a small part of the geographic range of a species or in species restricted to a narrow range of habitats. Gómez and Zamora (1992) have found a potential example in a high-elevation crucifer in the Sierra Nevada of Spain. The plant is primarily pollinated by worker ants despite the death of much of its pollen when in contact with the ants' integument. The abundant workers of *Proformica longiseta* can make up more than 80% of the visits to flowers of *Hormathophylla spinosa,* and experiments in which other pollinators were excluded show that the ants are effective pollinators. Nevertheless, these flowers are also visited by thirty-eight other insect species belonging to eighteen families, and it is not clear whether this plant is evolutionarily specialized for ant pollination. If so, these ants, or related species, would have had to have been predictably very abundant for a long period of the plant's past evolutionary history. These kinds of conditions must have been very rare in evolutionary time, because so few plants have been able to evolve the ability to specialize on the ubiquitous ants for pollination.

Conclusions

Combining the conclusions of this and the previous chapter, we find that specialization by hosts to one visitor species in nonsymbiotic mutualisms is limited by at least five aspects of these interactions: (1) the tendency for mutualistic resources to attract new visitor species, (2) the lack of predictability in the visitor assemblage, (3) the inability of hosts to evolve a selective sieve that permanently allows visitation by specialist visitors but not by thieves and cheaters, (4) the masking of mutualistic effects of any one visitor species by the other visitor species, and (5) the direct and indirect effects of the other interactions that affect a host's fitness. These influences on specialization will surely vary among populations over evolutionary time, providing opportunities for the evolution of different degrees of specialization in different populations. The coevolutionary process between any two species will be shaped in some cases by these geographic differences in specialization and outcome among populations.

PART IV
SPECIALIZATION AND
COEVOLUTION

CHAPTER TWELVE

Genetics of Coevolution

This chapter marks a transition. The arguments leading up to it have been devoted to the problem of how and why a wide range in specialization evolves in interactions between species and populations, creating a geographic mosaic in the structure of many interactions. The challenge confronting us now is how these differences in specialization and outcome among populations and species influence the process of coevolution.

A short definition of coevolution is 'reciprocal change in interacting species' (Thompson 1982). A more descriptive definition is given in the *Random House Dictionary of the English Language* (1987), which was the first general English-language dictionary to include a definition: "evolution involving a series of reciprocal changes in two or more noninterbreeding populations that have a close ecological relationship and act as agents of natural selection for each other, as the succession of adaptations of a predator for pursuing and of its prey for fleeing or evading." Although this definition uses an example of reciprocal adaptation between a pair of species, speciation in one or both species may also be an outcome of coevolution. Speciation may be a direct result of a particular form of coevolution or a fortuitous result of differential coadaptation of populations in different geographic areas. In either case, both adaptation and speciation require study as part of the process of reciprocal change.

The one-to-one interactions common until recently in mathematical models of coevolution and in many empirical studies of coevolution within local populations are the ideal extreme of the coevolutionary process. We consider the populations as pandemic units, and we track the average values for the traits as they are shaped by reciprocal selection (e.g., corolla depth and pollinator tongue length). Comparisons of cladograms in phylogenetic studies of coevolution are similarly built upon the view that the essence of reciprocal evolution is in the net effect of directional selection preserved in the means of traits found in different species. We track the distribution of fixed traits in the cladograms of interacting species to understand large-scale patterns in the evolution of arms races or mutualisms. As a result, we see only the part of coevolution that results in strong directional selection that has spread throughout the distribution of a species.

These are useful approaches to the study of coevolution, but they do not adequately incorporate the genetic dynamics of coevolution or the rich geographic structure of populations that shapes reciprocal evolutionary change. They do not take into account the differences in specialization among populations and the differences in outcomes of interactions among environments discussed in the previous chapters. As a result, studies of coevolution within local populations or comparisons of cladograms of species cannot fully represent the actual process or dynamics of coevolution.

In this chapter and the next two I explore the genetics and geographic structure of coevolution in interactions between pairs of species. My aim is to develop a more geographic view of the process of coevolution and of the relationships between specialization and coevolution. The final chapters then broaden the argument and consider the influences of asymmetries in specialization and geographic structure on coevolution involving more than two species. This chapter begins the argument by analyzing the genetics of coevolution in two interactions that exhibit geographic structure, producing different traits or outcomes in different populations: gene-for-gene coevolution between plants and pathogens, and mimicry in butterflies.

Gene-for-Gene Coevolution

There is a hierarchy of possible views of coevolution based upon the genetics and geographic structure of the interaction. The simplest possible view is that of reciprocal change within a local population driven solely by the evolution of major genes. Most explicitly genetic models of coevolution are based upon the assumption that major genes govern the coevolutionary process. There is no reason to believe that most of coevolution is driven by major genes, but it has been a convenient place to start. It must, however, be only the starting point. Defense and counterdefense driven by major genes, which can create discrete polymorphisms, must surely produce a dynamics in coevolution somewhat different from arms races governed solely by selection on quantitative traits. For example, the dynamics of frequency-dependent selection is likely to be different in a population divided into discrete morphs (e.g., for resistance alleles, color patterns, or prey preference) from that in a population exhibiting a continuous distribution for these characters. At present, however, we know most about the genetics of coevolution between pairs of species whose interactions are governed in whole, or to a large degree, by major genes. The two forms of interaction that seem best to show the effects of major genes are gene-for-gene coevolution between

Table 12.1. The expected compatibility between homozygous genotypes in a single-locus gene-for-gene interaction

	Host genotype	
Pathogen genotype	*RR*	*rr*
VV	Incompatible	Compatible
vv	Compatible	Compatible

Source: From Thompson and Burdon 1992. Reprinted with permission from *Nature* 360:121–25. Copyright 1992 by Macmillan Magazines Ltd.

Note: R is a dominant host gene conferring resistance to the pathogen, and *r* is a recessive host gene conferring susceptibility. *V* is a dominant pathogen gene conferring avirulence, and *v* is a recessive pathogen gene conferring virulence.

parasites and plants, and the evolution of Batesian and Müllerian mimicry complexes.

The hypothesis of gene-for-gene coevolution is based upon the view that for each gene causing resistance in a host there is a corresponding (matching) gene for avirulence in the parasite. According to this view of coevolution, a resistant (i.e., incompatible) reaction depends upon *both* the presence of a gene for resistance (*R*) in the host *and* the corresponding gene for avirulence (*V*) in the parasite. Several molecular models for gene-for-gene relationships have been proposed, but the one best supported by recent data is the elicitor-receptor model (Keen et al. 1990; de Wit 1992). Plants with the appropriate *R* gene have a receptor that reacts to the presence of a particular gene product (the race-specific elicitor) produced by parasites with the corresponding *V* gene. Although the word 'elicitor' has often been used historically for the parasite gene product that induces the host response, it unfortunately conveys the erroneous impression that the pathogen is actively producing a substance for the purpose of eliciting a host response. Instead, the elicitor is any substance produced by the parasite, probably as a result of its normal metabolism, that the host happens to be able to recognize as a result of having the appropriate receptor. The parasite evolves to be virulent (i.e., to produce a compatible response) through a new mutation that allows it to escape recognition by the host.

Combinations of homozygous hosts and parasites give three compatible and one incompatible outcome (table 12.1). An avirulence gene in a parasite cannot be detected if the host lacks the corresponding resistance gene. Similarly, a resistance gene in a host cannot be detected if the parasite lacks the corresponding avirulence gene. In diploid species, there are five other possible combinations of resistance and avirulence genes (*Vv RR, Vv Rr, Vv rr, VV Rr,* and *vv Rr*). These combinations, however, are generally phenotypically indistinguishable from other compatible or incompatible reactions, because resistance is usually dominant to susceptibility, and avirulence dominant to virulence.

A gene-for-gene relationship is inferred if phenotypic responses consistently correspond with theoretical Mendelian expectations. Formal genetic analyses of both the host and the parasite, however, have actually been performed for only a very small number of associations between species and often for only a few resistance and avirulence genes. These genetic analyses are often extremely difficult, if not impossible, to perform. Commonly one or both species seldom or never undergoes meiosis. Consequently, most genetic evidence for gene-for-gene interactions comes from a small number of well-studied cases. Gene for-gene interactions involving the formal analysis of at least a few genes for resistance and avirulence have now been determined in more than a dozen agricultural host/parasite associations and inferred in more than thirty others (table 12.2). Although resistance and avirulence are largely inherited as single genes in all the suggested cases of gene-for-gene interaction, it would be naive to think that these interactions are governed solely by single, matching major genes. These species also include genes in which expression of resistance or avirulence is controlled by two genes or is influenced by modifier genes. Moreover, although both resistance genes in hosts and avirulence genes in parasites commonly segregate independently, this is not always so. Tight linkage groups have been detected among both resistance and avirulence genes (Christ, Person, and Pope 1987; Hulbert, Sudupak, and Hong 1993).

In addition to associations in which the formal genetics have been studied, some other associations between plants and pathogens (including fungi, viruses, and bacteria) also appear to fit a gene-for-gene relationship. These interactions have been inferred to be gene-for-gene using the less rigorous criteria of compatibility reactions and studies of host genetics alone. These associations include a broad range of plant and pathogen taxa and life histories (table 12.2). The plants include grasses, annual and perennial herbs, shrubs, and trees. Most of these examples involve associations in which the pathogen is specialized for feeding on only one host species.

Although common in interactions between plants and pathogens, it is not yet clear to what extent gene-for-gene interactions occur in associations between other taxa. Such relationships have been postulated for interactions between plants and some phytophagous insects and nematodes (Parrott 1981; Sidhu 1981; Cox and Hatchett 1986; Saxena and Barrion 1987), but, in most cases, even less is known about the formal genetics of these associations than about fungal pathogens. The only well-studied example for interactions between plants and insects is that between wheat and Hessian fly. Twenty genes for resistance to Hessian fly are now known in wheat, and there are eleven

Table 12.2. Demonstrated and suggested gene-for-gene interactions between plants and their parasites

FUNGAL PATHOGENS	FUNGAL PATHOGENS
Basidomycetes Uredinales (rusts)	Pythiaceae
Avena—Puccinia framinis	*Glycine—Phytophthora megasperma*
Avena—Puccinia coronata	*Solanum—Phytophthora infestans*
Coffea—Hemileia vastatrix	Other fungi
Glycine—Phakopsora pachyrhizi	*Brassica—Plasmodiophora brassicae*
Linum—Melampsora lini	*Callistephus—Fusarium oxysporum*
Lycopersicon—Cladosporium fulvum	*Hordeum—Rhycosporium secalis*
Maize—Puccinia sorghi	*Malus—Venturia inaequalis*
Pinus—Cronartium quercuum	*Oryza—Pyricularia oryzae*
Triticum—Puccinia graminis	*Phaseolus—Colletotrichum lindmuth—*
Triticum—Puccinia recondita	*ianum*
Triticum—Puccinia striiformis	*Saccharum—Bipolaris sacchari*
Basidomycetes Ustilaginales (bunts and rusts)	*Solanum—Synchrytium endobioticum*
Avena—Ustilago avenae	BACTERIA
Hordeum—Ustilago hordei	*Capsicum—Xanthomonas campestris*
Triticum—Ustilago tritici	*Gossypium—Xanthomonas malvacearum*
Triticum—Tilletia caries	*Pisum—Pseudomonas syringae*
Triticum—Tilletia controversa	VIRUSES
Ascomycetes Erysiphales (powdery mildews)	*Capsicum—tobacco mosaic*
Hordeum—Erysiphe graminis	*Glycine—soybean mosaic*
Secale—Erysiphe graminis	*Lycopersicon—tobacco mosaic*
Senecio—Erysiphe fischeri	*Lycopersicon—spotted wilt*
Triticum—Erysiphe graminis	*Phaseolus—bean yellow mosaic*
Peronosporaceae (downy mildews)	*Solanum—potato virus x*
Lactuca—Bremia lactucae	INSECTS
Medicago—Peronospora trifoliorum	*Triticum—Mayetiola destructor*
Sorghum—Peronsclerospora sorghi	NEMATODES
	Solanum—Globodera rostochiensis

Sources: Burdon 1987; Hulbert and Bennetzen 1991; Hulbert, Lyons, and Bennetzen 1991; Janssen, Bakker, and Gommers 1991; de Wit 1992; Thompson and Burdon 1992 (reprinted with permission from *Nature* 360:121–25, copyright 1992 by Macmillan Magazines Ltd.); and Bakker et al. 1993.

fly biotypes that differ in their abilities to attack wheat varieties with these various resistance genes (Weller et al. 1991). Diehl and Bush (1984), however, have cautioned that the gene-for-gene view of this interaction may be partially an artifact of the experimental protocols used. Virulent fly biotypes have usually been obtained by 'purifying' strains over several generations on wheat varieties with different resistance genes, and then using a hybridization test in the F_4 generation to eliminate any heterozygotes. This process is likely to magnify single-gene effects.

Some other studies of interactions between insects and plants have grouped insects into biotypes, based upon the idea that each biotype may differ monogenically at least for virulence genes. As some of these insects have been studied in more detail, the genetics of virulence has turned out to be more complicated. For example, greenbug (*Schizaphis graminum*) biotypes are now known to be complexes of genotypes in which virulence for wheat is determined by the interaction of three loci: two duplicate genes that confer virulence when recessive and a modifier gene that is epistatic to one of the duplicate genes when dominant (Puterka and Peters 1989). Similarly, the biotypes of rice brown planthopper (*Nilaparvata lugens*), which were previously assumed to be distributed among rice genotypes in a gene-for-gene fashion, are now known to be a collection of polygenically determined genotypes. Each biotype is a mix of extreme genotypes for virulence selected by particular rice varieties from a more continuous range of virulence (Den Hollander and Pathak 1981; Claridge and Den Hollander 1982). The performance of insect larvae on different host plants often does not seem to take the form of discrete responses. Instead, there are often distributions of survivorship, growth rates, and pupal masses on different plants, and, at least in some insects, these different distributions are only partially correlated with one another (Thompson, Wehling, and Podolsky 1990). Hence, the genetics of attack by insects may make them generally unlikely to evolve in a gene-for-gene manner with their host plants, but the genetic studies are still too few to know.

Gene-for-gene interactions have been studied for only a few natural populations of plants and their parasites (all pathogens), but convincing evidence that such relationships can occur in nonagricultural settings now exists (Parker 1985; Burdon, Brown, and Jarosz 1990; Jarosz and Burdon 1991, 1992). Moreover, the results so far suggest that the evolution of these interactions is fully interpretable only in a geographic context. Although some mathematical models predict cyclical polymorphisms of resistance to virulence genotypes within local populations (Levin 1983), it is not yet clear whether gene-for-gene coevolution in natural populations proceeds in this way. Such polymorphisms can be produced by frequency-dependent selection acting against the most common local host genotype. Long-term maintenance of polymorphism in these and related models usually requires that the costs of resistance and virulence and the time delays in frequency-dependent selection be within a particular range of values (e.g., Bell and Maynard Smith 1987). Other values produce more-chaotic fluctuations in gene frequencies and the loss of polymorphisms.

In fact, chaotic fluctuations and loss of polymorphisms are becoming an increasingly common finding in recent genetic models of gene-for-gene coevolution. Such findings call into question just how often gene-for-gene coevolution is driven by frequency-dependent selection within local populations. For example, in Barrett's (1988) models, asymmetries in the relative

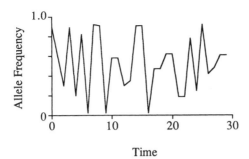

Fig. 12.1. Example of chaotic dynamical change in allele frequency in a host arising when natural selection exerted by a parasite is both frequency and density dependent. The mathematical model producing this pattern of change assumes a haploid host and a host density regulated solely by the parasite. Redrawn from May 1985.

costs of resistance and virulence can lead to rapid fixation of parasite genotypes, which results in coevolutionary arms races that rely upon the continued evolution of novel host genotypes rather than the cycling in frequency of current genotypes. Other recent models, incorporating at least two loci and resulting in three or more genotypes within a population, produce cycles that are even more complex and chaotic, and sometimes qualitatively different, from one-locus models (Seger 1988).

When ecological factors, such as changes in the abundance of host populations or different transmission rates, are built into gene-for-gene models, a whole range of outcomes becomes possible, and chaotic fluctuations are a common result (Anderson and May 1991; Frank 1991). The models of Anderson and May (1982; May 1985; May and Anderson 1990) combine epidemiology with genetics and show that epidemiological factors such as density dependence produce polymorphisms that are more likely to be cyclic or chaotic than the results predicted by purely genetic models (fig. 12.1). Frank's (1992, 1993) models, too, have indicated that ecological and demographic factors such as birth and death rates have at least as much effect on the coevolutionary dynamics of gene-for-gene interactions as genetic factors such as the pleiotropic costs of different resistance alleles or virulence alleles.

Together these mathematical models incorporating both genetics and ecology suggest that the trajectory of gene-for-gene coevolution in natural populations may depend to a large degree on the demographic structure and ecological setting of the interacting populations. Consequently, the local evolutionary dynamics of the interactions is likely to vary considerably among populations, thereby creating a geographic mosaic in the evolution of these relationships. Moreover, studies of natural populations are indicating that frequency-dependent selection by parasites may at least sometimes be weak or intermittent (Jarosz and Burdon 1991; Roy 1993) and that gene-for-gene coevolution is probably often best understood as a geographic process.

This geographic aspect of coevolution is evident in the results of studies of Burdon and colleagues on the interaction between plants and fungal pathogens in Australia. In the interaction between wilds oats, *Avena barbata,* and *Puccinia* rusts, for instance, the distributions of resistance in the plant populations to particular pathogen genotypes and the distribution of virulence in the rust populations vary greatly along a geographic gradient from north to south (Burdon, Oates, and Marshall 1983; Oates, Burdon, and Brouwer 1983). In this particular interaction it is not clear that coevolution is directly driving these geographic changes in distributions of resistance and virulence, but the evidence is convincing in the long-term studies of another interaction, that between an Australian wild flax, *Linum marginale,* and its associated rust fungus, *Melampsora lini. Linum marginale* is a congener of cultivated flax, *Linum usitatissimum,* which is the species that Flor (1942, 1956) used in developing the gene-for-gene concept. This wild flax is a short-lived perennial endemic to Australia. It occurs in a remarkably wide range of habitats, from coastal vegetation to marshes to subalpine. But throughout its range it occurs as small, localized populations, often a few hundred individuals or fewer. The rust fungus forms orange pustules on the leaves and significantly reduces longevity in its host (Jarosz and Burdon 1992). Variation in resistance and virulence occurs within populations, among populations within regions, and over large geographic areas (Lawrence and Burdon 1989; Burdon and Jarosz 1991; Jarosz and Burdon 1991).

In a survey of the resistance and virulence structure of ten populations throughout Kosciusko National Park in New South Wales, Jarosz and Burdon (1991) found little correspondence between the genotypes of host and pathogen populations. Although populations differed in the pathogen races attacking them and in the resistance genes they harbored, the distribution of genotypes and gene frequencies did not suggest that the interaction between these species is maintained within local populations by polymorphisms shaped by frequency-dependent selection. Some populations were susceptible to all the pathogen races attacking them, others were resistant to some pathogen genotypes but not all, and still others were fully resistant to pathogens that did not occur within the population at all (fig. 12.2). Jarosz and Burdon (1991) argue that annual bottlenecks in population size, periodic extinctions, and asexual reproduction work together to produce variation among demes in genetic structure and poor local matching of local resistance genes with virulence genes. Rather than being driven by frequency-dependent selection within demes, the interaction appears to evolve through a combination of natural selection, genetic drift, and gene flow among

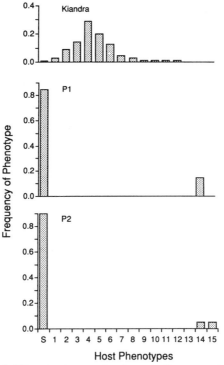

Fig. 12.2. Frequency of different resistance phenotypes in *Linum marginale* from three neighboring subalpine populations in Australia. Resistance is to the rust fungus *Melampsora lini*. Phenotype S is susceptible to all the rust races that were tested. Redrawn from Jarosz and Burdon 1991.

demes. The small size of the populations and the considerable differentiation among demes also allow for the possibility that differential extinction of demes (e.g., those that happen not to harbor resistance to any of the pathogen genotypes that attack them) may also be an important part of the process.

Interaction of Major Genes and Modifier Genes in Coevolution

Batesian and Müllerian mimicry are the other two kinds of potentially coevolutionary interaction known to involve major genes and to vary geographically. Unlike gene-for gene coevolution, however, the evolutionary genetics of mimicry is thought to be a two-stage process. Major genes produce a fairly close resemblance to a model in one step; then modifier genes, with smaller individual effects, refine that resemblance over time (Nicholson 1927; Turner 1987; Sheppard et al. 1985; Mallet 1989; Nijhout, Wray, and Gilbert 1990).

There is general agreement that both Batesian and Müllerian mimicry involve this two-stage process, but the evidence is based almost exclusively on studies of Lepidoptera, including swallowtail butterflies (Clarke and Sheppard 1960, 1963, 1971), *Heliconius* butterflies (Sheppard et al. 1985; Turner 1987; Mallet 1989; Nijhout, Wray, and Gilbert 1990), and *Zygaena* moths (Sbordoni et al. 1979).

In *Heliconius* butterflies, which form some of the most visually impressive mimicry complexes known, there are tremendous geographic differences in the color combination (black, red, yellow, and white) on which the species converge. *Heliconius melpomene* and *H. erato,* for example, occur together throughout Central and South America, and the color patterns on their wings vary in tandem over their geographic ranges. At least eleven different color patterns occur (J. Turner 1984), and the close resemblance in color patterns between these species is not due to interspecific hybridization. These two species come from different groups within the genus (Gilbert 1983; Mallet et al. 1990). The color patterns can differ markedly even over short distances, as occurs between the populations in several river drainages in Peru (fig. 12.3).

In *H. melpomene, H. erato,* and *H. cydno* alone, at least thirty-five genetic loci are known to affect these color patterns, but large changes in color and pattern between populations are often due to only a few loci. Some of these loci, however, may actually be tightly linked blocks of genes (supergenes) (Mallet and Barton 1989). The exact way in which the patterns are built up on the wings has recently undergone a major reinterpretation. The studies of Nijhout, Wray, and Gilbert (1990) indicate that the yellow and white portions of the pattern constitute the 'background' colors, on which the red and black patterns are expressed. This contrasts with the earlier view, in which black was thought to be the background rather than the pattern superimposed on the background. This change of view has reduced the number of phenotypic effects (e.g., color and the size, shape, and position of the pattern) resulting from the changes at single genes.

In the *H. erato* and *H. melpomene* populations in Peru, the color patterns are controlled by three unlinked loci in *H. erato* and by five loci—two linked pairs and a fifth locus possibly linked to one of the pairs—in *H. melpomene* (Mallet et al. 1990). Where populations of the different Peruvian color forms meet, they form a narrow hybrid zone about 10 km wide (Mallet 1989). Study of the color patterns within these hybrid zones has indicated that frequency-dependent selection for the more common local color pattern is probably quite strong, although other factors contributing to clines, such as epistasis, could also be involved (Mallet and Barton 1989; Mallet et al. 1990). Butterflies transferred experimentally across the hybrid zone suffered a 52% selective disadvantage over the resident morph, with most deaths occurring soon

Fig. 12.3. Geographic variation within Peru in mimetic resemblance between two species of *Heliconius* butterfly: *H. erato* (*left*) and *H. melpomene* (*right*). Black is shown in black, yellow as clear, and red as stippled. Redrawn from Mallet et al. 1990.

after release, presumably due to differential attack by predators (Mallet and Barton 1989). The ability of insectivorous birds to discriminate among color patterns on lepidopteran wings, avoiding after experience distasteful aposematic species and their mimics and thereby acting as agents of natural selection, has now been demonstrated repeatedly both in cage experiments and in elegantly designed field studies using a variety of species and color patterns (e.g., Jeffords, Sternburg, and Waldbauer 1979; Jeffords, Waldbauer, and Sternburg 1980; Brower 1984; Chai 1986).

The overall coevolution between *H. erato* and *H. melpomene,* then, is built up from all the local populations of these two species evolving toward different adaptive peaks (color combinations). Throughout their ranges *H. erato* is generally more common than *H. melpomene* (Gilbert 1983, 1991). Consequently, *H. erato* may often serve as the model and *H. melpomene* as the mimic, with little actual coevolution. At times, however, the numbers of both species can drop to low levels (Gilbert 1984) and they may both then serve as reciprocal models for one another. Hence, some local pairs of populations may exhibit coevolution, whereas others may not, or they may coevolve erratically. The overall process, however, creates a geographic mosaic in the coevolution of these species.

Elucidating the conditions favoring coevolution in mimicry is one of the oldest problems in evolutionary biology, but not until Gilbert's (1983) synthesis was there a comprehensive view of the ecological situations favoring coevolution in both Batesian and Müllerian mimicry. Gilbert argued that coevolution in mimicry depended upon two influences: the relative palatability and the relative abundances of the species (fig. 12.4). Müllerian mimics such as *Heliconius* butterflies are most likely to coevolve when they are equal in abundance and equal in unpalatability to predators. If one species is rarer than the other, the more (or most) abundant species is likely to be the model, with the rarer species converging on that color pattern. In contrast, in Batesian mimicry the likelihood of coevolution is high wherever the palatable mimic is as abundant or more abundant than the model. The reason is that when the palatable mimic is abundant relative to the model, the predators will more commonly encounter the mimic. The result may be a coevolutionary chase between model and mimic as natural selection favors in the model divergence from the color pattern of the mimic, while simultaneously favoring in the mimic tracking of the model's changes.

An additional difference between Batesian and Müllerian mimicry links palatability, population abundance, and genetics. Just as in gene-for-gene coevolution between pathogens and plants, the evolution of polymorphism is one of the outcomes of mimicry, but generally only for Batesian mimics (Ford 1964; Charlesworth and Charlesworth 1976; Turner 1980). In Müllerian mimicry, natural selection favors fixation of the most common genotype. Mutations producing phenotypes different from the model would often suffer lower fitness, because predators may not recognize and avoid these rare patterns. Consequently, almost all Müllerian mimicry complexes are locally monomorphic, except at hybrid zones between regions.

In Batesian mimicry, in contrast, the fitness of a mimetic form decreases as it becomes more common relative to the model: as the palatable mimic becomes common, naive predators would not learn as quickly to associate that particular color pattern with unpalatability. Hence, new rare mutations, producing either a cryptic morph or one that is mimetic of a different local mimicry complex, can be favored by natural selection, thereby producing a

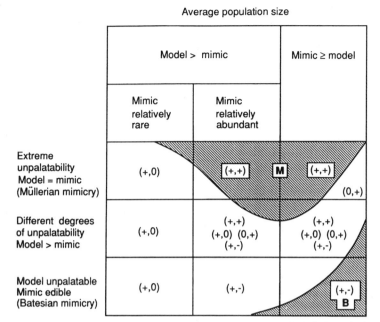

Fig. 12.4. Ecological conditions favoring coevolution in Batesian and Müllerian mimicry. The rows represent a gradient in palatability of two species, and the columns represent the relative abundance of these species. The net effect of the two species (mimic, model) is given in each cell. The relationship is uncertain for some combinations and will depend upon local ecological conditions. The shaded areas are the conditions under which coevolution is likely in Müllerian (M) and Batesian (B) mimicry. Redrawn from Gilbert 1983.

stable polymorphism within the mimic population. Rarely, however, does the polymorphism result in more than two morphs, one mimetic and the other cryptic. Highly polymorphic Batesian mimics, with multiple mimetic forms, occur within some swallowtail butterfly species (Clarke and Sheppard 1960, 1963, 1971), but they stand out as marvelously showy exceptions. The reason that this result is rare appears to be that maintenance of polymorphism for two or more models requires that the genes for the different mimetic morphs be closely linked; if they are unlinked, then natural selection favors either fixation or elimination of one of the mimetic morphs (Charlesworth and Charlesworth 1976).

    Together, the evolution of Müllerian and Batesian mimicry illustrates how superficially similar interactions can differ markedly in the ecological conditions favoring coevolution and in the genetic outcomes of the coevolutionary process. They also suggest that the process of coevolution in these interactions depends upon the relative population abundances of the species. Because population numbers often vary geographically, the probability of coevolution, and its outcome, are likely to vary among populations.

Polygenic Coevolution

The evolution of mimicry and gene-for-gene coevolution both involve major genes, but both undoubtedly involve polygenes as well. The theory of the evolution of mimicry explicitly includes polygenic inheritance as the second stage of the process, whereas both the theory and the study of gene-for-gene interactions have focused exclusively on major genes. As in the study of the evolution of mimicry, analyses of gene-for-gene interactions will undoubtedly benefit from the study of ways in which modifier genes influence gene-for-gene coevolution. It is certainly easy enough to find examples of resistance to parasites in both plants and animals spanning the entire spectrum from major single-gene effects to additivity across several to many loci (e.g., Burdon 1987; Osborn et al. 1988; Simms and Rausher 1989; Dobson, Hudson, and Lyles 1992; Fritz and Simms 1992). For example, Strong, Larsson, and Gullberg (1993) found evidence that resistance to gall midges in basket willow (*Salix viminalis*) is controlled by a combination of major genes and polygenes. Even highly selected crops such as wheat and maize include examples of resistance ranging from monogenic through oligogenic to polygenic (Khush and Brar 1991). Similarly, the ability of some insects to survive and develop on different plants is known to be determined by at least several genes (Thompson, Wehling, and Podolsky 1990).

The problem, however, is not simply to determine whether an interaction is governed by a gene-for-gene relationship or by more-complicated polygenic relationships. Rather, the problem is to understand the interplay between the mode of inheritance and the overall dynamics of coevolution. The mode of inheritance may have an important effect on the trajectory of coevolution. There are two general expectations in the pattern of coadaptation between pairs of species. One is directional selection leading to an arms race in antagonistic interactions or to increasing benefit to each partner in mutualisms. The other is a stable or fluctuating polymorphism of defenses and counterdefenses maintained by frequency-dependent selection, density-dependent selection, or both. These different expectations may be a consequence, at least partially, of differences in the mode of inheritance of defenses and counterdefenses: arms races in interactions governed mostly by polygenic inheritance, and stable or fluctuating polymorphisms in interactions governed by major genes.

Few models exist of the effect of mode of inheritance on the dynamics of coevolution, especially for polygenic coevolution, but the indication from models published thus far is that the underlying genetics can in fact have a powerful effect on coevolutionary dynamics. Seger (1992), for example, has argued that, at least for the simplest purely genetic models, species coevolving through changes in polygenic characters are less likely to cycle in gene frequencies than populations coevolving through changes in major genes. He suggests that the reason for this difference is that, with polygenic inheritance, rare phenotypes can increase only if the mean of the population is moving in the direction of that phenotype. That is, selection is more constrained, and therefore cycling less likely, than under major-gene coevolution.

The underlying genetics, however, is not necessarily a fixed constraint shaping the dynamics of coevolution. The genetics of interactions could itself be a result of natural selection favoring different modes of inheritance under different ecological conditions and in different populations. By favoring alleles with epistatic effects, or by favoring fixation of some loci but not others, selection could create some discrete polymorphisms from continuous distributions of characters. Alternatively, by favoring modifier genes, selection could produce more continuous distributions of phenotypes from discrete characters. I know of no models or experiments designed to ask specifically if the process of coevolution in natural populations could adjust the mode of inheritance of characters, but the results from artificial selection in agricultural crops suggest that such a change could possibly occur. The critics of gene-for-gene coevolution argue that the commonly observed gene-for-gene relationship between crop plants and fungal pathogens is partially an artifact of plant breeding, which magnifies major-gene effects. If the critics are right, then they are at the same time asserting that the process of selection—albeit in this case artificial selection—can reshape polygenic coevolution into gene-for-gene coevolution.

Conclusions

Studies of gene-for-gene interactions and mimicry suggest that there is geographic variation in the genetic structure and outcome of coevolution. Populations differ in their specialization to particular genotypes of other species, creating geographic differences in the number of coexisting genotypes and the direction of natural selection. The local genetic dynamics of coevolution may therefore involve only a subset of genes that are involved in the long-term coevolution of two species.

We are only at the very earliest stages of understanding how the mode of inheritance shapes coevolutionary interactions and, reciprocally, how the process of coevolution shapes the mode of inheritance of characters. We will need, however, to understand these relationships if we are to make progress in determining the conditions that favor different kinds of coevolutionary dynamics, from frequency-dependent cycling of characters to unidirectional arms races. Gene-for-gene coevolution, for example, may produce different evolutionary dynamics from more polygenic interactions between species. Interactions between plants and pathogens may be more likely to result in gene-for-gene coevolution than most other associations between species. Coevolution in other interactions probably often involves both major genes and modifier genes. Even superficially similar forms of interaction, as between Batesian and Müllerian mimicry, can create very different coevolutionary dynamics (e.g., selection for polymorphic vs. monomorphic populations). A more complete theory of the process of coevolutionary change is going to require a better combined understanding of the genetics shaping both specialization and coevolution in a geographic context.

CHAPTER THIRTEEN

The Geographic Mosaic Theory of Coevolution

Whether governed by major genes or polygenes, hypotheses and models of the rate and direction of coevolution are built mostly upon the combined effects of mutation and natural selection within local populations or upon the accumulation of fixed traits in species as mapped on phylogenetic trees. The arguments and examples developed in the previous chapters, however, suggest that this is an overly restrictive view of the coevolutionary process. In this chapter I develop more fully the argument for a geographic view of the coevolutionary process. It incorporates what we currently know about the geographic structure of specialization and adaptation and therefore bridges the gap between coevolution within local populations and the broad patterns found in the phylogenies of interacting species. I call this overall view the geographic mosaic theory of coevolution.

Toward Specific Hypotheses on the Coevolutionary Process

One of the reactions to the coevolution-is-everywhere attitude that dominated evolutionary ecology in the 1960s and 1970s has been an increasing recent tendency to dismiss coevolution as an important evolutionary process. This tendency has developed, I think, from three causes. First, convincing examples of coevolution have accumulated slowly, because it is extremely time-consuming and expensive to study interactions from the viewpoints of both (or all) species. In the interim, it has become cliché to hear at scientific meetings that there are no convincing examples of coevolution known for this kind of interaction or that kind of interaction. Meanwhile, the past decade, and especially the past five or so years, have produced a small number of studies that quite convincingly suggest coevolution. Nevertheless, it takes a while to change overall perceptions, and the impact of these studies is only now starting to have an effect on evolutionary biology.

Second, until recently we have had very few testable hypotheses on the process of coevolution. The general categories of specific coevolution and diffuse coevolution made few testable predictions beyond a general expectation of reciprocal evolutionary change. As a result, it was difficult to design studies in ways that specifically evaluated the coevolutionary process. That, too, is changing rapidly now.

Third, as the studies in the previous chapter suggest, we have usually failed to make adequate connections between the coevolution of local populations and the coevolution of species. Coevolution within local populations exemplifies only the most local level at which this process acts to shape interactions between species. The patterns that we see in cladograms are the other extreme: the traits have become fixed throughout the geographic ranges of species. The connection between the processes found in local populations and the patterns seen in cladograms is through the geographic structure of populations. Until that connection is made, and examined for its effect on different kinds of interactions, we can make little headway in understanding the actual dynamics of the coevolutionary process, and we are likely to underestimate its importance.

It is therefore no longer sufficient to think about coevolution as either specific or diffuse, and as either a local or a species-level process. In table 13.1 I have listed a variety of views of the coevolutionary process and how a geographic perspective leads to expectations different from those of a local perspective. These different views are not mutually exclusive. A particular interaction may shift among the different modes of coevolution in the same way that a particular population may shift between directional and stabilizing selection or be subject to a mix of frequency-dependent and density-dependent selection. The challenge for studies in coevolution is to understand how different ecological and genetic conditions and different configurations of population structure favor different modes and outcomes of the coevolutionary process. It is no longer good enough to lump all the different forms of coevolution together any more than it is good enough to lump all the modes of natural selection together. Different modes of natural selection are a result of different ecological conditions, and they produce different outcomes. We should expect the same to be true of coevolution.

The views of coevolution in table 13.1 are certainly not exhaustive. They are only some of the ways in which the processes and outcomes of reciprocal change vary among interactions. The various forms of multispecific coevolution, in particular, need much new thought and insight, but I think that any new specific hypotheses will come mostly from studies of the geographic structure of interactions.

The Geographic Mosaic Theory of Coevolution

The overall view of reciprocal evolutionary change that I wish to suggest builds upon the conclusions of the previous chapters and can be called the geographic mosaic theory of coevolution:

Table 13.1. The effects of a geographic mosaic view on expectations about how species coevolve

WITHIN-POPULATION VIEWS

- **Gene-for-gene coevolution.** Stable or fluctuating polymorphisms generally are driven by frequency-dependent and density-dependent selection.
- **Escalating evolutionary arms races.** Increasing defenses and counterdefenses are mediated by the rates of mutation and natural selection.
- **Genetic feedback.** Local population fluctuations of parasites and hosts are damped through increased host resistance driven by coevolution (Pimentel 1961, 1988).
- **Mutualism dependence in mutualisms.** Adaptation and specialization are driven by directional selection within local populations and limited by counterbalancing selection and pleiotropy.
- **Coevolutionary successional cycles.** Local succession after disturbance involves not only the sequential replacement of species but also coevolutionary change between pairs or groups of species. This reciprocal change occurs as local populations change over time in age structure and size structure, and the initial diversity of genotypes is sorted by natural selection (Turkington 1989). Each cycle of local succession could produce a different specific coevolutionary outcome in the interactions between a pair or a group of species.

GEOGRAPHIC MOSAIC VIEWS

- **Gene-for-gene coevolution.** Polymorphisms are maintained at the geographic level rather than locally and are driven by a combination of selection and genetic drift within local populations, differences among populations in the local combinations of resistance and virulence alleles, gene flow among populations, and extinction of populations (Burdon and Jarosz 1991; Jarosz and Burdon 1991; Thompson and Burdon 1992).
- **Geographically structured arms races.** Escalating defenses and counterdefenses are mediated by rates of mutation, natural selection, and genetic drift within local populations, differences in degrees of specialization among populations, and gene flow among populations.
- **Geographically structured mutualisms.** Degrees of mutual dependence between species are influenced by adaptation and the evolution of specialization within populations, differences in adaptation and specialization among populations, and gene flow among populations.
- **Coevolutionary successional cycles.** Local successional cycles produce different specific coevolutionary outcomes in the interactions between a pair or a group of species. This patchwork of coevolutionary successional cycles on the landscape scale creates an ever-changing geographic mosaic in the genetic dynamics of interactions between species.
- **Coevolutionary turnover.** Asymmetric competition among groups of species on archipelagoes creates repeated cycles of invasion, local coevolution and competitive exclusion of the resident population, subsequent evolution of the invader in the absence of competitors, and reinvasion by other populations of these species (Roughgarden and Pacala 1989).
- **Coevolutionary alternation.** Reciprocal evolution between one parasite (or grazer or predator) species and a group of victim species proceeds as the parasite alternates among species over evolutionary time. Alternation results because populations of the currently attacked victim species evolve greater defenses, while populations of the currently unattacked victim species lose their defenses (Davies and Brooke 1989a,b).
- **Diversifying coevolution**
  (a) **Symbiont-induced speciation.** Endosymbionts that cause at least partial reproductive isolation among host populations may coadapt so differently with host populations in different environments that these interactions sometimes lead to permanent reproductive isolation (Thompson 1987b).
  (b) **Speciation through control of gamete movement.** Speciation in plants may sometimes occur through coevolution with highly host-specific pollinators that differ among populations in preference for particular plant genotypes. A specific version of this view is plant and pollinator speciation occurring through pollinator-mediated sexual selection (Kiester, Lande, and Schemske 1984).

Table 13.1. Continued

GEOGRAPHIC MOSAIC VIEWS
- **Escape-and-radiation coevolution.** Ehrlich and Raven's (1964) five-step hypothesis of coevolution between plants and herbivores incorporates evolution of new adaptations in populations of plants free of their herbivores and subsequent specialization and adaptive radiation in the herbivores. The hypothesis has been used primarily to study species-level patterns of adaptive radiation, but it can also be applied to the geographic process of coevolution.

Note: References are to papers that either state the listed view explicitly or describe results for interactions that can be extrapolated to such a view.

1. Interspecific interactions commonly differ in outcome among populations. These differences result from the combined effects of differences in the physical environment, the local genetic and demographic structure of populations, and the community context in which the interaction occurs.

2. As a result of these differences in outcomes, an interaction may coevolve in some populations, affect the evolution of only one of the participants in other populations, and have no effect on evolution in yet other local populations. Still other populations may fall outside the current geographic range of the interacting species.

3. In addition, populations differ in the extent to which they show extreme specialization to one or more other species. Some populations may specialize on and sometimes coevolve locally with only one other species, other populations may specialize on and perhaps coevolve with a different species, and yet others may coevolve simultaneously with multiple species.

4. These interpopulational differences in outcome and specialization create a geographic mosaic in interactions, which is the raw material for the dynamics of coevolution. This mosaic also creates the possibility that the overall evolution of a species is a result of coevolution with several species, even though individual populations are specialized to only one or two of the species.

5. Gene flow among populations and extinction of some demes reshapes the geographic mosaic of coevolution as the adaptations and patterns of specialization developed locally spread to other populations or are lost. Characters evolved in some local populations will temporarily have the greatest effects on the overall direction of coevolution between a pair or group of species, whereas those evolved in other populations will contribute little.

6. The result is a continually shifting geographic pattern of coevolution between any two or more species. Much of the dynamics of the coevolutionary process need not result eventually in an escalating series of adaptations and counteradaptations that become fixed traits within species.

The geographic mosaic theory therefore suggests that the coevolutionary process is much more dynamic than is apparent from the study of individual populations or the distribution of characters found in phylogenetic trees. Adaptations appear and are lost. Some populations become highly specialized for the interaction as others remain or become less specialized. Some populations may fall outside the geographic range of the other species, lose some of their adaptations for the interaction, and then later be drawn back into the interaction. A few populations may temporarily become evolutionary 'hot spots' for the overall trajectory of coevolution between the species, whereas other populations act as evolutionary sinks. The overall course of coevolution between any two or more species is driven by this ever shifting geographic mosaic of the interaction.

The geographic mosaic theory therefore specifically incorporates as part of the coevolutionary process two observations that have sometimes been used as evidence against the importance of coevolution: lack of apparent coevolution in the interaction between a local pair of populations, and lack of biogeographic congruence in the distributions of some pairs or groups of species. To invoke these as evidence against coevolution is to take either a solely local view of coevolution (no coevolution in this population) or a species-level view (the interaction must occur in all, or almost all, populations), without linking them through the geographic mosaic. The geographic mosaic therefore offers a different approach from that of others, such as Jordano (1993), who argued that low biogeographic congruence of species limits the possibilities for pair-wise coevolution, or Roubik (1989), who asserted that exact congruence in the distribution of pairs of species (of plants and pollinators) in pristine habitat is strong evidence for coevolution.

According to the geographic mosaic theory, broad geographic congruence in the distributions of taxa is, by itself, evidence neither for nor against the importance of coevolution in the overall relationship between two or more species. I agree with others that the interaction must be sufficiently common for coevolution to be an important force in the evolution of species, but I do not think that the interaction must be strong and coevolutionary throughout all, or perhaps even most, populations. Patterns of geographic congruence and results from single local populations tell us something important about the particular geographic structure of an interaction, but they are only two of the factors contributing to the dynamics of the coevolutionary process. The following sections develop these points in more detail.

The Importance of the Geographic Structure of Interactions

Coevolutionary hypotheses based solely upon reciprocal interactions between local populations or upon traits that become fixed within species ignore three major aspects of the ecology of species and interactions that we know are true from empirical studies (Thompson 1994):

1. Many species have a geographic structure made up of local demes connected to varying degrees (Gill 1978; Harrison, Murphy, and Ehrlich 1988; Gilpin and Hanski 1991; Ehrlich 1992; Thomas and Harrison 1992). Local populations may go extinct, but the interaction may persist at the regional level. Alleles may appear within local populations through gene flow and disappear through genetic drift. As a consequence, individual populations sometimes include only subsets of the alleles involved in the evolution of an interaction. Even neighboring populations can differ significantly in the genes they harbor as a result of their interactions with other species (Burdon, Jarosz, and Kirby 1989; Burdon, Brown, and Jarosz 1990). These observations emphasize that any theory of coevolution between two species that is based only on mutation rates and selection intensities within local populations is missing a potentially important aspect of the structure and dynamics of the coevolutionary process.

2. Very few pairs of interacting species have identical geographic ranges. A population outside the geographic range of the other species may lose its adaptations for the interaction. Hence, meadow pipits and white wagtails on Iceland, where they are not parasitized by cuckoos, show less discrimination against eggs unlike their own than the same species in Britain, where they are parasitized (Davies and Brooke 1989a). Similarly, *Cecropia peltata* populations outside the range of their mutualistic ants have lost some of the adaptations for attracting ants. Populations on the neotropical mainland produce Müllerian bodies that are eaten by the mutualistic ants that live in their stems, but populations on the Caribbean Islands, where the ants are absent, either lack the hirsute pads (trichilia) that produce the Müllerian bodies or have only incompletely developed trichilia (Janzen 1973; Rickson 1977). The geographic ranges of species change over evolutionary time, and some populations outside the current range of an interaction may later be drawn into the relationship.

3. The outcomes of interactions and the degrees of specialization between pairs of species vary among physical environments, and they vary with the community context in which the interaction takes place. Resistance of hosts to their parasites, the competitive ability of species, and the degree of mutualism in interactions can all vary in outcome according to changes in the physical environment (e.g., Rice and Menke 1985; Bryant et al. 1987; Burdon, Jarosz, and Kirby 1989). Variation in outcome among environments can occur even if the initial genetic structure of the interacting populations is the same (Thompson 1988d; Wade 1990). The reason is that the reaction norms of genes important in interactions can vary among environments. These reaction norms result in 'interaction norms' of varying outcomes among environments, which is part of the raw material for the geographic differentiation of interactions (Thompson 1986c, 1988d). The presence of other species (e.g.,

competitors, co-mutualists) and selection on specialization can then modify or completely reshape the outcomes of interactions in different environments.

The long-term studies of Abrahamson and Weis on the tritrophic interactions between goldenrods (*Solidago altissima*), stem gall-making flies (*Eurosta solidaginins*), and the parasitoids and birds that attack the gall-feeding larvae provide additional evidence for the view that interactions vary in outcome and selection among environments. Gall size is determined by the genotype of the plant, the genotype of the fly, and the environment in which the plant is growing (Weis, Abrahamson, and McCrea 1985; McCrea and Abrahamson 1987; Weis and Gorman 1990). Parasitoids have greater success in attacking flies in small galls, thereby favoring flies that induce large galls. Birds, however, attack large galls more heavily (Abrahamson et al. 1989; Weis, Abrahamson, and Anderson 1992). Parasitoids and birds therefore place opposing directional selection pressures on the flies, and the net direction of selection in local populations depends upon the relative attack rates of these two enemies. Studies of sixteen populations have shown that selection pressures are highly variable both within and among populations. Parasitoids consistently place directional selection on the flies for large gall size in most populations and years, whereas birds place more sporadic directional selection on the flies for small gall size (Weis, Abrahamson, and Anderson 1992). The overall opposing patterns of attack currently result in stabilizing selection in about half the populations that have been studied. The combined effects of differences in selection on the flies, differences in plant genotype, and environment create the opportunity for a complex geographic pattern to the evolution of these interactions.

These three empirical observations—the subdivision of species into demes, the lack of congruence in geographic distributions of interacting species, and variation in outcome and specialization in different physical and biotic environments—together suggest that the theory of coevolution must be based upon the view that the geographic mosaic of interactions molds the overall pattern and dynamics of adaptation between interacting species. This mosaic, forged by the geographic subdivision of populations and natural selection acting within local populations, will then be modified by three important aspects of geographic structure: gene flow, genetic drift, and local extinction.

Gene flow. The movement of individuals between populations can spread novel, beneficial mutations among populations, allowing them to respond faster to other species than would be possible if they relied solely upon the development of new mutations only within their own populations. Differences in the geographic structures of two interacting species, which can affect gene flow among populations, may produce differences in their relative rates of evolution, magnifying or lessening differences in rates due to life histories. For example, large differences in the generation times of hosts and parasites are often cited as one of the major problems in the evolution of defense, because these differences create an asymmetry in arms races that favors parasites. For each host generation there may be ten or even a thousand parasite generations. Sexual reproduction (Hamilton, Axelrod, and Tanese 1990), somatic mutations in plants and other modular organisms (Whitham 1981), and the immune system of vertebrates are all potential ways in which long-lived hosts cope with short-lived parasites. There is no denying that the asymmetry exists, but if there is more gene flow between host demes than there is between parasite demes, then the asymmetry may be lessened. Price (1980, 1992) has argued that the lifestyles of many parasites make them more likely than other organisms to become differentiated into local, isolated demes. If so, parasites may experience less gene flow and more genetic drift (including loss of favorable alleles) than some host populations. Such differences in geographic structure of hosts and parasites are unlikely to eliminate the advantage of short generation time found in many parasite species, but they can at least change the odds to some extent.

Gene flow, however, can also have another, quite different, effect. It can homogenize populations, lowering their degree of adaptation to local conditions (e.g., Slatkin 1987; Briskie, Sealy, and Hobson 1992). Similarly, the coevolutionary fine-tuning of local mutualisms may be prevented to some extent by gene flow from other populations. These various, and sometimes conflicting, effects of gene flow together help to shape the geographic structure of coevolution.

Genetic drift. Inbreeding and the loss of genes by random genetic drift can have three effects on the geographic structure of evolving interactions. It can lower the degree of adaptation within local populations by eliminating favorable genes through chance, it can increase the degree of geographic differentiation among populations, and it can influence which alleles become established and spread during coevolution. Wade's (1990) experiments on competition between *Tribolium* beetles have shown that metapopulation structure and the pattern of genotype-environmental interaction (reaction norm) can create a complex pattern of adaptations among populations, which affects outcomes in competitive interactions with other species. He began a stock culture from a 'natural' population of *T. castaneum,* collected from a bag of heavily infested rice in Little Rock, Arkansas, and then subdivided the culture into twenty populations, each with twenty-five adults chosen at random. He maintained these populations separately for fourteen generations, beginning each generation with twenty-five adults. During the fourteen generations, the populations became differentiated due to genetic drift and possibly natural selection. Each population was then placed in a larger con-

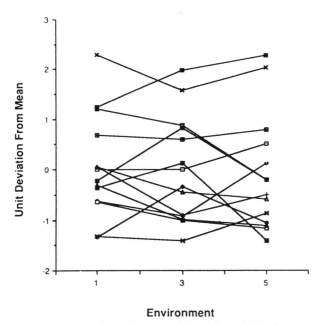

Environment

Fig. 13.1. Variation among populations in competitive ability of *Tribolium castaneum* in the absence of competition and in the presence of different genetic strains of a competitor. Environment 1 was free from interspecific competition, environment 3 included competition from strain b-SM of *Tribolium confusum,* and environment 5 included competition from strain b-I of *T. confusum.* Reprinted, by permission, from Wade 1990.

tainer, allowed to grow to a larger size, subdivided into replicates, and subjected to two physical environments (warm/wet, cool/dry). Each physical environment was in turn divided into competitive conditions (no competition, competition with strain b-SM of *Tribolium confusum,* or competition with strain b-I of *T. confusum*). Therefore, each population experienced six environments (2 physical × 3 competitive). After 52 days the populations were censused for one composite trait—rate of population increase in *T. castaneum*—which Wade took as a measure of competitive ability.

Wade found that a small number of generations of genetic drift followed by a short period of different treatments had created a mosaic of competitive outcomes. Populations of *T. castaneum* differed significantly in their competitive ability, depending upon both the physical environment and the genetic strain of the competitor (fig 13.1). Wade interpreted these results as an indication that, through population subdivision, natural selection would result in significant differences in coevolution among the populations.

Subdivision of populations into smaller, inbreeding demes can also influence which alleles become established in evolving interactions. Templeton

(1982) has shown, as a thought experiment, that natural selection for resistance to malaria could favor a balanced polymorphism for the A (susceptible) and S (resistant, sickle-cell) alleles when populations breed as large pandemic groups, but fixation for a recessive resistance allele (c) when populations breed as small, inbreeding demes. The S allele confers resistance to malaria when heterozygous with the A allele. In the homozygous state the S allele causes anemia, clotting, and other problems, which lower the ability of those individuals to survive. The A and S alleles are therefore maintained as a balanced polymorphism through the fitness advantage of the heterozygotes. In a large, panmictic population, the c allele would have little chance of becoming established, even though it has a somewhat higher fitness as a homozygote than the AS combination. The problem is that, in the absence of inbreeding, the c allele would have little chance of occurring as a homozygote when rare. Hence, if both the S and c alleles were introduced into a population at the same time, the S allele would increase in frequency faster than the c allele, because initially the c allele would almost always be in combination with allele A. As S increased, c would combine either with it or with A, neither of which would serve to increase the frequency of c.

In small, inbreeding populations, however, the c allele could occur as a homozygote even when in low frequency. If cc homozygotes had higher fitness than AS heterozygotes, then c could (although not inevitably) increase quickly in frequency within the local deme. The result could be different patterns of coevolution in different demes or fixation of the c allele over broader geographic scales if gene flow from the cc population to other populations increased sufficiently in future generations to allow cc individuals to become established in those other populations.

Extinction. The loss of whole demes could potentially change geographic patterns in allele frequencies so much that, in some interactions, the overall direction of coevolution would change. Imagine, for example, a polymorphism for resistance to parasites that is locally transient but geographically stable. The alleles may be transient locally despite frequency-dependent selection because the relative generation times and intrinsic rates of natural population increase of the interacting species are very different or because genetic drift sometimes eliminates alleles during the phase of the frequency-dependent cycle when they are rare in the population. During some generations, some of these alleles could be present in only one or two populations. More rarely, most of the genes important to the interaction could by chance be concentrated in one or two populations for several generations. These alleles could completely disappear if the one or two demes currently harboring them become extinct. At that point, an interaction governed by frequency-dependent selection at the geographic level could become one that is driven by directional selection on what is perhaps the one remaining major resistance allele and its modifiers. The result could be a new evolutionary race between host and parasite rather than a fluctuating polymorphism.

With a geographic view of coevolution, then, extinction becomes part of the process. Interactions may persist at the regional level even though one or the other species routinely goes extinct at the level of local populations (Fagerström 1988; Hassell, Comins, and May 1991; Jarosz and Burdon 1991). Local polymorphisms may be transient, but they may be retained regionally through gene flow among populations. Long-term reciprocal evolution in interactions becomes possible in interactions that may otherwise often lead to extinction of one of the species or loss of the interaction.

Long-term coevolution of competitors, for example, has always been a problem in coevolutionary theory at the level of local populations for two reasons, which are resolved with a geographic view. First, many local competitive hierarchies are highly asymmetric, making extinction of the poorer competitor possible and the long-term local persistence of the interaction often unlikely. Second, even with persistence, long-term local coevolution seems less likely in competitive interactions than in any other form of antagonistic interaction (Thompson 1982). In other antagonistic interactions, natural selection favors hosts or prey that decrease the chance of interaction with their enemies, but at the same time selection favors parasites, grazers, or predators that increase the chance that they encounter their hosts or prey. In each of these interactions, the interaction is held together because one species is pursuing the other in evolutionary time. But there is none of that in competitive interactions, unless the interaction involves a third species such as a predator that disproportionately attacks the better competitor. Even with selective predation or similar influences, selection will not favor an increased chance of interaction in either competing species (except under some rare conditions identified in some mathematical models). Once one of the species evolves away from the other, thereby decreasing the chance of interaction between the two, natural selection on the more competitive species decreases. Consequently, there is little selection for repeated bouts of reciprocal evolution. But in a geographic context, long-term coevolution of competitors becomes more likely as one species wins in some environments and the other wins in other environments.

Similarly, adoption of a geographic view provides a more realistic view of the long-term persistence of other interactions, such as those between parasites or parasitoids and hosts. Recent mathematical models by Hassell, Comins, and May (1991; Comins, Hassell, and May 1992) have shown that in the absence of geographic structure, the population dynamics of interacting parasitoids and hosts can often lead to extinction of either the parasitoid or both species. In their models, persistence in local populations in a homogeneous environment is possible (at least in populations with discrete generations) only when the parasitoids attack hosts nonrandomly or when the populations are regulated by density-dependent effects (Comins, Hassell, and May 1992). Dispersal among neighboring populations, however, can stabilize the interaction and lead to long-term persistence.

The Geographic Mosaic Theory of Coevolution and the Shifting-Balance Theory of Evolution

Geographic differences in interactions, formed by the interplay of natural selection, gene flow, genetic drift, and local extinction, create the conditions for a shifting-balance view of coevolution. Wright (1932, 1982) viewed the evolutionary landscape as a series of adaptive peaks and valleys that correspond to the interaction of alleles at multiple gene loci. Some gene combinations result in higher fitness than other gene combinations, and the combinations of high fitness may occupy positions far apart on the adaptive landscape, separated by deep valleys of low fitness. Through natural selection, populations move toward the peak to which they are nearest in their combination of genes. That local peak, however, may not necessarily be the peak of highest potential fitness. The population simply evolves through natural selection toward the nearest adaptive peak that does not require it to move through an adaptive valley of lower fitnesses. Different populations, then, could evolve toward different peaks, depending upon their current combination of genes. The result is a mosaic of populations with different suites of adaptations. A population with one group of characters—say, the suite of floral characters that attract particular pollinator species—could become saddled evolutionarily with a mediocre pollinator even though, with a change of multiple characters, it could exploit a better pollinator. It becomes stuck on the lower peak, because flowers with characters that fall in the valleys between the 'peak' combinations are selected against.

Adaptive landscapes and the associated geographic differentiation in populations they produce could result either because natural selection directly favors particular combinations of genes or because natural selection is constrained to certain gene combinations by the underlying genetic architecture (the genetic variance-covariance structure) of a species (Lande 1986; Felsenstein 1988; Riska 1989; Price, Turelli, and Slatkin 1993). Armbruster's (1990, 1991) studies of pollination and floral morphology in *Dalechampia* species are among the few that have tried to characterize adaptive landscapes for characters important in evolving interactions and to interpret the causes behind the topography of the landscapes. *Dalechampia* is a group of tropical plants that produce in their flowers triterpenoid resins that are collected by female bees, including euglossines, megachilids, and meliponines. Whether a bee contacts the stigmas and anthers while collecting resin depends upon

Fig. 13.2. Inflorescence of *Dalechampia scandens* showing gland (g), anthers (a), and stigmas (s). Reprinted, by permission, from Armbruster 1985.

bee size and three floral characters: size of the resin gland (which determines the kinds of bees that are attracted), the distance between the gland and the stigma, and the distance between the gland and the anther (fig. 13.2).

Through an analysis of forty-five populations of the twenty-five species of *Dalechampia,* Armbruster concluded that the adaptive landscape of these characters is probably best characterized as adaptive linear chains of peaks (an adaptive cordillera or ridge) that drop off on either side to valleys of lower fitness (fig. 13.3). The combination of small gland area and large stigma-gland or anther-gland distance results in low pollination rates, because it favors attraction of small bees (small gland area) but requires large bees (large stigma-gland and anther-gland distances) for pollination. Armbruster argues that the combination of very large gland area and small stigma-gland or anther-gland distance could also result in lowered fitness, for two reasons. First, if the cost of resin production increases linearly with

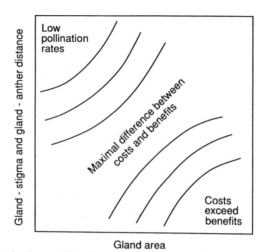

Fig. 13.3. Adaptive landscape of floral morphology in *Dalechampia*. The landscape is built upon the assumption that investment into resin production imposes a significant cost on a plant's carbon budget. Redrawn from Armbruster 1990.

gland area and the benefits of pollen deposition and dispersal reach an asymptote above a certain size (as the plant's stigmatic surfaces become saturated with pollen, or surrounding neighbors become saturated with that plant's pollen), then gland areas above a certain size will result in lower fitness as costs exceed benefits. Second, flowers with very large gland areas are more likely to be pollinated by a larger variety of less specific bees. Sharing pollinators in *Dalechampia* results in lowered pollination at least in some communities (Armbruster and Herzig 1984). Hence, evolution in floral morphology may be limited to movement along the diagonal cordillera of these character combinations.

As with *Dalechampia,* Benkman's (1993) studies of the North American red crossbills show that evolution of bill morphology in these species or subspecies can best be understood as complexes of characters grouped onto different adaptive peaks, resulting in populations of specialists (see chapter 8). Also, as in *Dalechampia* floral morphology, the peaks of crossbill morphology are distributed across the adaptive landscape. So long as the peaks do not form a continuous ridge, the species will become differentiated into populations of specialists.

In Wright's view, one of the major problems in evolution was to understand how populations shifted from the 'domain' of one adaptive peak to that of another, higher peak. This is where the geographic structure of populations becomes important. In trying to solve this problem, Wright imagined populations divided into demes of intermediate size. Through random genetic drift, one of the populations may change gene frequencies sufficiently that it moves into the domain of a higher adaptive peak. As individuals on the higher peak begin to produce more offspring than those on the lower peak, some of the individuals disperse to the demes on the lower peak(s). This infusion of alleles may change the frequencies sufficiently in these other demes to draw them across the adaptive valley into the domain of the higher peak, decreasing population differentiation as the populations converge on the higher peak.

There are two views of the relative importance of gene flow and natural selection in this final phase of convergence on the higher peak. Crow, Engels, and Denniston's (1990) model suggests that the shift occurs through a small amount of gene flow followed by strong natural selection in favor of a shift to the higher peak. Barton's (1992) model suggests that the shift occurs because gene flow swamps local selection. Whichever view is correct, the interplay of gene flow and natural selection will determine the final geographic structure of a species. One can extrapolate these results to interactions to explain how adaptations and patterns of specialization developed through coevolution in one population can spread through other populations. Gene flow from a population on a higher adaptive peak may change the trajectory of coevolution in a local population, moving it toward that, or some other, adaptive peak. Under these conditions, coevolution at the level of the local population is clearly the raw material from which the trajectory of coevolution is shaped at the geographic level.

Empirically testing such a view for interspecific interactions is as difficult as testing Wright's overall shifting-balance theory of evolution. For the general point I want to make here, however, it does not matter whether every step in Wright's theory is a common and important part of the evolution of species or interactions. What is important is that the shifting-balance theory of evolution provides a framework for understanding why the geographic mosaic of specialization and coevolution may be crucial for understanding the dynamics and trajectory of coevolutionary change. The shifting-balance theory illustrates one way that interactions could evolve through a combination of genetic drift, gene flow, and natural selection, making the overall dynamics of coevolution between two species very different from the patterns observed within local populations. Parts of the shifting-balance theory may be important in some interactions, and other parts in other interactions.

In the same vein, parts of the geographic mosaic of outcome, adaptation, and specialization will be important in some interactions, and other parts in other interactions. Hence, I am not trying to argue that the geographic mosaic is crucial to the evolution of all interactions or that it always governs the process of coevolution. Rather, my point is to argue that the geographic structure of interactions has the potential to be involved to some degree in the evolution of many relationships between species. Any full analysis of the coevolutionary process, and of evolving interactions in general, therefore requires a perspective that ranges from local populations, through the geographic structure of interactions, to the larger patterns in the phylogeny of species.

Using Geographic Structure to Resolve Diffuse Coevolution into Specific Coevolution

A geographic mosaic view of coevolution also suggests how reciprocal evolutionary change can shape interactions that appear at the species level to be asymmetric (one species interacting with several) but are one-to-one at the local level. It is not uncommon in studies of evolutionary ecology to find arguments suggesting that specific coevolution between two species seems unlikely because one of the species interacts with different but related species throughout its geographic range. From this species-level view of specialization, these kinds of interactions become lumped under the catchall of diffuse coevolution. I think this lumping masks the important relationships between specialization, the geographic structure of populations, and coevolution and preempts the search for specific ways in which coevolution can link the evolutionary histories of more than two species.

Some interactions between pollinators and plants appear to fit the pattern of one species interacting with groups of related species over a broad geographic range and provide the opportunity for studies that resolve diffuse coevolution into more-specific patterns of reciprocal change. For example, plants producing oils as floral rewards in addition to, or instead of, nectar and pollen occur in over two thousand species in ten families (Buchmann 1987). These flowers are visited by bees in fifteen genera distributed mostly over three families (Buchmann 1987). The bees use the oils for larval food, sometimes for nest lining, and possibly for adult food as well. Some of these oil-collecting bees are morphologically highly specialized for collecting oils from flowers, and at least some plant species appear to be pollinated solely or primarily by these oil-collecting bees (Cane et al. 1983; Simpson, Neff, and Seigler 1983; Vogel and Michener 1985; Steiner and Whitehead 1990).

Among the most specialized of these interactions are those between *Rediviva* bees and *Diascia* (Scrophulariaceae) flowers in southern Africa. *Diascia* includes about fifty species restricted to southern Africa (Buchmann 1987), and *Rediviva* includes at least eight species, all from Africa (Michener 1981). *Rediviva neliana* is a widespread species that appears to be the primary pollinator of twelve *Diascia* species (Steiner and Whitehead 1988, 1990), and the interaction creates the impression of being highly asymmetric in degree of specialization. But through extensive study of the geographic distribution of this interaction, Steiner and Whitehead (1990) have found that different populations of *R. neliana* appear to be adapted to different species of *Diascia*. Of twenty-two sites that they have studied over a wide geographic range, only three have more than one *Diascia* species present. Local populations of *R. neliana* have foreleg lengths that match the spur lengths of the flowers of the local *Diascia* species they visit. To reach the oil in the flowers, a bee must insert its unusually long forelegs into the corolla and down the long twin spurs (fig. 13.4). Mean spur length varies from 5.3 to 13.9 mm among *Diascia* populations and species, with most of the variation occurring among species. Foreleg length of *R. neliana* varies accordingly: the correlations between female foreleg length and *Diascia* spur length explain 91% of the variance in these traits. Hence, although at the species level it appears that *R. neliana* is a generalized visitor to *Diascia* species, at the population level these interactions are highly species-specific and seem to be fine-tuned.

Although Steiner and Whitehead (1990) interpret these interactions as asymmetric and diffusely coevolved, this seems to me to be a species-level interpretation of these associations. When viewed in a geographic context, this interaction, as they have described it, does not appear to be diffuse, even though it is one pollinator species interacting with twelve plant species. *Diascia* happens to be broken up into twelve recognized reproductively isolated taxa, whereas *R. neliana* is not, but that does not make the interactions any less specific at the level of local populations. If these populations are as mutually dependent as Steiner and Whitehead's results indicate, then individual populations of *R. neliana* are coevolving with individual populations of *Diascia,* some of which are recognized as separate species. The differences in spur lengths among *D. capsularis* populations are as great as some of the differences among other *Diascia* species visited by *R. neliana.*

Similar geographic correlations between foreleg length and floral morphology occur in another *Rediviva* species and its *Diascia* hosts (Steiner and Whitehead 1991). In explaining the evolution of these interactions, Steiner and Whitehead (1991) follow Darwin's ([1862] 1979, 1877) suggestion that the evolution of long-spurred flowers is a result of a coevolutionary race driven by selection for increased pollination efficiency. The stamens of most *Diascia* occur between the two spurs, and the bees come in contact with the anthers as they insert their forelegs into the spurs to reach the oil at their bases. Longer spurs would increase the chance of contact with the anthers and may also increase the amount of pollen transferred to the bodies of the bees. Selection for longer spurs may then select for longer forelegs in a coevolutionary cycle whose limits are set by genetic and physical constraints.

This coevolutionary hypothesis has been attractive to workers on pollination ever since Darwin first proposed it for long-spurred Madagascan orchids, but it was more than a hundred years before it received some direct

Fig. 13.4. *Rediviva neliana* bee visiting a flower of *Diascia capsularis.* Note the close fit between foreleg length in the bee and spur length in the flower. Reprinted, by permission, from Steiner and Whitehead 1990.

experimental support. Using the hawkmoth-pollinated orchid *Platanthera bifolia* in Sweden, Nilsson (1988) experimentally constricted the orchid's spurs at fixed distances from the entrance and recorded both pollinia removal and receipt during visitation by pollinators. Pollinia removal and receipt were both reduced in short-spurred plants, although the effect was greater on pollen receipt than on removal. For the long-spurred Madagascan orchids that led Darwin to his coevolutionary hypothesis, coevolution may be a complex mix of specific and multispecific coevolution within local populations. There are about sixty species of hawkmoths in Madagascar, and at least six of these

Fig. 13.5. The extraordinarily long-spurred Madagascan orchid *Angraecum arachnites* and the equally impressively long proboscis of the hawkmoth *Panogena lingens*. Photographs courtesy of Anders Nilsson. See Nilsson et al. 1987.

have proboscises longer than 7 cm (Nilsson et al. 1987). During a several-week study in central Madagascar, Nilsson and his colleagues found that one hawkmoth species, *Panogena lingens,* was the primary pollinator for six long-spurred orchid species. That hawkmoth, however, has a bimodal distribution in tongue lengths and widths. Only the long- and slender-tongued morph is able to probe the long, thin spurs of *Angraecum arachnites* effectively (fig. 13.5), and pollination of this plant species was restricted during the study to that one form of this one moth species (Nilsson et al. 1985). Now, it may be that the moth *P. lingens* is dimorphic, but it is also possible that there are two closely related species adapted to flowers with spurs of different lengths and widths. Consequently, the interaction may involve either a number of orchid species interacting with one morphologically varied hawkmoth population or a combination of orchids and sibling species of moths with different degrees of local specialization. As was the case for oil-collecting bees and their plants, understanding the structure of coevolution between these orchids and moths may require a more thorough analysis of the taxonomy of the species and the geography of the interactions.

All these interactions, from the mutualisms between pollinators and flowers to the competitive interactions described earlier between *Tribolium* species and the antagonistic interactions between *Linum* and *Melampsora* and between crossbills and seeds, point to the need to develop a more geographically based view of the process of coevolutionary change. Studying solely the dynamics of coevolution within single populations or, alternatively, lumping populations together into a species-level analysis of coevolution ignores the intermediate levels of population structure, which may be the most active and dynamic levels of reciprocal evolutionary change.

Conclusions

This chapter has developed the geographic mosaic theory of coevolution. The specific parts of the hypothesis are outlined point by point in the chapter, but overall the hypothesis suggests that the process of coevolution often depends upon the subdivision of populations into demes, geographic differences in outcome, adaptation, and specialization, and the additional combined effects of genetic drift, gene flow, and extinction. Studies of coevolution within local populations may tell us little, when taken alone, about the coevolutionary dynamics of an interaction. In fact, some forms of coevolution, such as gene-for-gene interactions, may require a metapopulation structure to be maintained over long periods of evolutionary time. Consequently, we require a geographic, and perhaps shifting-balance, view of the process of coevolutionary change. What we think of as diffuse coevolution sometimes comes from failure to consider the geographic structure of an interaction in which individual populations are highly specialized. Understanding the differences in expectations resulting from local and geographic views of coevolution will be one of the most important challenges in coevolutionary studies.

CHAPTER FOURTEEN

Diversifying Coevolution

The geographic mosaic of interactions is likely to be important not only for reciprocal adaptation of interacting species but also for speciation driven by interactions. The ways in which coevolution can influence speciation are among the least understood aspects of reciprocal evolutionary change (Thompson 1982). Nevertheless, we are getting closer to having specific, testable hypotheses. In this chapter I consider a range of ideas on how coevolution may result in speciation, driven by geographic differences in specialization in interactions.

If gene flow between populations is sufficiently restricted, speciation may result as a fortuitous consequence of populations adapting to different local ecological conditions as the interactions evolve in different ways. There are, however, some interactions in which the evolution of reproductive isolation can be a direct result, rather than a chance consequence, of an interaction between two species. In these relationships, one species exerts significant direct control over either the movement of gametes in the other species or the success of matings among subgroups in the other species. These two kinds of control over mating in another species may result in 'diversifying coevolution', producing speciation in one or both of the interacting species (Thompson 1989, 1990). Two kinds of interaction may best fit the conditions of diversifying coevolution: those involving pollinators and plants showing extreme reciprocal specialization and those involving hosts and maternally inherited intracellular symbionts.

In both kinds of interaction, one species controls which hosts successfully mate with whom. The mechanism of control, however, is different. Maternally inherited symbionts, at least in some species, can control which matings result in viable and fertile offspring, whereas pollinators directly control the pattern of movement of a plant's gametes. Interactions between pollinators and plants are, in fact, unique: in no other form of interspecific interaction does one species so directly control the movement of gametes of another species. Consequently, these interactions may be more likely to produce diversification in host populations than any other form of interaction.

Speciation in Pollinator-Plant Interactions through Sexual Selection and Control of Gamete Movement

The mathematical models of Kiester, Lande, and Schemske (1984) have indicated that the geographic structure of populations and sexual selection may both play vital roles in inducing speciation through coevolution between pollinators and plants. In their models, they assume that the pollinator and plant species are specific to each other, and that coevolution involves normally distributed quantitative traits in both species: some continuously varying character such as flower size, scent, or shape in the plant, and preference for that character in the pollinator. The pollinator's preference may be absolute, such that it always prefers a certain value of the character regardless of its frequency in the population, or relative, such that it develops a 'search image' for a particular size, scent, or shape based upon its relative abundance in the plant population.

In large populations of the pollinator and plant without geographic structure, the pollinator and plant populations coevolve in a particular, unvarying direction over time. If instead the pollinator and plant populations are subdivided into small, geographically isolated demes in which genetic drift occurs, then the populations may rapidly diversify if the pollinator's preferences are absolute. The role of genetic drift is to shift the mean pollinator preference and the mean value of the plant character away from the deterministic direction of evolution that would occur in single large populations. (In the terminology of Wright's adaptive landscape, genetic drift moves one or more of the small populations from the domain of one adaptive peak to that of another peak.) Divergence among populations then proceeds at an arithmetic rate through time as the plant characters and pollinator preferences in the different populations stabilize around different means.

Moreover, sexual selection in the pollinator population can make the rate of diversification in small populations exponential rather than arithmetic. In some pollinator species only one sex visits flowers, and which flowers are visited may be a result of sexual selection. For instance, male euglossine bees collect and store floral fragrances from orchids and use these fragrances for attracting mates. Sexual selection could favor in different populations males that gather different fragrances, which would place strong directional selection on both pollinator preference and plant characters. Following the logic of Fisher's ([1930] 1958) argument for the appearance of extreme traits through runaway sexual selection, sexual selection in pollinators based upon their choice plant characters may rapidly drive both plant characters and pollinator preference in different directions in different populations.

This view of plant speciation driven by coevolution with pollinators requires extreme specialization in both the pollinators and the plants. Although specialization driven by sexual selection seems possible in euglossine bees, euglossines often visit more than one plant species. Hence, only some of these plants are likely candidates for pollinator-driven speciation. Another way, however, in which sexual selection may drive plant speciation is through floral mimicry of female insects (Kiester, Lande, and Schemske 1984). The flowers are pollinated as male insects attempt to copulate with them. This kind of floral mimicry demands a high degree of specialization for a particular pollinator, and the particular pollinator species that is mimicked could vary among populations.

The other major route to extreme specialization in these interactions is through pollination by floral parasites. Here the specialization is driven not so much by sexual selection as by the parasitic lifestyle of the insects. Fig wasps, yucca moths, and globeflower flies all show a high degree of reciprocal extreme specialization (Addicott 1986; Bronstein 1987; Pellmyr 1992). These interactions may, in fact, provide the most-plausible conditions for reciprocal speciation through coevolution, but the actual effects of gene flow mediated by the pollinators have not yet been studied in detail.

Symbiont-Induced Speciation

Hypotheses on symbiont-induced speciation rely similarly upon extreme specialization and geographic differentiation. Complete or partial reproductive isolation maintained by cytoplasmically inherited symbionts is now known from about a dozen taxa (table 14.1). Similar unidirectional incompatibility has been shown among other taxa, although cytoplasmically inherited symbionts have not been directly investigated or shown unequivocally to be the cause (Boller et al. 1976; Overmeer and Van Zon 1976; Irving-Bell 1983; Noda 1984). In cases of symbiont-induced reproductive isolation between populations, the isolation is asymmetric: all crosses are successful except for matings between females from uninfected populations and males from infected populations. Treatment of individuals with antibiotics that kill the symbiont—or, in one case, ingestion of bacteria known to produce tetracycline (Stevens and Wicklow 1992)—has been shown to break down this partial reproductive isolation, at least under laboratory conditions. In fact, treatment with antibiotics is the usual experimental treatment used to demonstrate the role of symbionts in reproductive isolation.

In other species, infection results in complete parthenogenesis. For example, some species of *Trichogramma* wasps that are fully parthenogenetic become sexual after treatment with antibiotics (Stouthamer, Luck, and Hamilton 1990; Stouthamer et al. 1990; Stouthamer and Luck 1993). In yet other species, the presence of the symbionts causes different degrees of distortion of the sex ratio, biasing it toward production of female progeny (e.g., Werren,

Table 14.1. Examples of taxa in which cytoplasmically inherited symbionts (parasitic or mutual-istic) have been implicated, usually through antibiotic treatment, in reproductive isolation be-tween host populations or as the cause of parthenogenetic reproduction in host species

Hosts	References
Drosophila melanogaster	Hoffmann 1988
Drosophila paulistorum	Somerson et al. 1984; Ehrman et al. 1989
Drosophila pseudoobscura	Powell 1982
Drosophila simulans	Hoffmann, Turelli, and Simmons 1986; Turelli and Hoffmann 1991; Rousset, Vautrin, and Solignac 1992; Boyle et al. 1993
Culex pipiens complex mosquitoes	Barr 1980; Awahmukalah and Brooks 1985
Ephestia cautella moths	Kellen and Hoffman 1981
Hypera postica weevils	C. Hsiao and T. H. Hsiao 1985; T. H. Hsiao and C. Hsiao 1985
Tribolium confusum beetles	Wade and Stevens 1985; O'Neill 1989
Nasonia wasps	Richardson, Holmes, and Saul 1987; Breeuwer and Werren 1990
Trichogramma wasps	Stouthamer, Luck, and Hamilton 1990
Porcellio dilatatus isopods	Legrand et al. 1986

Source: Updated from Thompson 1987b.

Skinner, and Huger 1986; Ebbert 1992). Recent studies of the ribosomal RNA of reproductive-incompatibility bacteria and parthenogenesis-inducing bacteria have indicated that these symbionts all belong to a monophyletic group in the alpha subdivision of proteobacteria, and they are now collec-tively assigned to the genus *Wolbachia* (Rousset, Vautrin, and Solignac 1992; Stouthamer et al. 1993). The bacteria affect host reproduction by altering the early stages of mitosis. Incompatibility bacteria disrupt incorporation of paternal chromosomes in fertilized eggs, and parthenogenesis-inducing bac-teria prevent segregation of chromosomes in unfertilized eggs (Stouthamer et al. 1993).

Because treatment with antibiotics can sometimes restore sexual reproduc-tion or reproductive compatibility, speciation through acquisition of these symbionts is certainly not inevitable. Nevertheless, if reproductive isolation is maintained for a long period of evolutionary time, speciation is at least a possibility. Moreover, recent studies have shown that mating between indi-viduals from some populations results not just in unidirectional cytoplasmic incompatibility but in bidirectional incompatibility, indicating the existence of different incompatibility strains (e.g., Breeuwer and Werren 1990; O'Neill and Karr 1990; Rousset, Vautrin, and Solignac 1992). This bidirectional in-compatibility could enhance the possibility of symbiont-induced speciation.

At least three different kinds of symbiont involvement in host speciation have been proposed, all based upon the observation that symbionts can cause reduced viability or fertility in hosts (Thompson 1987b). In one version, a cytoplasmically inherited symbiont causes reduced viability or fertility in male hybrids in crosses between host populations, and speciation may by driven by this hybrid inferiority (Powell 1982). In other interactions, a parasite may cause sterility in males in some populations, thereby favoring the evolution of parthenogenetic sibling species (Stefani 1956, 1960; White 1978). In a third version, a noncytoplasmically inherited parasite (e.g., *Plasmodium* and its associated disease, malaria) may have more of a detrimental effect on some genotypes, or populations, than others, and again speciation may be driven by selection against hybrids (Wheatley 1980; Steiner 1981).

Of these three versions, this last one is probably the least likely route to speciation in host populations. It involves only a difference in the mean response of host populations to parasite attack. The other two versions, however, either involve cytoplasmically inherited symbionts that cause complete or almost complete failure of reproduction in one of the two sexes in interpopulational crosses or eliminate sexual reproduction altogether. The result is a direct and immediate effect on host population structure.

Most hypotheses of symbiont-induced speciation that build upon asymmetric failure in reproduction during interpopulational crosses rely specifically upon selection for reinforcement genes that prevent individuals from different host populations from mating with one another (reviewed in Thompson 1987b). Individuals that mate with others from their own population are favored by natural selection, because they exhibit higher viability or fertility than hybrids between the populations. Selection for such reinforcement genes has long been a controversial topic in evolutionary biology. There is still no consensus on the conditions under which selection would favor genes that prevent mating with individuals from neighboring populations rather than genes that decrease inferiority of the hybrids (Templeton 1981; Paterson 1982; Barton and Hewitt 1985; Butlin 1987; Coyne, Orr, and Futuyma 1988; Harrison 1990) or on the specific role that asymmetric reproductive isolation may play in speciation (Ehrman and Wasserman 1987; DeSalle and Templeton 1987).

An alternative is to consider the effects on host speciation of geographically structured coevolution (Thompson 1987b). Populations may come to differ in their interactions with a symbiont species because the interaction produces a different outcome in different environments. The interaction in, say, a colder, wetter, or more infertile environment may be mutualistic, whereas the same interaction in a different environment may be antagonistic. Even given the same environmental conditions, the populations may come to differ if genetic drift moves the populations into the domain of different adaptive peaks. Consequently, the interactions may quickly coevolve in completely different ways in different populations, building up different blocks of alleles. When the host populations come into contact, they may differ in so many genes that selection against hybrid inferiority may not be feasible. Speciation, then, may result as a fortuitous consequence of differential coevolution between hosts and symbionts in different populations.

These two views of symbiont-induced speciation, one based upon reinforcement genes and the other on a fortuitous consequence of adaptation to different physical and biotic environments, are versions of the old arguments about the evolution of reproductive isolation that extend all the way back to Darwin and Wallace. The potential difference, however, is that coevolution with symbionts may speed the process of differentiation between host populations in two ways. First, unlike responses of populations to different physical environments, coevolution involves evolutionary responses of symbionts and hosts specifically to one another. Salinity does not respond to evolutionary changes in organisms to cope with salt in the environment, but symbionts may respond quickly to evolutionary changes in host populations directed at coping with the symbionts, thereby inducing further changes in the host population. This is the same argument that has been used to develop the view that the evolution of recombination and sex is more likely to be a result of variation in interactions with parasites than variation in the physical environment (Hamilton, Axelrod, and Tanese 1990). Consequently, overcoming male sterility may not be simple in hybrids between host populations if many coevolved genes in the host and symbiont are involved. Second, the interaction between the symbiont and host directly affects host reproduction and is the specific cause of hybrid inferiority. Hence, much of the focus of coevolution within local demes may involve selection on how the symbiont affects host reproduction.

Coevolution between symbionts and hosts in geographically structured populations may not often lead to speciation. But because cytoplasmically inherited microbial symbionts appear in so many taxa and sometimes have such strong control over host reproduction, it is a possibility that is worth much more study.

Symbiont-induced parthenogenesis within populations presents a potential route to speciation different from symbiont-induced reproductive isolation between sexual populations. The symbiosis may produce one or more parthenogenetic 'species' rather than separate sexual species. No examples are yet known of populations or species that have been rendered permanently asexual as a result of coevolution with their symbionts. In the species that have been studied in recent years, parthenogenetic individuals often compose less than 5% of the population, and treatment with antibiotics has restored sexual reproduction (e.g., Stouthamer and Luck 1993). There is, however, geographic variation in the proportion of parthenogenetic individuals among populations. On Kauai, Hawaii, Stouthamer and Luck (1993) found fully parthenogenetic populations of the parasitoid *Trichogramma pretiosum*. Although these populations could be rendered sexual with antibiotics, there is the possibility that a population fully parthenogenetic over a long period of time would remain parthenogenetic even after antibiotic treatment if evolutionary changes in host genetics had occurred during that time.

The problem will be in finding such examples among parthenogenetic populations. Restoration of sexual reproduction with treatment with antibiotics indicates that parthenogenesis had been maintained by the symbiont. Lack of restoration of sexual reproduction after treatment, however, is more difficult to interpret. It could mean that parthenogenesis always had a genetic, rather than cytoplasmic, basis in that population, or it could mean that parthenogenesis induced by the symbiont led to genetic changes in the host permanently favoring parthenogenesis. Comparative studies of the mitotic processes in the parthenogenetic and the related sexual species would be needed to propose that symbionts had played a role in speciation.

Alternatively, the biology of these interactions may in some way be intrinsically different from those involving reproductive isolation between populations. Symbionts that induce sex ratio distortion within populations often persist at low levels, commonly around 10%, over long periods of time rather than sweep through the whole population (Ebbert 1992). Symbionts involved in reproductive isolation between populations have also been found in some cases to occur in only some individuals within populations, although it is not clear if these are maintained as stable polymorphisms (Stevens and Wade 1990). In *Drosophila simulans* in California, infection by its *Wolbachia* symbiont has rapidly reached almost 100% of individuals in populations into which it has recently spread (Turelli and Hoffmann 1991).

Little, however, is known about how these interactions between endosymbionts and hosts evolve. Parasite-induced parthenogenesis has been reported in twenty-eight insect species (Ebbert 1992), but the effects on host fitness have been studied in only a few species: *Drosophila melanogaster* (Counce and Poulson 1966), *Drosophila willistoni* (Ebbert 1991), and *Trichogramma* spp. (Stouthamer and Luck 1993). In these species the interaction not only results in the death of males but also causes reduced viability and/or fecundity in females. In addition, Ebbert's careful study of the interaction in multiple populations showed that these outcomes of the interaction (which together constitute what she calls the interaction phenotype) in *D. willistoni* vary both with the host and with the spiroplasma line. Moreover, her crosses between lines suggest that coevolution has happened in some of the laboratory cultures.

Similar rapid evolution of outcome has been found in several other kinds of interaction between endosymbionts and hosts. In 1966 a laboratory strain of *Amoeba proteus* became infected with a virulent endosymbiont, now called x-bacteria (Jeon 1991), that killed most of the amoebae. Some, however, survived and within a few years the outcome of the interaction had changed so completely that the amoeba population was now in fact dependent upon the bacteria for survival (Jeon 1972). This dependence has been demonstrated experimentally through nuclear transplantation between the symbiont-dependent population and a symbiont-free population (Jeon and Jeon 1976) and by selective removal and reinsertion of the x-bacteria from cells (Lorch and Jeon 1980). Removal of the bacteria results in loss of cell viability; reinsertion restores viability. The symbiont-dependent amoebae have persisted in the laboratory, and Jeon (1991) has argued that they could possibly be considered a separate species, although he suggests that for now they are best considered a variant strain.

Two studies of interactions between the bacterium *Escherichia coli* and different plasmids have also shown decreased negative effects on host fitness over time. Bouma and Lenski (1988) cultured *E. coli* with the antagonistic plasmid pACYC184 in the presence of antibiotic and found by the end of five hundred generations that the *E. coli* population had evolved to ameliorate the deleterious effects of the plasmid. The plasmid, however, had not evolved during the course of the experiment. Modi and Adams (1991), however, recorded coevolution between a population of *E. coli* and a different plasmid in a glucose-limited environment after approximately eight hundred generations. The changes in both the bacterial population and the plasmid contributed to the decreased deleterious effect of the plasmid on host fitness.

These examples of amoebae, bacteria, and other symbionts are from laboratory populations in simple environments. Populations in natural environments are undoubtedly more genetically heterogeneous. Nevertheless, each of these studies indicates how quickly interactions between endosymbionts and hosts may evolve, and sometimes coevolve, under some conditions. Occasionally, these changes in outcome may result in speciation and, very rarely, even new forms of life, as has been suggested by the endosymbiotic theory of the eukaryotic cell (Margulis and Bermudes 1985; Bermudes and Margulis 1988; Margulis and Fester 1991; Price 1991b).

Asexuality, Coevolution, and Speciation

Other than symbiont-induced speciation, relationships between specialization, coevolution, sexuality, and speciation are mostly unexplored. When one species in an interaction is obligately asexual, the relationship between specialization and coevolution takes on a special meaning. As with all obligately asexual organisms, species limits are somewhat arbitrary. What matters for the relationship between specialization and coevolution is the number of species with which an asexual clone interacts. By lumping all the clones together, it may appear that one asexual species has a broad range of hosts when in fact it is a group of highly host-specific clones.

Leaf-cutter attine ants include about 190 species, all of which cultivate fungus gardens (Stradling 1991). The ants collect leaves, fruits, and flowers from an extraordinarily wide range of plant taxa, but all these plants are simply substrate for fungal growth in the ants' highly specialized gardens. According to current taxonomy all attine ants garden the same single fungus species, which over the past hundred years has received several names (Cherrett, Powell, and Stradling 1989). The problem in sorting out the taxonomy of this fungus is that it does not form sporophytes. Although some sporophytes have been reported in both the laboratory and the field, most, if not all, are probably contaminants (Stradling 1991). The fungus is completely dependent upon the ants. It occurs only in attine nests, loses in competition to other fungi and bacteria in the absence of cultivation, and relies upon its ant gardeners to disperse it to new nests. A queen inoculates a new nest by actively collecting mycelia before leaving her mother's nest on her nuptial flight. The mycelia are carried in a pouch called the infrabuccal chamber and regurgitated onto the floor of the new nest (Hölldobler and Wilson 1990). The interaction is reciprocally obligate and undoubtedly highly coevolved. Yet by current usage in evolutionary biology, it would commonly be dubbed diffuse coevolution, because this one fungus interacts with 190 ant species.

But there is nothing diffuse or multispecific about this interaction. The fungus is not dispersing among the nests of even several of the 190 attine species. Rather, each fungus colony is actively carried generation after generation to new nests of the same attine species. It is unknown whether after all these millennia of cultivation the fungus colony of one attine species could even survive and grow normally in the nests of other attine species. The interaction between attines and this fungus surely goes back to the early days of attine evolution, before they radiated into so many species. The development of reproductive isolation among attine populations has not made the ant-fungus interaction any more diffuse. Instead, there now appear to be 190 versions of this one-to-one interaction, each coevolving separately. It may not even be fair to argue that the fungus is one species if at least some of these clones are able to survive only in the nest of particular attine species.

Similar situations arise in interactions involving rust fungi, algae, and bacteria. For example, the endosymbiotic dinoflagellate (zooanthellae) *Symbiodinium microadriaticum* has been reported as inhabiting a wide range of marine hosts, including corals, jellyfish, anemones, and giant clams. Yet individuals taken from different hosts differ in chromosome number and volume, suggesting that these symbionts may be at least a species complex, with greater specificity than is implied by its single Latin binomial (Blank and Trench 1985).

The relationship between coevolution and speciation in interactions involving asexual species will continue to be difficult to understand because there are no readily definable limits to an asexual species. Nonetheless, these examples of diversification in associations involving asexual species suggest that the coevolutionary process may be involved in the diversification of these kinds of interactions just as it may be in the diversification of interactions involving sexual species.

Coevolution of Speciating Congeners

The relationships between specialization, geographic structure, coevolution, and speciation take on yet a different meaning when the taxa are very closely related competitors (sister taxa) that are incompletely, or only recently completely, reproductively isolated from one another. Galápagos finches, *Plethodon* salamanders, and dozens of avian species pairs in montane New Guinea all provide examples of pairs of species descended from the same parental species and now thought to be coevolving as competitors (Grant 1986; Diamond 1986b; Hairston, Nishikawa, and Stenhouse 1987). Most hypotheses on the evolution of competition between sister taxa are based upon divergence while populations are geographically isolated (allopatric) followed by coevolution resulting in increased character displacement or increased competitive ability when the populations come back into contact. The divergence that occurs while the populations are allopatric can be considerable or very little. In some cases, populations could diverge so much in use of habitat or resources that, when they come back into contact, competition between them is already greatly reduced and little or no coevolution occurs (Connell 1980). Alternatively, the populations could diverge while in allopatry mostly in characters affecting reproduction, and divergence in habitat or resource use then occurs through direct competition after the populations come back into contact (Lack [1947] 1983; Diamond 1986b). Diamond (1986b) has called these the complete allopatric model and the partial allopatric model of ecological segregation. It is only the partial allopatric model that includes coevolution as an important part of the process of divergence and speciation.

Because both reproductive character displacement and ecological character displacement can be involved in the evolution of sister taxa, this form of coevolution is both unique and difficult to analyze for any one pair of species. Divergence in habitat use could result either from direct competition for resources or from a combination of interspecific competition and intraspecific selection on reproductive behaviors. Divergence mediated by reproduction rather than resource competition could result as selection acts to minimize reproductive contact with the other species (driven by low fitness of hybrid offspring) or as runaway sexual selection favors individuals that preferentially search and choose mates in different habitats. Diamond (1986b) has addressed the problem of the role of competition and coevolution between sister taxa by analyzing 154 closely related pairs of bird species in New Guinea in which at least one of the species is montane. His method offers promise for similar studies of how commonly coevolution occurs between closely related congeners. The 154 sister pairs in New Guinea exhibit different degrees of geographic differentiation and overlap, from completely allopatric populations through species with sharply abutting geographic ranges to species that overlap completely. By comparing each of these geographic relationships between pairs of closely related species, it is possible to study the role that coevolution may play in shaping these taxa.

Diamond divided the pairs of species into six categories, depending upon their degree of geographic overlap. He found that spatial segregation, especially altitudinal segregation, was the major form of ecological segregation among pairs of closely related species. Only with significant geographic overlap do pairs begin to differ in body size or food. Moreover, he found that divergence in these birds generally follows the partial allopatric model rather than the complete allopatric model. That is, most divergence between pairs occurs after species come back into contact. For ten of the nineteen pairs of closely related species that occur both in allopatry and in partial sympatry, altitudinal ranges differ more markedly in sympatry than in allopatry. At least eight of the pairs have sharply abutting altitudinal range limits in sympatry, suggesting that they have reciprocal influences on one another. Finally, the species that have achieved broad geographic overlap are those in which ecological differences between them have become genetically fixed—that is, those in which removal of one species of the pair would no longer result in a broadened use of habitat or resources by the other species. Some of that genetic difference occurs through differences in body size. Only pairs that differ in body weight by a ratio of 1.79 or more overlap in use of habitat.

Diamond emphasized that his findings apply only to New Guinea birds in which at least one species is montane. Other taxa at other places may differ in how and when ecological segregation and specialization occur. New Guinea has a simple geography, with a long central chain of mountains covered with similar vegetation and surrounded by coastal lowlands. The patterns of divergence and ecological segregation among pairs of species and the importance of competition in shaping patterns of specialization may be different in areas in which the habitats are more ecologically diverse and more divergence occurs while populations are allopatric. Nevertheless, his results suggest that under some ecological conditions coevolution between pairs of closely related congeners is a common part of the process of speciation and the evolution of specialization.

Further evidence of coevolution between competing congeners comes from character displacement of myzomelid honeyeaters on Long Island off the northeastern coast of New Guinea (Diamond et al. 1989). Long Island erupted about three centuries ago in a massive volcanic explosion that may have surpassed in volume that of Krakatau. The explosion collapsed the central caldera, creating in its place a doughnut-shaped island covered by several meters of ash. Since then the island has been recolonized by taxa noted for overwater dispersal. Nine pairs of congeneric bird species occur on Long Island and its neighbors, including two small-island specialist species of honeyeater, *Myzomela pammelaena* from the northern Bismarcks and *M. sclateri* from the southern Bismarcks. Although eight of the pairs are known to occur in sympatry elsewhere, these two species are sympatric nowhere else in their geographic ranges and appear to have encountered each other for the first time on Long Island. They are also the only species that have diverged in size from their relatives found elsewhere. The larger *M. pammelaena* is even larger on Long Island and its neighbors than elsewhere, and *M. sclateri* is smaller. What makes this study different from others that have compared sizes of species in allopatry and in sympatry as a way of studying character displacement through coevolution is that the date of first contact between these species and the ancestral size of both species can both be inferred.

An alternative explanation for the divergence in size found in these two species is that the small number of founders of each of these species could have just happened by chance to be larger (for *M. pammelaena*) and smaller (for *M. sclateri*) than the means for these species on other islands and that no further change has occurred in these populations since colonization. That possibility cannot be eliminated, but it would be a rigidly contrariant view to the possible role of competition. If competition, however, has molded the sizes of these species, it could have happened in either of two ways. The two species could have colonized the islands at or near the mean sizes found on other islands and diverged through natural selection to larger and smaller sizes over subsequent generations. Alternatively, one species could have colonized first and diverged from its ancestors due to selection based upon the available resources on Long Island. When the other species arrived, only those colonists whose sizes were most different from the first species may have been successful. As a result, most of the difference in size could have occurred at the time of colonization.

Both of these competitive processes—character displacement and differential colonization—seem to have been involved in the evolution of another group of congeneric birds on islands, the Galápagos finches (*Geospiza*).

Through their long-term study of these birds, the Grants and their colleagues have shown that competition and coevolution have shaped the evolution of these congeners (Grant 1986; Grant and Grant 1989). Because these birds have moved back and forth among the islands of the Galápagos archipelago over evolutionary time, reconstruction of the evolutionary history of these interactions is more difficult than on New Guinea or Long Island. Nonetheless, the evidence for competition and coevolution is compelling. Evidence for differential colonization of islands comes from comparisons of the distribution of species among islands. Some combinations of species are rare or do not occur, and these missing combinations are generally those that are similar in beak size and diet. Evidence for character displacement comes from the greater differences in characters found in populations in sympatry than in allopatry. What makes the arguments compelling, however, is that the Grants and their colleagues have been able to tie the differences found in these species to competition for food. In the cases they have studied of differential colonization, there is never more than one species associated with the peak density of resources, and the species using the most-abundant resources on each island varies among islands. In cases of character displacement, they can show that the differences found in populations in sympatry compared with those in allopatry are not due to differences in the range of available resources, which has been a major criticism of many past studies of character displacement. Hence, as for the birds of New Guinea, Grant (1986) has concluded that competition and coevolution have played an important role in the process of differentiation and speciation in these birds.

Whether coevolution of competing congeners ever drives populations to extreme specialization in their interactions with yet other species is unknown. Natural selection through competition seems more likely to act in a way that favors the use of different habitats or kinds of resources rather than a particular single food species or mutualist (e.g., pollinator, dispersal agent). Nevertheless, among pairs of species that are already highly specialized to interact with only a few other species, competition may push such species to even more extreme specialization. Populations of some Galápagos finches rely heavily upon a single species of *Opuntia* cactus, and the use of this and other food resources is governed by competition with congeners (Grant and Grant 1989).

Although most studies suggesting coevolution between congeners in natural populations have found evidence of increased specialization in use of habitats or character divergence in sympatry, Hairston's long-term study of *Plethodon* salamanders in the mountains of the southeastern United States has indicated increased competitive ability in at least one zone of contact (Hairston, Nishikawa, and Stenhouse 1987). *Plethodon glutinosus* and *P. jordani* are very closely related and form hybrids in a few areas within their geographic ranges, but most populations seem to be reproductively isolated in their zones of contact (Hairston et al. 1992). *Plethodon jordani* occupies the higher elevations in both the Great Smoky Mountains and in the Balsam Mountains, but the zone of overlap is much narrower in the Smokies than in the Balsam Mountains. Through removal experiments and transplantation experiments, Hairston, Nishikawa, and Stenhouse (1987) concluded that the reason for the narrow zone of overlap in the Smokies was greater interference mechanisms (partially through aggressive behavior in at least *P. glutinosus*) among individuals in these populations than in the Balsam Mountains. The selection pressures driving these differences between the mountain ranges are unclear. Neither food nor microhabitat appears to be limiting in the tests that have been conducted.

The results for the Galápagos finches and *Plethodon* salamanders suggest that the competitive interaction between two congeners can result in different evolutionary outcomes in different zones of contact. As with all the other kinds of coevolution between pairs of species considered in this chapter, these results argue for a broad geographic perspective in understanding how coevolution shapes the evolution of competing species. The local outcomes, whether coevolutionary or not, are only part of the larger-scale evolutionary dynamics of interactions between pairs of species.

Conclusions

Speciation may sometimes be a specific consequence of the geographic structure of coevolution. Potential examples span a wide range of interactions, including those between some pollinators and plants, maternally inherited symbionts and their hosts, mutualistic fungi and their hosts, and competitors. Each of these examples of how coevolution may shape speciation emphasizes the need for a geographic view that goes beyond reciprocal adaptation in local populations.

CHAPTER FIFTEEN

Asymmetries in
Specialization and Coevolution

At La Selva, in the lowland tropical forest of Costa Rica, the cycad *Zamia skinneri* is attacked by just one major herbivore species, a lycaenid butterfly whose larvae feed locally on only this one cycad species (Clark and Clark 1991). A few languriid beetle species sometimes nibble at the male cones and scrape the surfaces of the thick sporophylls, often at sites already damaged by the lycaenid larvae, but during seven years of study of this cycad population, the Clarks never found these beetles to cause anything more than minimal damage. The interaction between the cycad and the lycaenid larvae is essentially one-to-one, and each has a major effect on the fitness of the other. This is the kind of rare, pair-wise association between species that sets the stage for coevolution in its simplest form. It is also the kind of highly specific association that many evolutionary ecologists hope to find in studying coevolution.

In fact, a common assumption in many arguments and studies of coevolution is that a reciprocally highly specific interaction is actually a prerequisite for coevolutionary change. By this way of thinking, a plant species whose flowers are visited and pollinated by only one host-specific bee species is a good candidate for coevolution, but a plant pollinated by social bees, which also visit other plants at other times of year, is not. For example, in discussing the role of rodents in the pollination of southern African *Protea* species, Wiens and Rourke (1978) argued that coadaptation could not occur because rodent-pollinated proteas do not flower throughout the year.

In this chapter I discuss why it is wrong to assume that reciprocal extreme specialization is a prerequisite for coevolution, and I consider how different kinds of asymmetry in specialization lead to different forms of coevolution. Toward the goal of replacing the general concept of diffuse coevolution with more testable hypotheses, I discuss three specific hypotheses on reciprocal change between groups of species: coevolutionary alternation, successional cycles of coevolution, and coevolutionary turnover. All are based upon the view that specialization is evolutionarily dynamic rather than a dead end.

The Fallacy of Extreme Specialization as a
Prerequisite for Coevolution

Undoubtedly, a one-to-one relationship provides the most-convenient conditions, and the easiest opportunity, for observing coevolution. Nonetheless, there is no reason to assume that two species not fully specific to one another cannot undergo reciprocal evolutionary change. Species need not be specific to one another year-round for coevolution to be possible. An interaction that occurs between a pair of species only during part of each year may greatly affect fitness in both of them and thereby drive coevolution between them, even though they interact with other species at other times of the year.

Imagine, for instance, a community in which plants come into flower at different times of the year. A bee population visits the flowers of the different plant species as they come in bloom. At some times during the year many plants may be in bloom, but at other times very few may be. There may be a time during each year when the bee population has few options and predictably relies upon one plant species for nectar. That short period of time each year could impose stringent selection on the bee population to be highly effective at harvesting nectar from that one plant species. At the same time, that plant species may evolve in morphology or blooming time to take advantage of the availability of this one bee species during this window in time. As a result, the bee and plant species may undergo specific coevolution even though the bee population visits many other plant species at other times of the year.

The sequence of visiting different plants at different times of the year may impose some conflicting selection pressures on the bee population. The outcome may be a suite of traits in the bees that is a compromise between the need to be effective at harvesting nectar from the one species available during the lean period and the need to harvest nectar from other plants throughout the rest of the year. But as long as the bee population must predictably rely heavily upon only one plant species for part of each year, selection during that period may be crucial in molding the overall morphology and foraging tactics of these flower visitors.

The same logic applies to all interactions. Reciprocal extreme specialization is not a requirement for coevolution. Species may interact with many other taxa and yet have much of their evolution driven by their relationships with only one or a few other species.

The Catchall of Multispecific Coevolution

One of the most promising developments in the study of evolving interactions in recent decades has been the increasing number of studies designed to explore the middle ground between species pairs and whole communities. Ideas and experiments on guilds (Root 1967; Simberloff and Dayan 1991), component communities (Root 1973), keystone predators (Paine 1966), keystone mutualists (Gilbert 1980), strong versus weak links in food webs (Paine 1980), tritrophic-level interactions (Price et al. 1980), parasite-mediated outcomes of competition (Price et al. 1986), and other indirect effects have all been aimed at finding how different clusterings of relationships between species affect the ecology and evolution of interactions. Many of these concepts have been used solely to search for ecological patterns, but some have been used in searches for what can be called the evolutionary units of interaction (Thompson 1982). Each of these concepts emphasizes in one way or another that many interactions are multispecific and asymmetric.

As the number of interacting species increases, the problem of what constitutes reciprocal change also increases. In a group of interacting taxa, not every species will change in response to each change in every other species. Rather, there may be bouts of alternating, episodic change. Two pollinator species using the flowers of one plant species interact directly with the plants and indirectly with each other. As more plants and pollinators are collected into the community over evolutionary time, the connections get even more complicated. The result can be a guild of hummingbird species that interacts with a guild of hummingbird-pollinated plants. A tremendous amount of reciprocal change may occur between and within these guilds over evolutionary time. To ignore this kind of coevolution because it is not possible to pick out individual pairs of species undergoing reciprocal change would be to ignore one of the potentially most important aspects of evolution and the origins of community organization.

Currently, these multispecific associations are generally lumped together into what has variously been called diffuse coevolution, guild coevolution, or multispecific coevolution. Multispecific coevolution emphasizes that the evolutionary unit of an interaction may involve more than a pair or trio (in tritrophic interactions) of species. The problem with multispecific coevolution for some evolutionary biologists is in where to draw the line short of some Gaia-like extreme of oxygen-producing plants and oxygen-consuming animals. But this is no more a problem than that of setting the limits of the effects of a gene with all its main, pleiotropic, epigenetic, and heterochronic ramifications. Some effects are much more important and tractable by experimental study than others.

Part of the problem is that the broad designation of multispecific coevolution covers such a wide range of interactions. Consequently, it may hide important differences in the process and outcome of reciprocal change. Consider, for example, some of the diverse kinds of interaction that we would currently bundle together under the general title of multispecific coevolution.

1. One parasitic species coevolving with several hosts: one species of cuckoo parasitizing the nests and mimicking with different color patterns the eggs of several avian hosts, each of which has evolved defenses against this cuckoo species.
2. Several highly host-specific parasites attacking one host species within a local community: the insect fauna restricted to a population of one milkweed (*Asclepias*) species within that community.
3. Several host-specific parasites attacking one host species but with different subsets of parasite species attacking different host populations: the insect fauna associated with a milkweed species throughout North America.
4. One major grazing herbivore coevolving with many plant species: bison and grasses on the Great Plains of North America.
5. Many grazing herbivores coevolving with many plant species: ungulates and grasses in the Serengeti Plain.
6. Interactions among a group of competing species over a broad geographic range in which local interactions may often involve extinction of populations: lizard species distributed over islands in the Caribbean.
7. Several potentially mutualistic visitors coevolving with several hosts: a guild of large frugivorous birds visiting lauraceous plants producing large-seeded fruits within a community; a guild of hummingbirds visiting plants with long-tubed corollas.

To corral all these interactions together as one form of coevolution can only lead to confusion and to lack of explanatory and predictive power in hypotheses about outcomes of reciprocal evolution. At the very least, hypotheses of multispecific coevolution will need to be partitioned into more specific hypotheses that take into account two important components of these interactions: (1) the origins and effects of asymmetry (e.g., one species coevolving with two or more other species) and (2) the effects of the geographic structure of specialization within populations.

The Origins of Asymmetries in Coevolution

Asymmetry in the degree of specialization within an interaction can arise in at least two ways: accumulation of unrelated species and differential rates of speciation. As an interaction grows and spreads within and among communities, asymmetries can develop as additional, unrelated species accumulate more on one side of the interaction than on the other. In addition, some of the interacting species may undergo more speciation than the others. As a result, an interaction between a pair of species may become a much more multispecific interaction over evolutionary time.

The asymmetries in coevolution between woody plants and plant-inhabiting ants appear to have arisen from these dual effects of accumulation of unrelated species and differential speciation. Janzen's pioneering study of coevolution between ants and acacias suggested that a small number of *Pseudomyrmex* ants have coevolved with swollen-thorn acacias in tropical and subtropical America (Janzen 1966, 1967b). The evolution of specialization in acacias and acacia-ants epitomizes what is probably a common pattern as interacting taxa diversify, collect additional species, and shift around geographically over evolutionary time. *Pseudomyrmex* is the largest of three genera of ants that compose the Pseudomyrmecine. Of the approximately 230 species in this subfamily, at least 37 are thought to be obligate to plant domatia (Ward 1991). The plants used as hosts are distributed over twenty genera in about fourteen families. Many of the domatia-inhabiting species tend coccids on their hosts, although those inhabiting bull's-horn acacias do not. Species such as *P. ferruginea* are clearly mutualistic with their host plants (Janzen 1967b), whereas *P. nigropilosus* is a parasite that offers no protection to the plants it inhabits (Janzen 1975).

Through a preliminary phylogenetic analysis, Ward (1991) has estimated that obligate domatia inhabitation may have arisen at least twelve times in the Pseudomyrmecine ants. Some may have colonized plants that already had associations with other Pseudomyrmecines, and some have apparently switched onto different plant genera or families over evolutionary time. The result is a complex mix of associations and degrees of specialization to particular plant species. Some of these associations are probably long-standing, with particular ant-plant pairs highly coevolved, and others the result of recent colonizations.

Similar analyses of specialization are now emerging for other mutualisms in which the plants provide a domicile and food rewards in return for the protection afforded by the guarding ants from herbivores or encroaching vines: *Macaranga* trees and *Crematogaster* ants in Southeast Asia (Fiala, Maschwitz, and Pong 1991), *Cecropia* trees and *Azteca* ants in the Neotropics (Davidson and Fisher 1991; Longino 1991), and *Leonardoxa* and their associated ants within the tribe Myrmelachistini in tropical Africa (McKey 1991). In all these cases, the phylogenetic studies of the plants and the ants are still incomplete. But even the partial analyses now available indicate that very similar associations among species can evolve different degrees of asymmetry as speciation and geographic distributions vary and shift among the interacting species.

The plant genus *Cecropia* includes over 100 species in the Neotropics, most of which have a suite of traits that appear specifically adapted for the maintenance of resident ant colonies: hollow stems used by their mutualistic ants as domatia; a thin area in the internodal wall lacking latex ducts and used as an entrance hole by founding queens; and glycogen-rich Müllerian bodies at the base of each petiole used by the ants as their primary food

Table 15.1. Specialization in the interaction between *Cecropia* plants and their obligate attendant ants at Reserva Tambopata and Estación Biológica de Cocha Cashu, Peru

Cecropia spp.	No. of plants	*Azteca ovat- iceps*	*Azteca xan- thochroa*	*Campo- notus balzani*	*Pachy- condyla luteola*
Light demanding					
membranacea	53	49	21	30	0
engleriana	6	83	17	0	0
polystachya	7	57	29	14	0
Shade tolerant					
tessmannii	31	0	0	10	90
species A	17	0	100	0	0
ficifolia	24	0	48	52	0

Source: Data from Davidson and Fisher 1991.
Note: The values are the percentages of plants inhabited by each of the ant species. Seedlings
sometimes contain foundresses of other ant species, but only the listed species are successful at
establishing colonies.

(Longino 1991). Of the approximately 150 named species and subspecies of
Azteca ants, at least 6 live obligately in *Cecropia.* There are marked differences in nest structure and behavior among these 6 obligate *Cecropia* nesters,
which Benson (1985) and Longino (1991) have used to argue for two or more
independent origins of this mutualism. The taxonomy of these ants, however,
is still in flux, and the geographic patterns in specialization are only now
being studied in some detail. In addition, there are other ants that live in
Cecropia, and some of these also appear to be host specific. In southeastern
Peru, there are 4 known species of *Cecropia* specialists: 2 *Azteca* species, a
Camponotus species, and a *Pachycondyla* species (Davidson and Fisher
1991). Established colonies of *Azteca xanthochroa* have been found in 5 of
the 6 local *Cecropia* species, *Camponotus balzani* in 4, *Azteca ovaticeps* in
3, and *Pachycondyla luteola* in 1. There is then a distribution in the degree
of local host specialization by the ants, but that specialization may be due
at least partially to the preference of the ants for particular habitats. *Azteca
ovaticeps* is found only on the 3 fast-growing, light-demanding *Cecropia*
species and never on the 3 slow-growing, shade-tolerant species (table 15.1).
    Although the number of sampled plants is still small for some species, this
complex pattern of host specificity in the ants also seems to cause a complex
pattern of specificity in the plants. Of the three light-demanding *Cecropia*
species, two are associated at Cocha Cashu in southeastern Peru with three
ant species, and the other with two ant species (table 15.1). The three shade-
tolerant species show greater specificity, associating with one or two ant species. This small group of six plants and four ants therefore gives an almost

complete range of possible degrees of specialization and asymmetry. The most reciprocally specific, at least locally, are *C.* species A and *A. xanthochroa,* and *C. tessmannii* and *P. luteola.* Davidson and her colleagues have argued that at least this latter interaction may be the result of selection on *C. tessmannii* favoring this relatively large ant species (Davidson et al. 1991; Davidson and Fisher 1991). The Müllerian bodies of *C. tessmannii* average three times the dry weight of those of the *Cecropia* species that normally harbor the smaller, *Azteca* ants. Moreover, *P. luteola* queens, whose nutrition and egg production appear to depend upon Müllerian bodies, typically reject the smaller Müllerian bodies of other species, either because they differ in size or perhaps because they differ in chemical composition.

At the other extreme are the interactions between *Leonardoxa* and its Myrmelachistini ants, in which the total number of species and the degree of asymmetry in both the plants and the ants is small. *Leonardoxa* has five species, two in the Zaire basin and three closely related species in the Lower Guinea coastal forest. All species have extrafloral nectaries. The two species from the Zaire basin have not been studied, but they lack any features other than extrafloral nectaries to attract ants and probably do not maintain a specialized relationship with any one ant species (McKey 1991). Of the three species from the Lower Guinea coastal forest, one has only extrafloral nectaries and attracts a variety of ants. The other two species, however, have extrafloral nectaries and swollen internodes that are domiciles for specialized ants. *Aphomomyrmex afer* lives within *Leonardoxa letouzeyi* but has also been collected, although rarely, in areas outside the range of *Leonardoxa. Petalomyrmex phylax* is a morphologically highly specialized species and appears to be restricted throughout its range to *Leonardoxa africana* (McKey 1991). These two ants are monotypic genera, and the only African representatives of the tribe Myrmelachistini. They may be sister species that have undergone considerable morphological differentiation in the course of coevolving with their respective *Leonardoxa* hosts. If *Leonardoxa* had undergone more extensive speciation, the relationships could possibly have appeared much more asymmetric, like those between *Cecropia* and its ants.

The pattern of specialization between ants and ant-fed plants is probably similar to that found in ant-protection mutualisms, involving a similar combination of coevolution, differential speciation rates, and switching of partners. Even more so than ant-protection mutualisms, these interactions may commonly involve the gathering of new, unrelated species into the interaction over evolutionary time. There are two routes along which plants have evolved to capture the by-products of ant activity for nutrition. Some tropical epiphytes not only have specialized cavities in their rhizomes, stems, roots, or leaves that are used as nesting sites by particular ant species but also appear to have evolved the ability to absorb the nutrients in the ants' debris

(Benzing 1970; Huxley 1978; Rickson 1979; Rico-Gray and Thien 1989). Some Neotropical epiphytes, distributed over at least eleven angiosperm families, grow on the carton nests of arboreal ant species, forming ant gardens (Ule 1901; Kleinfeldt 1978; Davidson 1988). Both types of interaction have appeared sporadically among plant families, leaving the distinct impression that new taxa have tapped into a preexisting interaction over evolutionary time. Ant-fed plants with nesting cavities are distributed across at least four to five angiosperm families and one fern family (Janzen 1974c; Huxley 1980, 1986; Thompson 1981b; Benzing 1991; Jebb 1991). Ant-garden plants are known from at least eleven families (Davidson 1988; Benzing 1991).

In the tropical Pacific, the main ant inhabitants of the plant cavities are several species of *Iridomyrmex,* whose distributions among plant species appear to be related more by habitat and elevation than by specialization to particular plants. Any new plant mutant with the appropriate cavity would probably have the potential to be incorporated into this interaction. The list of ant inhabitants from the Neotropics, for both ant-cavity and ant-garden plants, is longer, but *Azteca, Camponotus,* and *Crematogaster* are among the most commonly observed genera. At least some of the plants appear to be obligately associated with ant nests (Davidson 1988), but neither the plants nor the ants appear to show extreme specialization. The ants use chemical cues in choosing seeds to carry back to their carton nests, and they will even carry back bits of porcelain coated with the correct compounds (Davidson et al. 1991). Hence, as in many interactions between ants and other species (Hölldobler and Wilson 1990), tapping into an ant's communication system may be the first major step in a plant species' becoming incorporated into this interaction.

Although the number of species involved in these interactions is small, the number of plant families is probably too large for us to sort out how the interaction first began and which taxa are newcomers. This is a recurrent problem in many interactions, and it is at the root of much of the angst among evolutionary ecologists studying coevolution. One response is simply to sweep it all under the rug of diffuse coevolution and leave it at that. But rather than viewing it simply as a problem in studies of coevolution, we should view these complex taxonomic distributions of interacting participants as one of the important evolutionary outcomes of an initially reciprocal interaction. Interactions commonly grow through the accumulation of new taxa, and that is especially true for some kinds of mutualism (Thompson 1982). These complex associations show us one of the ways in which asymmetries in specialization develop over evolutionary time. They also provide us with clues about how coevolution influences the organization of communities by providing foci for the evolution of species that exploit preexisting, coevolved interactions.

The following sections examine three hypotheses on coevolution between groups of species that go beyond the catchall concept of multispecific coevolution. These hypotheses all require additional testing, but each of them shows that it is possible to replace the concept of diffuse coevolution with specific, testable views on how one species may coevolve with several other species through patterns of specialization that change over evolutionary time.

Hypothesis 1: Coevolutionary Alternation

Asymmetries in specialization need not simply increase over evolutionary time. They can vary over time specifically as a result of a process that can be called coevolutionary alternation. Imagine a parasite that initially attacks one host. The host evolves defenses against that parasite, which in turn results in one of three evolutionary outcomes: local extinction of the parasite, the evolution of counterdefenses in the parasite, or a shift onto a new, less-defended host. The first outcome ends the interaction, the second leads to an arms race between the two species, and the third sets the stage for coevolutionary alternation. If the parasite shifts to a new, poorly defended host, selection for defense will begin to decrease on the original host. As time proceeds, the original host may become poorly defended as the new host becomes better defended, and natural selection will then begin to favor those parasites that attack either the original host or yet another host. In this way, the parasite may come to alternate among several hosts and to coevolve with all of them (fig. 15.1). Time lags in the evolution of defenses and counterdefenses—and time lags in their loss—would produce a complex pattern of specialization in the parasite populations and a complex distribution of defenses among hosts.

Davies and Brooke (1989a,b) have suggested that brood parasitism by cuckoos (*Cuculus canorus*) in Britain follows just this kind of coevolutionary cycle. These cuckoos currently have four main hosts in Britain: meadow pipits (*Anthus pratensis*), dunnocks (*Prunella modularis*), reed warblers (*Acrocephalus scirpaceus*), and, to a lesser extent, pied wagtails (*Motacilla alba yarrellii*). In addition, they occasionally parasitize at least ten other bird species. The cuckoos are genetically polymorphic for the color of the eggs they lay. In Britain there are four different egg morphs, which correspond to the four major host species. Three of these egg morphs match their host eggs very well. The exception is the egg morph laid in dunnock nests, which is not mimetic (fig. 15.2).

Not only is the mimicry of eggs remarkably good for three of the morphs, but the cuckoos also use some behaviors to lessen the chance that a host will discover it has been parasitized. A cuckoo waits until a host female has begun to lay her eggs before laying her small (relative to cuckoo body size) egg in

Hypothetical Points in Evolutionary Time during Coevolutionary Alternation

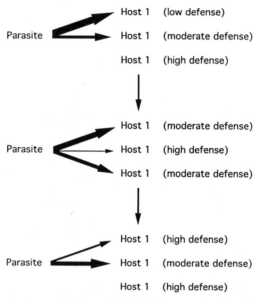

Fig. 15.1. The process of coevolutionary alternation between a parasite and several hosts. Only one of many possible patterns of shifts among hosts over evolutionary time is illustrated. The thickness of the arrows indicates the proportion of parasite individuals attacking a particular host. The proportion will vary among populations and, over time, within populations. Based upon the results of Davies and Brooke (1989a,b) for coevolution between cuckoos and their hosts.

the nest, usually removes one host egg before laying her own, and accomplishes it all in remarkably short time—about 13 seconds—during the afternoon, when she is least likely to be observed by the host. In an elegant series of experiments Davies and Brooke showed that all these behaviors increase the chance that a host will accept the cuckoo egg. They placed eggs of different colors in the nests and found that the major hosts, except the dunnock, either ejected eggs that were the wrong color and shading or abandoned the nest altogether. They also provided strong evidence that ejection behavior was an evolved response: meadow pipits and pied (white) wagtails in Iceland, where cuckoos do not occur, show less discrimination against eggs of the wrong color and shape than do birds of the same species in Britain (Davies

Fig. 15.2. Mimetic eggs laid by cuckoos (*Cuculus canorus*). Cuckoo populations in Britain have evolved eggs that mimic three of their four major hosts. *Top row:* cuckoo eggs from reed warbler mimics ('gentes'), meadow pipit mimics, and pied wagtail mimics. *Middle row:* model cuckoo eggs representing each of the three mimetic types plus a fourth representing a redstart egg (a suitable but currently rarely used host in Britain for which a mimetic type occurs in Finland). *Bottom row:* Eggs of the current favorite British hosts—reed warbler, meadow pipit, pied wagtail, dunnock. Reprinted from N.B. Davies and M. de L. Brooke, 1989, "An Experimental Study of Co-evolution between the Cuckoo, *Cuculus canorus*, and Its Hosts," *Journal of Animal Ecology* 58:207–24, by permission of Blackwell Scientific Publications Ltd.

and Brooke 1989a). Moreover, species in Britain with no known history of parasitism by cuckoos and species that are unsuitable as hosts (e.g., seed eaters) seldom reject experimental eggs of the wrong color and shape. Hence, only populations of the cuckoos' normal hosts seem to maintain high levels of discrimination against eggs that are unlike their own.

The exception is the dunnock, which provides the clue to the process of coevolutionary alternation. Although this a major host in Britain, it does not generally reject poorly matched eggs, and no egg morph in the cuckoo populations matches the shape and color of dunnock eggs. Davies and Brooke (1989b) have argued that the dunnocks have only recently become a common host of cuckoos in Britain. Although they are now a major host, only about

2% of dunnock nests are parasitized. Consequently, it could take thousands of generations for a gene favoring rejection of eggs unlike their own to spread and reach a high frequency in these populations. This evolutionary time lag would favor cuckoos that parasitize dunnocks rather than hosts that have already evolved high levels of egg discrimination. Parasitism of dunnocks should continue to be favored until they develop high degrees of discrimination, after which selection should favor cuckoos that attempt to lay in the nests of yet other hosts.

This is where another evolutionary time lag comes in. Once cuckoos have shifted onto a new host, the old hosts will retain their recognition abilities for a while until those genes have been lost. Some of the currently seldom-used hosts in Britain exhibit stronger rejection behaviors than some of the currently preferred hosts, which may indicate just this kind of time lag. Hence, the range of potential new hosts available to cuckoos will depend at any moment in evolutionary time upon how long ago each of its old hosts was abandoned.

The territorial breeding system of European cuckoos may add some additional twists to how these time lags affect the rate at which host shifts occur. In a five-year study of a population at Dungeness Bird Observatory near Kent, Riddiford (1986) found that territory-holding pairs of male and female cuckoos expelled nonterritorial females from the habitat of the preferred hosts but tolerated them at the periphery of their territories. Hence, nonterritorial females were denied access to the currently preferred hosts and were forced to lay elsewhere. If this combination of territorial and nonterritorial females is common in European cuckoo populations, then it has at least two implications for the evolution of host shifts. First, even after most cuckoos have shifted onto a new host, nonterritorial females may have little option but to continue using the old host and even some nonpreferred hosts. Hence, the rate at which old hosts lose their rejection behaviors will be slower than would occur if old hosts were completely abandoned for some time. Second, the territorial system may fortuitously allow a faster shift to new hosts than might occur if all the cuckoos were restricted to a single host at any moment in evolutionary time. As the currently preferred hosts evolve rejection behaviors, individuals forced to use nonpreferred hosts could gradually become the focus of selection for another host shift in cuckoo populations. These territorial and competitive effects are probably not an essential part of coevolutionary alternation. Instead, they influence the rates of evolution of defense in hosts and shifts among hosts by the cuckoos.

The potential for coevolutionary alternation may vary among taxa of brood parasites, because the effect of brood parasites on hosts varies. Cuckoos and cowbirds are two major avian groups that include species specialized for laying their eggs in the nests of other avian species. The number of hosts with which these two groups of brood parasites coevolve varies both among species and among populations. In South America the shining cowbird *Molothrus bonariensis* has been recorded from the nests of over 170 host species, whereas the screaming cowbird *M. rufoaxillaris* appears to be mostly host specific to a cooperatively breeding host, the bay-winged cowbird *M. badius* (Friedmann, Kiff, and Rothstein 1977; Mason 1986). These numbers, however, are snapshots of single moments in evolutionary time.

Once a brood parasite finds a host nest, it usually lays a single egg in it, often removing one of the host's eggs in the process. Depending upon the species of brood parasite, the parasite chick, soon after hatching, kills the host's chicks or develops alongside them. Either way, host fitness is decreased. Consequently, natural selection should favor hosts that are alert to brood parasites searching for nests, recognize eggs and chicks of brood parasites in their nests, and adopt behaviors that minimize the chance of being parasitized or minimize the effect if they are parasitized. In turn, selection should favor brood parasites that are sneaky when laying their eggs, and it should often favor eggs that mimic those of the hosts.

The coevolutionary asymmetries in these interactions seem to develop because of constraints in the evolution of host defenses and evolutionary time lags and shifts among hosts. Ejection of a brood parasite's eggs imposes both costs and constraints on hosts, which may influence both the evolution of host defense and specialization in the parasites. Avian hosts with large bills can eject parasite eggs by grasping them in their bills and lifting them out of their nests, but this is not easy for smaller hosts. In Norway, large-billed hosts are grasp ejectors, whereas medium-billed hosts generally attempt to puncture a parasite's egg in order to get a grip for ejection, and small-billed hosts must simply abandon their nests (Moksnes, Røskaft, and Braa 1991). The host species with the smallest bills therefore suffer from the greatest costs of defense, although puncture ejection can also impose important costs. The eggshells of many brood parasites are thicker than those of other similarly sized birds (Spaw and Rohwer 1987). When northern orioles attempt to puncture the thick-shelled eggs of North American brown-headed cowbirds, they sometimes damage their own eggs (Rohwer, Spaw, and Røskaft 1989).

Consequently, the evolution of defense against brood parasites should vary with both host size and the cost to host fitness of accepting a parasite's egg. Brown-headed cowbirds differ from European cuckoos in that unlike cuckoos they do not eject the hosts' eggs or nestlings. Hence, the cost for a small host of accepting a cowbird egg is less than the complete loss in reproduction that results from accepting a cuckoo's egg. Selection should always favor in cuckoo hosts either ejection of parasite eggs or abandonment of the nest, but not so for cowbird hosts. The high cost of attempted ejection or nest abandonment may, in fact, favor acceptance in small species attacked by cowbirds (Rohwer and Spaw 1988). Damage to her own eggs during attempted ejection or nest abandonment may result in a greater loss in lifetime fitness for the host bird than simply rearing the cowbird along with her own off-spring. In comparing the proportion of small- to medium-billed Norwegian species that reject cuckoo eggs with the proportion of similarly sized North American species that reject cowbird eggs, Moksnes, Røskaft, and Braa (1991) found that more cuckoo hosts attempt to eject eggs than cowbird hosts.

This difference in the cost of defense against cuckoos as compared with cowbirds may partially explain why European cuckoos are subdivided into genotypes that are host specialists, producing eggs that closely mimic their hosts, whereas brown-headed cowbirds are not. Studies of other cuckoo species suggest that these differences in patterns of specialization may not be restricted to these two species alone. In Australia, bronze-cuckoo species specialize on only a few hosts. Horsfield's bronze-cuckoo (*Chrysococcyx basalis*) most commonly parasitizes fairy-wrens (*Malurus* spp.), although it also parasitizes thornbills (*Acanthiza*), robins (*Petroica*), and several other genera of small passerines throughout its range; the shining bronze-cuckoo (*C. lucidus*) most commonly parasitizes thornbills (Brooker and Brooker 1989). Local populations use more than one host, but the range of hosts is narrow. During four years at Gooseberry Hill on the eastern outskirts of Perth, Horsfield's bronze-cuckoos parasitized both splendid fairy-wrens and western thornbills, whereas shining bronze-cuckoos parasitized yellow-rumped thornbills and western thornbills.

In considering the high degrees of host specialization found in bronze-cuckoos, Brooker and Brooker (1989) dismiss rejection by hosts as a cause and argue instead for a combination of two selective forces. First, the birds may be restricted to hosts whose eggs approximately match the sizes of their own eggs, so that incubation takes about as long as a host egg. Second, among the hosts with appropriately sized eggs, competition among cuckoos may have favored even greater degrees of host specificity. The competition they envisage takes the form of a bronze-cuckoo selectively removing the egg of any other bronze-cuckoo she encounters when laying an egg in a nest. Severe competition among cuckoos is certainly a possibility in the Gooseberry Hill population. During four years of study, Horsfield's bronze-cuckoo parasitized on average 24% of splendid fairy-wren nests and 12% of western thornbill nests. During the previous eleven years, these cuckoos had parasitized on average 17% of splendid fairy-wren nests. Similarly, shining bronze-cuckoos parasitized an average of 26% of yellow-rumped thornbills and 8% of western thornbill nests. With these high percentages of parasitism, bronze-cuckoos searching for nests must commonly come upon nests that have already been parasitized. There is some weak evidence from studies of brood parasitism of cuckoos, although not these species, that a female may preferentially remove a cuckoo egg rather than a host egg before laying her own egg in a nest. Brooker and Brooker argue that if this occurs in bronze-cuckoos, it could favor females that lay eggs only in the nest of species whose eggs their own mimic. They use the same argument to explain the evolution of polymorphism in egg color within other cuckoo species.

This 'incubation time and competition' argument for the evolution of specialization in bronze-cuckoos arises from the observation that the host populations near Perth do not eject the eggs of these brood parasites. Moreover, the hosts do not abandon parasitized nests any more frequently than unparasitized nests (Brooker and Brooker 1989). Nevertheless, hosts could be more involved in the long-term evolution of specialization than implicated by this hypothesis. Fairy-wrens and thornbills are among the smallest of Australian birds. Ejection may not be a viable evolutionary option for the current hosts of these bronze-cuckoos, just as it does not seem to be a viable option for the small hosts of cuckoos in Europe. Parasitism of larger hosts, which could eject eggs, may have been selected against over evolutionary time. In fact, bronze-cuckoos lay eggs that are much smaller than one would predict from their body size, approximating instead the size of the eggs found in their much smaller hosts (Brooker and Brooker 1989). The incubation time and competition argument seems insufficient to explain why these bronze-cuckoos are so specialized to only the very smallest of Australian birds.

That leaves the question of why these small hosts do not abandon parasitized nests. Until more is known about the behavior of hosts in other populations throughout Australia, it is not possible to know if this is an invariant behavior common throughout fairy-wren and thornbill populations and species or if these are newly parasitized host populations that have not yet responded. It would be somewhat surprising if these birds continued over a long period of time to exhibit no response to brood parasitism. Rejection behaviors in populations faced with brood parasitism seem to evolve readily in passerines (Rothstein 1990). Among reed warblers (*Acrocephalus* spp.), the clamorous reed warbler (*A. stentoreus*) accepts unlike eggs in unparasitized populations in Australia (Brown et al. 1990), but the reed warbler (*A. scirpaceus*) and the great reed warbler (*A. arundinaceus*), which are parasitized in Europe, reject eggs that appear unlike their own. Parasitized populations of meadow pipits (*Anthus pratensis*) and white wagtails (*Motacilla alba*) in Britain are more likely to reject unlike eggs than unparasitized populations in Iceland (Davies and Brooke 1989b). Similarly, American robins (*Turdus migratorius*) in southern Manitoba, where they are attacked by brown-headed cowbirds, show higher rates of rejection than robins in northern Manitoba, which are outside the geographic range of these cowbirds (Briskie, Sealy, and Hobson 1992). And magpies (*Pica*), which are parasitized by great spotted cuckoos (*Clamator glandarius*) only in parts of their geographic range in Europe, show a gradient in their response to brood parasitism, depending upon how long the association has been in place: they reject both mimetic and nonmimetic eggs in areas of ancient sympatry, show some rejection, especially of nonmimetic eggs in areas of recent sympatry, and accept both mimetic and nonmimetic eggs in populations beyond the range of the cuckoos (Soler and Møller 1990). Evolutionary time lags in defense are probably a common aspect of coevolution between brood parasites and their hosts (Davies and Brooke 1989a,b; Rothstein 1990). The tendency of passerine hosts to evolve rejection behaviors only when faced with brood parasitism and the time lags associated with the gain and loss of these behaviors are precisely what may drive coevolutionary alternation, with parasites changing local hosts over evolutionary time as hosts develop defenses.

Coevolutionary alternation is a powerful view of asymmetric coevolution that may well extend beyond brood parasitism to other interactions between parasites and hosts, grazers and hosts, and predators and prey. It provides a clear view of how one species can coevolve with at least several other species, sometimes simultaneously and sometimes in an alternating sequence of recurring interactions that may take thousands of years to cycle. Moreover, it highlights the inadequacy of studying any pair-wise interaction, or any group of current interactions, in isolation from the historical community context in which it has developed.

Hypothesis 2: Successional Cycles of Coevolution

On a shorter time scale than coevolutionary alternation, the process of succession may produce minicycles of coevolution replete with asymmetries. The patch dynamics of species (Thompson 1978; Pickett and White 1985) also creates a patch dynamics of interactions (Thompson 1982, 1985b). The repeated process of disturbance and succession within communities guarantees that most populations are almost always a mix of subgroups experiencing different outcomes in their interactions with other species. The outcomes of interactions within a newly created light gap in a forest, a plowed field, or on a piece of bare rock in a rocky intertidal zone can be quite different from the outcomes among the same species five or fifty years later (Tilman 1988; Bazzaz 1990a).

Succession has been studied primarily as an ecological process, but it is just as much a genetic process. Within any one patch, the ages, sizes, and genetic composition of the interacting species can all change over time. Each successional patch can become a microcosm of natural selection, sorting out genotypes and favoring individuals that interact in different ways at different stages of succession. Small differences in soil moisture, density, and neighbors can have large effects in the distribution of genotypes within populations even over a small number of generations (e.g., Zangerl and Bazzaz 1984). The frequency with which new patches become available, the size of the patches, and the length of time that any patch remains intact will all influence how interactions between species are shaped by natural selection, genetic drift, and gene flow among patches.

Few studies have begun to address the question of how interactions may change during succession due to the sorting of genotypes. The results, however, of studies on interactions between white clover, *Trifolium repens,* and grasses within pastures suggest how succession may provide the setting for asymmetric coevolution among competitors. Although the studies do not show coevolution (they were designed to test evolution in *Trifolium* but not in the grasses), they suggest how a population could cycle over time in the number of species with which it interacts. By extrapolating the results as a thought experiment, however, this interaction suggests how cycles of succession could also be cycles of asymmetric coevolution. In 1979, Turkington and Harper published the results of a series of experiments testing how well *T. repens* grew when planted next to any one of four grass species in Wales: *Lolium perenne, Holcus lanatus, Cynosurus cristatus,* and *Agrostis tenuis.* For these studies, they collected *T. repens* plants growing next to each of these grass species from a pasture that had been established approximately a hundred years ago. The plants were multiplied vegetatively in the greenhouse, then placed back into the field in all possible combinations of clover type (e.g., clover originally found next to *L. perenne*) and site of origin (i.e., each of the four grass species). Most clover types survived best or grew best when grown next to the grass species with which they were originally associated. Turkington and Harper interpreted these results as evidence for selection among *T. repens* genotypes during succession. No formal genetic analysis was done to evaluate these *T. repens* types, but the results pointed to the possibility of this sorting of genotypes within the pasture over time.

Ten years later, Turkington (1989) constructed a more complicated version of this experiment in the same pasture, but this time with only three of the grass species. As in the earlier experiment, he demonstrated a home-site advantage for *T. repens* plants, except for those taken from stands of *H. lanatus.* Similar results were obtained for yet another experiment, in which *Trifolium* plants from pastures in Switzerland, Italy, France, and England were grown in Wales next to the grass species with which they were found in their native pastures and also next to other grasses (Evans et al. 1985). After two years, all five clover populations showed highest yields when grown next to their matching grass species.

In between these two experiments, Aarssen and Turkington (1985) conducted similar trials on plants from a pasture in western British Columbia.

The pasture had many of the same species as occurred in Wales. The major difference was that the Welsh pasture was about one hundred years old and grazed by sheep, whereas the one in British Columbia was about forty years old and grazed by cattle. Again, they found a home-site advantage for *T. repens*. A follow-up experiment, however, by Evans and Turkington (1988) suggested that these differences, at least in British Columbia, were due not to selection for different genotypes depending upon neighbors but rather to phenotypic plasticity. When grown under common garden conditions, *T. repens* plants collected from different grass neighbors retained their differences for at least four months. But after twenty-seven months, the differences had disappeared. They argued, therefore, that transplant experiments using clones of perennials may require much longer periods of preconditioning to standardize the plants than are often allowed in such experiments.

The results of Evans and Turkington (1988) call into question the genetic interpretation of all the other experiments done in both Wales and British Columbia. Moreover, in yet another set of experiments, they found that the species of neighboring grass could affect the interaction between *T. repens* and its root-nodulating symbiont *Rhizobium leguminosarum* (Chanway, Holl, and Turkington 1989; Thompson, Turkington, and Holl 1990). Hence, there is another species whose genetic or phenotypic responses can affect the outcome of the grass-*Trifolium* interaction over time. Turkington's (1989) current thinking is that the results from Wales probably show genetic differentiation, whereas those from British Columbia do not. A formal genetic analysis of the Welsh population is the next logical step. For now, the results indicate that the outcome of the interaction among these perennial neighbors appears to be molded over time within these pastures through phenotypic plasticity, differential survival and growth of *Trifolium* genotypes depending upon the grass neighbor, or both.

A more recent study in France has reinforced a genetic interpretation of the evolution of these interactions during succession. Lüscher, Connolly, and Jacquard (1992) tested combinations of *Trifolium* and *Lolium* from a cattle-grazed pasture in central France in a longer-term experiment that eliminated the methodological problems discovered in earlier studies. They addressed the problem of carryover effects by preconditioning the plants for eighteen months, then ran the experiment testing combinations of *Trifolium* and *Lolium* for another seven months. They addressed the problem of the effect of *Rhizobium* by eliminating the native *Rhizobium* and inoculating the *Trifolium* with a nonnative strain. After eliminating these complicating factors, they found that *Trifolium* genotypes still grew better when grown next to the *Lolium* genotype that had been its natural neighbor in the pasture than when grown next to other *Lolium* genotypes.

The mechanisms by which these evolutionary changes occur are unknown.

It may not be as simple as differential survival of *Trifolium* alone. Aarssen and Turkington (1985) suggested that, during succession, natural selection could actually favor grass genotypes with decreased competitive ability relative to *Trifolium*. If the interaction between *Trifolium* and *Rhizobium* enriches the surrounding soil with nitrogen, then grass individuals next to thriving *Trifolium* plants would fare better than other grass individuals. That is, natural selection on the grasses during succession would favor genotypes that are competitive enough to survive and grow next to *Trifolium* but not so competitive as to kill off neighboring *Trifolium*. Under these conditions, *Trifolium* and the grasses would coevolve during succession as the initially inferior *Trifolium* increased its competitive ability and each of the grass species reached an intermediate level of competitive ability.

If differential survival of *Trifolium* genotypes or coevolution between *Trifolium* and grasses occurs commonly during succession in pastures, then the number of grass species with which *Trifolium* interacts and the distribution of genotypes among individuals could be very different in a fifty-year-old pasture from those in a one-year-old pasture. The *Trifolium* population in a fifty-year-old pasture could be mostly a collection of, say, *Lolium* specialists, whereas the population in a one-year-old pasture would reflect the distribution of genotypes that initially colonized the field from the seed bank and from neighboring fields. The process of coevolution would not necessarily result in any long-term directional selection on the interacting competitors. Instead, it would create successional cycles of coevolution in which the distributions of competitive abilities and the number of species with which any one species associated changed over time.

Hypothesis 3: Coevolutionary Turnover

Moving to a broader geographic scale, some forms of the 'taxon cycle' among groups of species on islands may produce a form of multispecific coevolution involving continually mixing groups of competitors. When Wilson (1959, 1961) introduced the concept of the taxon cycle, he did so in an attempt to explain patterns in the distribution of ant species in New Guinea and other islands of Melanesia. He argued that the Melanesian ant fauna originated from a subset of ant species that had been displaced from the inner rain forests into the marginal lowland forests, savannas, and seashore habitats of southeastern Asia. These ants, characterized by the large colonies common to ants in impoverished habitats, eventually colonized Melanesia, expanded into the inner rain forests of those islands, and diversified in species. This cycle of displacement to marginal habitats, colonization of new islands from those habitats, expansion into more-favorable habitats, and diversification could have happened repeatedly among the islands.

Roughgarden and his colleagues have used the taxon cycle to construct a particular version of multispecific coevolution, called coevolution-invasion turnover (referred to here by the shorter term 'coevolutionary turnover') (Roughgarden 1983b; Rummel and Roughgarden 1985; Roughgarden and Pacala 1989). During this process, larger groups of species distributed among islands are linked through bouts of colonization, coevolution, and extinction between pairs or smaller subgroups of species. Coevolutionary turnover is the only form of coevolution that explicitly uses extinction of local populations as part of the process of reciprocal evolutionary change over broad geographic scales. The hypothesis was devised to explain the distribution of *Anolis* lizards on islands of the northern Lesser Antilles. The islands in this archipelago each have only one or two anole species living in native habitats. According to the hypothesis, the equilibrium number of anoles on these islands is only one species. When only one species occurs on an island, the population evolves toward an intermediate size, which maximizes population density. That population, however, is subject to invasion by larger anoles, which successfully outcompete their smaller congeners. During an invasion by a larger species, both populations evolve toward smaller size. Ultimately, the resident species is driven to extinction, the invader evolves toward intermediate size, and the cycle begins again when the island is invaded by a larger anole (fig. 15.3).

In this view of coevolution, the interactions are between pairs of species, but over time, all the anole species are involved in various combinations. Coevolutionary turnover successfully explains the general pattern of anole distribution in the northern Lesser Antilles (Roughgarden and Pacala 1989), although an alternative view—character displacement—is still a possible explanation for some of the pattern (Losos 1992). The alternative view, first suggested by Williams (1972), is that these islands are in an early stage in the accumulation of anole species. As islands accumulate additional species, the species will diverge through character displacement, with some becoming larger and others smaller. Roughgarden and Pacala (1989) have argued that this alternative view is refuted by three observations. First, St. Maarten includes two anole species that are similar in size. If the anole fauna evolves through character displacement toward smaller and larger sizes, then these two lizard species should be recent colonists. In fact, these two lizards are differentiated morphologically and biochemically from populations on other islands, suggesting that they have been evolving with one another for some time, possibly in the direction predicted by coevolutionary turnover. Second, the fossil evidence from throughout these islands suggests that the anoles on these islands have been evolving toward smaller size. There is no evidence of evolution toward larger size, which should occur if stable character displacement has occurred on these islands. Finally, the character displacement

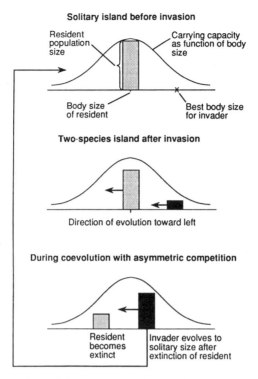

Fig. 15.3. The process of coevolutionary turnover of competitors on islands. The horizontal axis is some measure of body size, and the vertical axis is the number of individuals in each population. Redrawn from Roughgarden 1983b.

hypothesis assumes that invading species are similar in size to resident species, and that one population evolves to larger size as the other evolves to smaller size. Experimental studies, however, have shown that species of a size similar to that already on an island cannot successfully invade it. Only larger species are capable of invading the islands that already have a resident anole species.

Roughgarden and Pacala (1989) have argued that the coevolutionary turnover version of the taxon cycle may be a rare form of coevolution, restricted to small groups of islands and the kinds of ecological conditions and life histories found in species such as the anoles of the northern Lesser Antilles. The hypothesis does not, for example, apply readily to the anoles of the southern Lesser Antilles, where the species tend to be separated more by habitat, and the interactions among the species may differ from those in the northern Lesser Antilles. Character displacement seems to be the better explanation for the patterns of distribution of species and sizes of some other taxa on archipelagoes, including Galápagos finches (Grant 1986) and *Cnemidophorus* lizards (Case 1979). Moreover, mathematical models of character displacement suggest that coevolution during the taxon cycle may not always take the route suggested by the anoles on the northern Lesser Antilles. Taper and Case (1992) have evaluated three of the major mathematical approaches to the problem of character displacement and their application to the taxon cycle. They found that these approaches differ in their predictions about the role of body size in driving the taxon cycle. The coevolutionarily stable community model predicts a cycle as described by Roughgarden and colleagues in which larger populations replace smaller populations. The evolutionarily stable strategy (ESS) model and the quantitative genetic recursion model, however, predict that smaller anoles will be more successful invaders than larger anoles and that both species should evolve toward larger, not smaller, sizes.

In comparing the models, Taper and Case (1992) note that none adequately captures the biology, genetics, and ecology of these competitive interactions. For instance, in the coevolutionarily stable community model populations are monomorphic and there is no intraspecific frequency-dependent selection (i.e., there is no competition within the population, and selection for use of resources is all interspecific); in the evolutionarily stable strategy model populations are almost completely monomorphic and haploid; and in the quantitative genetic recursion model heritability of traits is constant. Nevertheless, all three approaches support the possibility of the taxon cycle itself. Hence, the general concept of a taxon cycle driven by coevolution appears to be robust, although the differences in predictions among models suggest that additional studies are needed to evaluate precisely how different forms of competition, life history, and genetics influence the patterns of coevolution during the cycle.

Conclusions

Coevolutionary alternation, successional cycles of coevolution, and coevolutionary turnover are all specific, testable views of how groups of species, rather than pairs of species, can coevolve as patterns of specialization change over evolutionary time. They all are based upon the view that specialization is evolutionarily dynamic rather than a dead end. Some or all of these hypotheses may require extensive modification after additional testing, but they are very important at this stage in the development of coevolutionary theory because they suggest processes that go beyond the catchall concept of diffuse coevolution. They show how we can develop specific hypotheses on coevolution between groups of species by focusing on the pattern of change in specialization and the sequence of events that occur during the coevolutionary process. As we explore the relationships between specialization and coevolution in greater detail, yet other specific forms of the coevolutionary process will undoubtedly be discovered. Each of these views helps us to understand better that there is often nothing at all diffuse about the reciprocal interactions among many groups of interacting species.

CHAPTER SIXTEEN

Pushing the Limits of Coevolution

Interactions among larger numbers of species over even larger geographic areas begin to push the boundaries of what we can practically study as coevolution. In this final chapter, I consider two kinds of complex, multispecific interaction that may involve reciprocal evolutionary change but may also be at the limits of what we can understand about the coevolutionary process. These two kinds of interaction are the process of escape-and-radiation coevolution proposed by Ehrlich and Raven (1964) and the pervasive and complex mutualisms between vertebrates and the fruits and seeds they eat.

Escape-and-Radiation Coevolution

In 1964 Ehrlich and Raven introduced the most complex form of coevolution so far conceived, involving both adaptation and speciation and creating asymmetries in the patterns of specialization. That paper has undoubtedly spawned more work on coevolution than any other single article. Their hypothesis was initially proposed as a partial explanation for the evolution of interactions and diversity in butterflies and plants. Unfortunately, it has also become the most misinterpreted form of coevolution. Parallel speciation of interacting species is one of the commonly cited expectations arising from Ehrlich and Raven's view, whereas asymmetry is probably the more common result of the process as they envisioned it.

Ehrlich and Raven's hypothesis, which can be called escape-and-radiation coevolution, has five steps:

1. Plants produce novel chemical compounds ('secondary compounds,' 'allelochemicals') through mutation and recombination.
2. The new compounds reduce the palatability of these plants to phytophagous insects.
3. Plants with these novel compounds undergo evolutionary radiation in species into a new adaptive zone in which they are free of their former herbivores.
4. A new mutant or recombinant appears in an insect population that is able to overcome the novel plant compounds.

5. These insects enter a new adaptive zone and radiate in species onto the plants containing the novel compounds.

The result of this process should be clusters of species of plants and insects that show this sequence of plant escape and radiation of species followed by insect colonization and radiation of species. The process of escape-and-radiation coevolution differs fundamentally from diversifying coevolution (Thompson 1989, 1990). In diversifying coevolution, one species directly influences speciation in another species by controlling the movement or survival of gametes. In escape-and-radiation coevolution, plant speciation occurs during periods in which the plants have escaped from insect attack. The insects do not control the movement or survival of gametes. Although there have been some attempts to test the Ehrlich-Raven hypothesis, none so far has been fully convincing (Berenbaum 1983; Smiley 1985; Thompson 1986c). The problem is that a complete analysis requires detailed phylogenies of both the plants and the insects and a thorough understanding of the biogeography, chemistry, and ecology of the interactions.

In addition, there have been dozens of purported tests of the hypothesis by some systematists who have equated the Ehrlich-Raven hypothesis with 'parallel cladogenesis' of species. One possible result of the five-step process of escape-and-radiation coevolution is close similarity in the sequence of speciation events in the plants and insects (or any interaction between host and parasite lineages). Some systematists have searched for such parallels in the speciation of interacting taxa and have used them—or the lack of them in some cases—as the basis for forming conclusions about the Ehrlich-Raven hypothesis. Miller (1987a), for example, wrote that stepwise coevolution, "as defined by Ehrlich and Raven (1964) and others," could be equated with parallel cladogenesis or association by descent. To do so, however, is to ignore the central point about coevolution—the idea that an interaction can evolve through reciprocal evolutionary change. Parallel cladogenesis can occur without any reciprocal evolution occurring at all. Commensals may speciate in parallel with their hosts without having any effect on adaptation or speciation of the hosts.

The interpretation of escape-and-radiation coevolution using systematics alone as the tool for evaluating the hypothesis became clouded even more with Jermy's (1976, 1984, 1993) introduction of 'sequential evolution' as an alternative hypothesis to the evolution of insects and angiosperms. It was not Jermy's hypothesis that clouded the issue but rather the use of the hypothesis as an alternative to coevolution. Jermy argued that phytophagous insects occur at very low population densities compared to the available biomass of their host plants and are therefore unlikely to be important agents of natural selection on plant populations. Instead, plants adapt and speciate over evolutionary time as a result of selection mediated by other influences, such as climate, soil, and competition, which together affect the biochemical diversity found among plant species. Insects diversify in response to this plant diversity without having any evolutionary influence on the plants.

Probably for most evolutionary ecologists, the data to the contrary are overwhelming. Although Jermy discusses the effects of insects on plant fitness, his overall premise equates low insect population densities with weak selection. There is, however, no such simple connection between number of plants attacked or the overall amount of each plant that is eaten and the intensity of natural selection. What matters is whether insects show differential attack among plant genotypes, and whether that differential attack results in plant genotypes that differ in fitness. Hail storms may kill 50% of a plant population, while insects kill only 5%. But if the death caused by hail is random among plant genotypes, whereas death caused by insects is nonrandom, then only insect attack would be the agent of natural selection. Studies have now shown differential attack of insects among plant genotypes for a large number of crop plants and for plants in natural communities as well (e.g., Gould 1983; Berenbaum, Zangerl, and Nitao 1986; McCrea and Abrahamson 1987; Maddox and Root 1987; Fritz and Price 1988; Anderson et al. 1989; Rausher and Simms 1989; Simms and Rausher 1989; Simms and Fritz 1990). This does not mean that insects are necessarily the major and only agents of natural selection on plant populations, only that the premise of 'sequential evolution', which attempts to preclude the likelihood of coevolution between plants and phytophagous insects, is faulty. Undoubtedly, the diversification of plants and insects involves a mix of coevolution at some times and in some species and sequential evolution, otherwise known as phylogenetic tracking, at other times and in other species. As Strong, Lawton, and Southwood (1984) summed up the problem using a marvelously visual metaphor, what we are searching for are coevolutionary vortices in an evolutionary stream.

The development of the idea of sequential evolution led to confusion, however, when some systematists began using parallel cladogenesis of interacting taxa as a way of differentiating between escape-and-radiation coevolution and sequential evolution. For example, Shields and Reveal (1988) argued that the relationship between the butterfly genus *Euphilotes* and their host plant genus *Eriogonum* in western North America is "one of sequential evolution rather than coevolution," because *Eriogonum* probably originated and underwent some diversification before the appearance of *Euphilotes*. It may well be that *Eriogonum* began to diversify before it was colonized by these butterflies, but that tells us nothing directly about the subsequent evolution, or coevolution, of these interactions.

A proper test of escape-and-radiation coevolution must involve an analysis of each of the five steps of the process. Although several attempts have been made, Berenbaum's (1983) analysis on coevolution between the plant family Umbelliferae and its insect herbivores comes closest to a full test. She concluded that these interactions provided evidence for the hypothesis, but subsequent evaluation suggested much more uncertainty in whether these interactions actually follow the process of escape-and-radiation coevolution (Thompson 1986c). The uncertainties in the analysis stem primarily from our lack of a consistent classification of genera within the Umbelliferae and an incomplete understanding at that time of the geographic patterns of specialization in the insects feeding on umbellifers. To evaluate fully Ehrlich and Raven's hypothesis will require continued refinement of our understanding of systematics, evolutionary genetics, the geographic structure of populations, and the ecology of interactions.

Coevolution and the Diversification of Mutualisms

A similar combination of approaches is needed to understand the role of coevolution in the growth and diversification of mutualisms. The tendency for short-term mutualisms to collect new species over evolutionary time creates a network of coevolved and non-coevolved participants. Disentangling that network will require intense work on the phylogeny, paleobiology, evolutionary ecology, physiology, and geographic structure of these interactions. Even so, we will probably be able to sort out only part of how these networks have evolved.

Studies of coevolution between pines and corvids (jays, nutcrackers, and related species) suggest some of the opportunities we have, and some of the limitations we face, in understanding how reciprocal evolution shapes these kinds of geographically variable, complex mutualisms. A number of pines in western North America, Europe, and Asia appear to have coevolved mutualistically with pinyon jays (*Gymnorhinus cyanocephalus*), Clark's nutcrackers (*Nucifraga columbiana*), and some related corvids. The cones of these pines are open rather than serotinous, and the large, wingless seeds are easily extractable by the birds (Vander Wall and Balda 1977; Tomback 1982). The corvids bury the seeds in small caches for later use, and seedlings germinate from those caches that the birds do not use later.

There is some evidence that coevolution between pines and corvids may have begun as a pair-wise interaction or as an association between a few related pine and corvid species but that additional species were collected over time, and perhaps others lost, as geographic distributions of species changed. Current views of pine systematics suggest that two or more groups of pines may have independently evolved traits for dispersal by corvids. The pines now involved in these interactions include several Eurasian species in the subsection *Cembrae,* the North American pinyon pines in the subsection *Cembroides,* and several other, perhaps unrelated, pine species (Tomback 1983; Vander Wall 1990). Not all the pines in these subsections, however, are involved in mutualisms with the most specialized of the corvid species. Of the pinyon pines of North America, which comprise eleven recognized species, only several are mainstays of the diets of the most specialized conifer-caching corvids. Some other pinyon pines occur in Mexico, south of the geographic range of pinyon jays and Clark's nutcrackers. Hence, phylogenetic analyses have suggested that the current interactions between pines and corvids may have involved convergence of several pine species from two or more lineages and that pine species within these lineages differ in the extent to which they interact mutualistically with the most-specialized corvids.

The seed-caching corvid species vary in the degree to which they are specialized for extracting and utilizing pine seeds. In North America there are four corvids that regularly cache conifer seeds: pinyon jays, Clark's nutcrackers, scrub jays (*Aphelocoma coerulescens*), and Steller's jays (*Cyanocitta stelleri*) (Marzluff and Balda 1992). Of these four, pinyon jays and Clark's nutcrackers have life cycles, behaviors, and morphological adaptations highly specialized to exploitation of pine seeds. Pinyon jays specialize on pinyon pines, especially *P. monophylla* and *P. edulis.* Clark's nutcrackers feed primarily on pines at higher elevations but readily eat pinyon pine seeds whenever they find them. But unlike these two specialized corvids, neither Steller's jays nor scrub jays have the specialized morphologies that enable them to open closed cones. Hence, they are more opportunistic feeders on pinyon pines. Marzluff and Balda (1992) have taken this spectrum of specialization in corvids and hypothesized that the interaction began as a close relationship between pinyon jays and pinyon pines (or their ancestors) and came to involve Clark's nutcrackers and the other two jays as the geographic distributions of the pines and corvids overlapped.

Marzluff and Balda's (1992) view of the growth of these interactions over evolutionary time is built upon the assumption that the most-specialized species in these interactions are the oldest participants. Although this is a tempting assumption for this interaction, it is one that has to be viewed with caution. Both the pines and the jays and nutcrackers have undoubtedly shifted their geographic ranges considerably over tens of thousands of years. Species of pinyon pines that may have been the mainstays of these corvids fifteen thousand years ago may not be the same species that are the mainstays today. Similarly, although pinyon jays may be the corvids that are now most completely dependent upon pinyon pines throughout their geographic range, they may not have been the initial selection agent on pinyon pines for dispersal by corvids. The interaction could have arisen initially between, say, an ancestor of Clark's nutcracker or even of scrub jays and one or more of the pinyon pines and was later co-opted and refined by an ancestor of pinyon jays. There is no way of knowing for certain, even if an unassailable phylogeny of these corvids were available. However the relationship between corvids and pines began, this complex association of species highlights the interplay of specialization, geography, speciation, collection of new participants, and potential loss of old participants that must be responsible for so many of the asymmetries and complex patterns of coevolution found in interactions.

Coevolution within Complex Webs of Interaction

There are more-complex interactions that may undergo coevolution but that permit specific hypotheses on only parts of the process. The intricate network of relationships between many birds and the flowers and fruits they visit has spread, diversified, collected so many unrelated taxa, and been rearranged so much over the millennia that the best we can do is to catch glimpses of the coevolutionary process. Studies of the interactions between hummingbirds and flowers throughout Central and South America, for example, have shown that each local community includes a spectrum of birds, from those using only a few plants to those using many, and a spectrum of plants, some pollinated by only one species and some pollinated by at least several species (Feinsinger 1978, 1980; Snow and Snow 1980; Stiles 1975, 1985). The same birds a hundred kilometers away or several kilometers higher in elevation may visit yet other flowers of phylogenetically unrelated plants. In addition, a number of bird species migrate thousands of miles and feed on different flowers in different parts of the world. The evolution of these interactions therefore occurs at a geographic scale beyond that at which all studies have been made.

The relationships between birds and fruits offer similar challenges. About half of the 281 wholly or partially terrestrial angiosperm families include at least some species with fleshy fruits, and about a third of the 135 terrestrial bird families include frugivores, as do a fifth of the 107 terrestrial mammal families (Fleming 1991). Many frugivores eat a broad range of fruits, and in many environments, there seems to be little mutual dependence between individual bird and plant species (e.g., Herrera 1984; Malmborg and Willson 1988). Fecal and stomach samples of small frugivores in Mediterranean scrublands and North American temperate forests commonly contain two or more fruit species: 45–75% of samples from six Mediterranean species and 5–39% (Herrera 1984) and 3–39% from eleven in North America (White and Stiles 1990). These interactions have grown over evolutionary time, accumulating birds and plants from a large number of phylogenetically unrelated orders and families.

In addition to the complexity introduced by the sheer number of terrestrial species involved in these interactions, avian migration has made these interactions a process that acts almost on a global scale. Although other animal taxa include migratory species, none approaches the degree to which migrating birds have linked large numbers of unrelated taxa in geographically separated communities. The evolution of frugivory, and to some extent nectarivory, in migrating birds has linked birds and plants both within and between continents, tying North America to Central and South America, and Europe to Asia and Africa. Fruit production in some temperate environments appears to be timed to avian migration (Thompson and Willson 1979; Piper 1986; Willson 1986; Skeate 1987). In the temperate deciduous forests of eastern North America, many wind-dispersed plants and ant-dispersed plants disperse their seeds in early summer or midsummer, whereas most fleshy fruited plants do not ripen their fruits until late summer or fall. Among the herbs in these forests, there is a temporal separation even within plant families between the early-maturing ant-dispersed seeds and the late-maturing bird-dispersed seeds (Thompson 1981a). Many birds rely upon the abundance of fall-ripening fruit to replenish their energy reserves during migration. Some avian species even change their foraging repertoires during migration (Moore and Simm 1985; Loria and Moore 1990).

Although long-distance migration by birds may have made the evolution of interactions between birds and flowers and fruits the most complex problem in the relationship between specialization and coevolution, it is important to try to disentangle what we can. These interactions involve too large a proportion of terrestrial species and they are too central to terrestrial community organization to be ignored or relegated to the general category of diffuse coevolution. The explosion of work on these associations over the past few decades has shown that it is possible to discern both pattern and process in the evolution of these interactions. What the results so far suggest is that the evolution of specialization and the scale of coevolution in these relationships must be analyzed from a broad community, geographic, and phylogenetic perspective.

From that perspective it is becoming increasingly evident that sugar composition of floral nectar and fruit pulp is one of the major organizing influences on specialization and coevolution between birds and the flowers and fruits they visit. Hummingbird-pollinated plants secrete predominantly sucrose-rich nectars, whereas passerine-pollinated plants secrete glucose and fructose-dominated nectars (Baker 1975; Baker and Baker 1983; Martinez del Rio, Baker, and Baker 1992; Stiles and Freeman 1993). Passerine-dispersed fruits similarly contain mostly glucose and fructose sugars, whereas mammalian-dispersed fruits commonly contain higher proportions of sucrose (Martinez del Rio, Baker, and Baker 1992). These patterns are apparent even within individual plant genera and families. For example, hummingbird-pollinated plants in the genera *Erythrina* (Fabaceae), *Puya* (Bromeliaceae), *Campsis* (Bignoniaceae), and *Fushia* (Onagraceae) all secrete sucrose-dominated nectars, whereas passerine-pollinated plants in the same genera have very little sucrose in their nectars; mammalian-dispersed fruits in the Rosaceae have significantly higher proportions of sucrose than avian-dispersed fruits (Martinez del Rio, Baker, and Baker 1992).

The sugar constituents found in flowers and fruits match fairly well the sugar preferences of the avian groups that visit them. Hummingbirds prefer sucrose over glucose and fructose (Stiles 1976; Martinez del Rio 1990b), and passerines generally seem to prefer glucose and fructose over sucrose, although there is considerable variation for sugar preference among passerine species (Martinez del Rio et al. 1988; Martinez del Rio, Baker, and Baker 1992). These differences in sugar preference among avian species appear to result from at least two differences in the digestive physiology of avian species (Martinez del Rio 1990a; Martinez del Rio, Baker, and Baker 1992). Some avian families—thrushes, starlings, and thrashers—lack intestinal sucrase and therefore cannot use sucrose. In addition, the extremely fast passage rates of food through the guts of some frugivorous passerines may prohibit efficient use of sucrose as an energy source because it cannot be hydrolyzed fast enough as it passes through the gut. (Sucrose must first be hydrolyzed to glucose and fructose before it can be absorbed by the intestine, whereas ingested glucose and fructose can be absorbed directly.) Relative to passerines, hummingbirds can hydrolyze sucrose very quickly. Nevertheless, it is easier, using current knowledge of digestive physiology, to understand the preference of passerines for glucose and fructose than it is to understand the preference of hummingbirds for sucrose. They all provide the same amounts of energy, and hummingbirds can assimilate all three sugars with equal efficiency.

The combined studies of sugar concentrations in flowers and fruits and avian preferences, however, have begun to suggest some specific hypotheses about the ways in which sugar constituents have shaped the evolution of these interactions. The hypotheses are based upon the phylogeny and geography of the interactions:

1. Sucrose preference evolved in hummingbirds because hummingbird-pollinated flowers are mostly derived from large bee- and butterfly-pollinated flowers, which secrete sucrose-rich nectars. Hence, these birds evolved specifically to use sucrose-rich flowers. This strong preference and adaptation for sucrose did not develop among avian taxa in other parts of the world, because large bee- and butterfly-pollinated flowers are less common in Australia and Africa, and, in the Old World, bird-pollinated flowers have been derived from flowers with lower proportions of sucrose (Martinez del Rio, Baker, and Baker 1992).

2. Specialization in coevolution between hummingbirds and flowers is mediated in part by nectar volume and, to a lesser extent, total sugar concentration. Although nearly all hummingbird flowers appear to have converged on sucrose-rich nectars to attract hummingbirds, plants pollinated by hermit hummingbirds (Phaethorninae) have higher sugar volumes and concentrations than most plants pollinated by other species of hummingbirds (Stiles and Freeman 1993).

3. The evolution of frugivory has demanded the evolution of fast digestive rates to void undigestible seeds. These fast digestive rates have favored the evolution of preference for fruits that can be rapidly assimilated, such as those containing glucose and fructose rather than sucrose. This preference has, in turn, selected for fruits with yet higher proportions of glucose and fructose rather than sucrose as well as other rapidly assimilated constituents such as free amino acids rather than polypeptides (Levey and Grajal 1991).

4. Although lack of sucrase is phylogenetically restricted in birds to the lineage that includes thrushes, starlings, and thrashers, these taxa make up a high proportion of frugivores in communities worldwide. These birds may therefore be responsible for the maintenance of low sucrose concentrations in bird-dispersed fruits (Martinez del Rio, Baker, and Baker 1992).

All these hypotheses will require continued rigorous evaluation as our phylogenetic understanding of both plants and birds improves. Each of them provides an avenue of research for decomposing diffuse coevolution into more-specific patterns of relationship between specialization and coevolution.

Other patterns in the evolution of these interactions still require further work from a combination of phylogenetic, geographic, and ecological perspectives in order to construct more specific coevolutionary hypotheses. Floral and fruit size, color, and shape all influence these relationships. For example, fleshy fruits adapted for avian dispersal have converged worldwide on black and red as the most common colors (Willson and Thompson 1982; Willson 1983; Wheelwright and Janson 1985; Willson, Irvine, and Walsh 1989). There is only weak evidence at best, however, that avian frugivores actually prefer black and red over other colors; hence the causes of the worldwide convergence on these two colors are not yet clear. Willson and Whelan (1990) proposed thirteen nonexclusive possible reasons for the predominance of these two colors—some based upon avian foraging, others based upon defense against enemies, and still others based upon physiological constraints or costs—but none has yet received strong support. Geographic variation in the relative frequencies of fruit colors appears to be related not so much to differences in the available suite of avian dispersers as to the relative frequencies of fruits adapted to mammalian, rather than avian, dispersal (Willson, Irvine, and Walsh 1989). There is also geographic variation in the frequency of avian-dispersed plants with bicolored fruit displays (Willson and Thompson 1982; Herrera 1987b; Willson, Irvine, and Walsh 1989). Some of these bicolored displays are temporal, changing, for example, from green to red to black such that a plant has a display of both red and black fruits; other bicolored displays are morphological, contrasting, for example, black fruits with red peduncles. Under at least some ecological conditions, certain kinds of bicolored displays result in faster fruit removal than monocolored displays (Willson and Melampy 1983; Willson and Whelan 1990). Moreover, there are geographic differences in the frequency of polymorphic fruits—that is, differences in the frequencies of plant populations in which different individuals differ in the color of the fruits they produce (Willson, Irvine, and Walsh 1989). These geographic patterns in the frequency of bicolored and polymorphic fruits are only now becoming evident, and the causes of the patterns are unknown.

Because the interactions between birds and flowers and fruits involve so many species, there is a tendency among some ecologists to refer to them as diffuse, loose, or sloppy. Thinking of these associations in that way is counterproductive, creating the impression of a free-for-all, or at least discouraging the search for ways in which pairs or small groups of species may coevolve amid interactions with other species. Although these relationships involve networks of species, the challenge is to understand how such networks result in specialization and coevolution among groups of species, creating geographic patterns in the evolution of these mutualisms.

Studying Coevolution in a Community Context

The complex networks of interaction that result from escape-and-radiation coevolution and the even more complex multispecific associations between birds and flowers and seeds and fruits are a marked contrast to the couplets or small groups of species that have often been the focus of studies on evolving interactions. This is not because we as biologists are somehow unaware of the community context in which a pair of species interacts. Rather, it more often simply reflects practical limits to what can be studied in depth given limited time and money. The problem that we will always face, though, is how to interpret these studies of pair-wise interaction if we are to use them as tools for understanding the evolution of interactions. To toss them all aside as inadequate is both self-defeating and wrongheaded. We will never be able to study entire biological communities in such depth that we could place a species pair in the context of all the other direct and indirect interactions that influence them.

Studies of species pairs in fact tell us a great deal about the ways in which the genetic, social, and demic structure of populations and the influences of different physical environments can shape interspecific interactions. These studies are tools for unraveling how natural selection, genetic drift, and phylogenetic history can mold the evolution of specialization, create geographic structure within interactions, and adjust the trajectories of coevolution. The mistake is to view these studies as more than tools and to believe that a short-term study of a pair of species, studied over a small part of their overall geographic ranges and considered apart from all the other interactions affecting them, tells us how that particular interaction has evolved and is evolving.

Studies of pair-wise interactions alone are insufficient for understanding the evolution of interactions in general and the coevolutionary process in particular. We have enough studies now to know that the interpretation of interactions between pairs of species can be terribly misleading when separated from the community, and sometimes geographic, context in which the interaction takes place. Detailed studies of the mechanisms of competition between pairs of species, for instance, are important, but only when analyzed in the context of how those mechanisms affect individuals in the context of the communities in which they operate. The elegant results of Feinsinger and his colleagues on competition among plants for hummingbird pollinators at Monteverde, Costa Rica, clearly show the importance of a hierarchical approach to the study of evolving interactions. Under highly controlled experimental conditions, strong competition among plants sharing the same pollinators is evident, and the mechanisms of competition can be readily deciphered (Feinsinger and Tiebout 1991). When placed within their natural communities, however, the competitive mechanisms penetrate only very weakly to population-level processes and community patterns, at least at the level of local neighborhoods (Feinsinger, Tiebout, and Young 1991).

Similar differences in interpretation between detailed experimental studies of pairs of species and broader studies of those same interactions in the context of natural communities are evident in other recent studies, including the associations between the tropical herb *Calathea ovandensis* and its bee pollinators (Schemske and Horvitz 1988) and that between the temperate herb *Lithophragma parviflorum* and its pollinating seed-parasitic moth *Greya politella* (Thompson and Pellmyr 1992). Interpretation of the evolution of these and other interactions may differ even more when studied over larger spatial scales (Thompson 1988d, 1994; Thompson and Burdon 1992; Wiens 1989; Feinsinger, Tiebout, and Young 1991). We are only now, however, beginning to come to grips with the scale of interactions. Until we do, we will not know.

Conclusions

Each of the forms of coevolution considered in this and the previous chapter suggests that asymmetries in specialization and geographic structure are a common and important part of the process of coevolution. In some cases, these forms of reciprocal change also illustrate that it is possible to replace untestable views of diffuse coevolution with testable hypotheses about how groups of species rather than pairs of species specialize and coevolve. These are not likely to be all the possible ways in which asymmetries and geographic structure influence the evolution of specialization and the process of coevolution, but they suggest that it is possible to dissect both pattern and process within the complexity of evolving interactions.

SYNTHESIS

The Geographic Mosaic
in Evolving Interactions

We are entering what I think will be the most exciting decade yet—perhaps ever—in the study of evolving interactions. We now have the tools for understanding very precisely the patterns of specialization and the processes of coevolution that occur within and among populations. Studies during the past decade have increasingly revealed hidden specialization within populations and species once thought to be locally and geographically homogeneous in their interactions with other taxa. Analyses of the phylogeny, genetics, and ontogeny of organisms all suggest that such specialization to other species is often evolutionarily dynamic rather than a phylogenetic dead end. Through genetic and ontogenetic mechanisms, species and local populations can even evolve to specialize on, and potentially coevolve with, more than one other species simultaneously. Clustered genetic loci that affect resistance to two or more very different parasites, preference hierarchies that can be genetically tailored in some species to the composition of local communities, local polymorphisms that subdivide populations into groups of specialists, and ontogenetic compartmentalization of specialization to different species are just some of the ways in which organisms have evolved to cope with the problems of specialization and coevolution in natural communities that change continually in species composition over evolutionary time.

These evolutionary dynamics of specialization are not random. The chapters in Part 3 have suggested that there are ecological patterns to the evolution of specialization found in organisms. Parasitism, grazing, predation, competition, symbiotic mutualisms, and nonsymbiotic mutualisms all show dynamic patterns of specialization that are molded in different ways both locally and geographically by natural selection through the community context in which the interactions occur. Differences among populations in the outcomes of interactions add to the geographic structure and dynamics of these interactions. Natural selection on the relationship between a pair or group of species may favor directional selection in one population and polymorphism in another. It may favor escalation of antagonism in some environments and reduced antagonism or mutualism in others. The result is that most interactions exhibit an evolutionarily dynamic geographic mosaic, formed by the combined interpopulational differences in outcome, adaptation, and specialization.

The most important conclusion from this book is that this geographic mosaic in the interactions between species often drives the coevolutionary process and the evolution of interactions in general. Studies of neither local populations alone nor species as a whole can capture the dynamics of evolving interactions. Associations between species are molded by local conditions, and these associations are then often reshaped over evolutionary time as some populations become extinct and others exchange genes. Hence, the outcomes within local populations—whether escalating arms races or intricate mutualisms—are often only the raw material for the patterns that develop and the processes that take place over larger geographic scales.

Therefore, a hierarchy of spatial scales shapes specialization and the coevolutionary process–from local populations, through clusters of demes that differ in their adaptations, to the fixed traits that characterize species. Just as much of the dynamics of coevolution cannot be seen by looking at only a local population, much of it will also be missed by looking only at the fixed characters that differentiate species and serve as the tools for systematics. Most of the action of the coevolutionary process (and most of the evolution of interactions in general, whether coevolutionary or not) occurs at the combined local and geographic scales, reshaping variable characters that shift in mean and variance over evolutionary time.

Many of the characters that result from the coevolutionary process will not necessarily ever become fixed throughout all the populations of a species. Searching the cladograms of interacting species for evidence of coevolution without concomitant ecological study of local and geographic variation will therefore underestimate the importance of coevolution, because it will focus on the fixed characters. It is the equivalent of making conclusions about the importance of natural selection using only studies of sustained directional selection while ignoring studies of frequency-dependent selection, density-dependent selection, stabilizing selection, and disruptive selection.

We will have a complete theory of evolving interactions only when we have fully incorporated into our thinking a geographically dynamic view of interactions. Toward that end, I have tried in this book to develop the major components of a geographic mosaic theory of coevolution. Moreover, I have tried to show some of the ways in which a geographic view of specialization and coevolution changes our expectations about the process and outcome of reciprocal change. A geographic perspective on coevolution relies less on long-term stable interactions within populations than a local view does. A geographic perspective also provides the possibility of a shifting-balance view of some associations between species, as outcomes, adaptations, and specialization change among populations. Different populations may specialize in different ways and coevolve toward different adaptive peaks. When that occurs, the overall trajectory of coevolution will depend upon the mix of natural selection, genetic drift, and gene flow that shapes each population and species. Some isolated populations may become stuck on one adaptive peak, while other groups of populations shift from one peak to another. This geographic view of interactions means that species do not need to be either identical in their geographic ranges or extreme in their specialization to one another throughout their geographic ranges for coevolution to occur. It is the differences among the populations, including those outside the range of the interaction, that provide the grist for the coevolutionary mill.

This geographic view of coevolution also helps in disentangling diffuse coevolution into more specific, testable hypotheses. It is no longer sufficient to think of coevolution as either specific or diffuse. A number of specific hypotheses on reciprocal change involving more than two species have appeared during the past decade, and I have tried to show that most of them are related in that they rely in one way or another upon the geographic structure of interactions. An immense amount of work, however, still needs to be done to understand how the different combinations of components in the geographic mosaic of interactions shape the process of coevolution in yet other ways.

There are eight areas of research that I think will be the most fruitful over the coming years in developing a geographic view of the evolution of specialization, the process of coevolution, and evolving interactions in general. The eight areas fall into two groups.

Life histories and the evolution of interactions:

1. Modes of interaction: How do natural selection and other evolutionary processes mold different kinds of interspecific interactions (e.g., parasitism vs. predation, symbiotic vs. nonsymbiotic mutualisms) under different ecological conditions (e.g., different population, community, and geographic structures)?
2. Evolutionary genetics: How do modes of interactions differ in the kinds of genetic change they elicit in populations, and in turn, how does the genetics of interactions shape and constrain the trajectory of coevolutionary change?
3. Molecular genetics: How does the genetics of organisms allow them to specialize and coevolve simultaneously with multiple species?
4. Specialization during ontogeny: How does natural selection mold life histories in ways that compartmentalize specialization and sometimes coevolutionary interactions during an organism's lifetime?

The hierarchical structure of evolving interactions:

5. Population structure: How do the genetics of interactions and mode of interaction influence the local structure of populations? How does local population structure, in turn, influence the genetics, mode, and evolutionary outcome of interactions?
6. Community context: How does the local organization of communities influence the evolution of interacting pairs or small groups of species? And, in turn, how does the local (co)evolutionary dynamics of interactions shape the structure of local communities; that is, what are the ripple effects of coevolution throughout the rest of a community?
7. Geographic structure: How do the demic structure of populations (e.g., metapopulation structure) and interdemic variation in the outcome of interactions influence the coevolutionary process? That is, what different forms can the geographic mosaic of coevolution take?
8. Coevolution and speciation: How do different kinds of interaction and geographic structure in populations influence speciation as a direct or indirect outcome of the coevolutionary process?

Studies of the geographic mosaic in evolving interactions are likely to continue to provide new insights into the relationships between specialization and the coevolutionary process. Although I have developed the arguments in this book primarily as a way of studying and understanding the coevolutionary process, most of the arguments apply equally to the evolution of interspecific interactions in general, whether coevolutionary or not. The geographic mosaic view of interactions is a way of understanding the evolutionary and coevolutionary processes that create such intricate patterns of life within the entangled bank.

EPILOGUE

Specialization, Coevolution, and Conservation

For our understanding of evolution, the real tragedy accompanying the destruction of natural communities is the loss forever of specialized and highly coevolved interactions. These are the relationships between species that are probably disappearing fastest, yet they are precisely the ones that could tell us the most about the evolutionary consequences of particular ways of interacting. We have detailed mathematical models that predict the long-term evolutionary outcomes of every sort of pair-wise interspecific interaction, but these will remain an untested academic exercise if the most specialized interactions have disappeared or have lost the community context in which they were formed. There is no way of adequately simulating the vagaries of ecology, genetics, and evolutionary constraints that have shaped particular interactions over tens of thousands or millions of years. Without these interactions as touchstones, we have no way of testing our intuition or our models.

Take away the few ridiculously long-spurred orchids in Madagascar and elsewhere and the long-tongued moths that pollinate them, and we lose some of the few opportunities we have remaining to understand how selection for pollinator efficiency can sometimes drive interactions to a coevolutionary extreme of specialization. Take away the highly specialized interactions between yuccas, figs, or globeflowers and their pollinating floral parasites, and we lose another perspective on how interactions between flowers and insects can lead to extreme specialization. Take away the yucca moths' host-specific relatives in the genus *Greya,* and we are robbed of a chance to understand how very similar interactions between flowers and insects can lead to different evolutionary results under different ecological conditions. The extreme specialists show us the limits of specialization in evolution in general and in coevolution in particular. They are the jewels in the crown of biodiversity. They give us a point of comparison by which we can evaluate and interpret other interactions and see what is possible given a different set of genetic and environmental conditions.

These extreme, and sometimes highly coevolved, specialists are the reason that concerns over maintenance of biodiversity need to be tied to our concerns for preservation of the few remaining tracts of intact natural communities. The most broadly convincing arguments for the maintenance of biodiversity have been those based upon ethics, esthetics, economics, and ecosystem services, as Ehrlich and Wilson (1991) have emphasized. Human curiosity about the world should in itself be a convincing additional argument, but as May (1990) has pointed out, astronomers have used this argument more successfully for studying the diversity of stars than biologists have used it for studying the much greater diversity of organisms. Yet the stars will be there a hundred years from now, whereas many species will not. Disappearing even more rapidly than species are the highly specialized interactions and the community contexts in which they have been formed.

Just as species are storehouses of genetic diversity, these interactions in intact communities are storehouses of information on how evolution shapes relationships between species under different ecological conditions. They are another important reason for a landscape, rather than a species, view of conservation. We are going to need the information on interactions in natural communities as we try to manipulate and control interactions in the agricultural, pastoral, silvicultural, suburban, and even urban environments we create. Attempts at biological control of weeds and pests, breeding schemes for plant resistance to pathogens based upon a gene-for-gene view or a 'durable resistance' goal, and protocols for coping with the evolution of virulence in diseases can all draw upon knowledge of the evolutionary outcomes of interactions in natural populations if we allow at least some of them to remain intact. We are only now starting to take advantage of what we can learn from these evolved interactions. Although gene-for-gene coevolution has for decades been the paradigm for research on interactions between crop plants and fungal pathogens, the first studies of gene-for-gene interactions in natural communities were begun only in the past decade.

Each remnant of the remaining, partially intact communities is precious. Our current understanding of patterns in the evolution of specialization is based primarily upon a very small subset of terrestrial plants and animals, some freshwater species, and an even smaller proportion of marine species. There are precious few studies on some taxonomic groups and communities. For example, the meiofauna of marine soft sediments includes a diverse array of copepods, nematodes, turbellarians, and other taxa whose interspecific interactions are virtually unknown (Watzin 1985). There are over 1200 tree species in the rain forests of Australia, but the first quantitative study of the pollination biology of an Australian rain forest tree was published only in 1986 (Crome and Irvine 1986).

Eliminating the extreme specialists leaves us with a natural world of dandelions, cheatgrass, starlings, and other opportunistic species. But even that is no guarantee against further extinctions. Two of the most abundant polyphagous and opportunistic animals in North America, both capable of literally blackening the skies, have become extinct during the past one hundred years: the passenger pigeon in the east and the Rocky Mountain grasshopper (*Melanoplus spretus*) in the west. Passenger pigeons fed on the seeds of a number of forest trees. The Rocky Mountain grasshopper, which fed on many crop plants as well as native plants, was considered one of the greatest impediments to settlement of western North America until its sudden decline in the 1890s and extinction around 1902 (Lockwood and DeBrey 1990). Both species had life histories apparently adapted for movement over wide geographic areas, and both were driven to much lower numbers by human activity before suddenly collapsing to extinction. The passenger pigeon was driven to low numbers through relentless hunting, and the numbers of Rocky Mountain grasshopper collapsed possibly through destruction of riparian habitat that may have been important for breeding. Whatever the exact causes of their demise, the loss of these two species illustrates graphically how driving populations below critical thresholds in size or geographic distribution can trigger the rapid extinction of a whole species. As we chip away at the numbers of extreme specialists, eliminating one local population after another, the story repeats itself, but the loss is quieter and less likely to be noticed.

If, as I have tried to argue in this book, much of the evolution of interactions occurs at a geographic scale—molding the raw material of specialization and coevolution within local populations into broader patterns—then this loss of one local population after another through human destruction will drastically change the evolutionary dynamics of interactions and rob many species of their evolutionary future. Our empirical studies and mathematical models of species interactions have become sophisticated enough now to leave little question that the regional, and even global, persistence of many interactions probably relies heavily upon metapopulation structure. Whether 'many' means tens of thousands or hundreds of thousands does not really matter. What are likely to go first are the extreme specialists. A few Yellowstone Parks scattered here and there will not make up for the loss of the geographic complexity in specialization, defense, and mutualism that must certainly maintain the long-term persistence of many of these interactions.

The loss of extreme specialists and of the geographic structure among the remaining species could also take with it some of the long-term economic rationale for preserving biodiversity. We argue, quite rightly, that natural populations are storehouses of genes with potential importance to human agriculture and medicine. These genes include the defenses that plants and animals have developed in their evolutionary battles against their own enemies. Specialists harbor many unique forms of these defenses. If these genes have been maintained through evolutionary time by their movements among populations—being favored here for a time until an enemy develops a counterdefense, and then there—they will be lost as we eliminate the geographic structure that has maintained them.

I realize that in writing this epilogue, I am speaking to the already converted. My reason for making these points at all is that the intellectual and economic value of extreme specialists, highly evolved interactions, and the geographic structure of interactions have not been part of the usual lexicon of arguments about the need for conservation and the preservation of biodiversity. We base our arguments most commonly upon the preservation of species and the maintenance of ecosystem functions, and we use the sizes of individual populations and the genetic variation they harbor as the currency of how well or poorly we are doing. But between species and ecosystems are the interspecific interactions that create the true richness of biodiversity, and between the local populations and the number count of whole species is the geographic structure that partially shapes those interactions.

About a year ago, I participated in a scientific workshop that was held to give input into a broader public document on strategies for preserving biodiversity. The meeting included ecologists, evolutionary biologists, and biological resource managers. I commented that in reading the preliminary draft of the document, I was struck that no mention was made of interspecific interactions as part of biodiversity, and that no mention was made of the importance of preserving interactions as part of an overall strategy of preserving biodiversity. The response I received from one of the members was this: "If we worry about the species, the interactions will take care of themselves." That is precisely the view that I think we need to change both for the welfare of our societies and for the intellectual opportunities that long-evolved interactions hold for our understanding of the history of life and the process of evolution. Specialization and coevolution have made the entangled bank more than a field of weedy species. Only by making the extreme specialists, the highly coevolved interactions, and the complex geographic structure of interactions a major part of our increasingly landscape-oriented view of conservation can we prevent, or at least slow, the loss of much of our opportunity for understanding how life evolves.

LITERATURE CITED

Aarssen, L. W., and R. Turkington. 1985. Biotic specialization between neighbouring genotypes in *Lolium perenne* and *Trifolium repens* from a permanent pasture. *Journal of Ecology* 73:605–14.

Abrahamson, W. G., J. F. Sattler, K. D. McCrea, and A. E. Weis. 1989. Variation in selection pressures on the goldenrod gall fly and the competitive interactions of its natural enemies. *Oecologia* 79:15–22.

Abrams, P. A. 1986. Adaptive responses of predators to prey and prey to predators: The failure of the arms race analogy. *Evolution* 40:1229–47.

———. 1987. On classifying interactions between populations. *Oecologia* 73:272–81.

Addicott, J. F. 1986. Variation in the costs and benefits of mutualism: The interaction between yuccas and yucca moths. *Oecologia* 70:486–94.

Addicott, J. F., J. Bronstein, and F. Kjellberg. 1990. Evolution of mutualistic life-cycles: Yucca moths and fig wasps. Pages 143–61 in F. Gilbert, ed., *Insect life cycles: Genetics, evolution and co-ordination.* Springer-Verlag, London.

Alexander, A. J. 1961. A study of the biology and behavior of the caterpillars, pupae and emerging butterflies of the subfamily Heliconiinae in Trinidad, West Indies. Part I. Some aspects of larval behavior. *Zoologica* 46:1–24.

Allee, W. C., O. Park, A. E. Emerson, T. Park, and K. P. Schmidt. 1949. *Principles of animal ecology.* W. B. Saunders, Philadelphia.

Allison, A. C. 1982. Coevolution between hosts and infectious disease agents, and its effect on virulence. Pages 245–68 in R. M. Anderson and R. M. May, eds., *Population biology of infectious diseases.* Springer-Verlag, New York.

Amadon, D. 1943. Specialization and evolution. *American Naturalist* 77:135–41.

AnanthaKrishnan, T. N. 1984. *Biology of gall-forming insects.* Arnold, London.

Anderson, B. W., and R. D. Ohmart. 1978. Phainopepla utilization of honey mesquite forests in the Colorado River valley. *Condor* 80:334–38.

Anderson, M. D., P. R. K. Richardson, and P. F. Woodall. 1992. Functional analysis of the feeding apparatus and digestive tract anatomy of the aardwolf *Proteles cristatus. Journal of Zoology, London* 228:423–34.

Anderson, R. M., and R. M. May. 1979. Population biology of infectious diseases. Part 1. *Nature* 280:361–67.

———. 1982. Coevolution of hosts and parasites. *Parasitology* 85: 411–26.

———. 1991. *Infectious diseases of humans: Dynamics and control.* Oxford University Press, Oxford.

Anderson, S. S., K. D. McCrea, W. G. Abrahamson, and L. M. Hartzel. 1989. Host genotype choice by the ball gallmaker *Eurosta solidaginis* (Diptera: Tephritidae). *Ecology* 70:1048–54.

Andrews, E. A. 1891. A commensal annelid. *American Naturalist* 25:25–35.

Armbruster, W. S. 1985. Patterns of character divergence and the evolution of reproductive eco-types of *Dalechampia scandens*. *Evolution* 39:733–52.

———. 1990. Estimating and testing the shapes of adaptive surfaces: The morphology and pol-lination of *Dalechampia* blossoms. *American Naturalist* 135:14–31.

———. 1991. Multilevel analysis of morphometric data from natural plant populations: Insights into ontogenetic, genetic, and selective correlations in *Dalechampia scandens*. *Evolution* 45:1229–44.

———. 1992. Phylogeny and the evolution of plant-animal interactions. *Bioscience* 42:12–20.

Armbruster, W. S., and A. L. Herzig. 1984. Partitioning and sharing of pollinators by four sym-patric species of *Dalechampia* (Euphorbiaceae) in Panama. *Annals of the Missouri Botanical Garden* 71:1–16.

Arnold, H. L. 1968. *Poisonous plants of Hawaii*. Charles E. Tuttle Co., Rutland, Vt.

Arnold, S. J. 1981a. Behavioral variation in natural populations. I. Phenotypic, genetic and envi-ronmental correlations between chemoreceptive responses to prey in the garter snake, *Tham-nophis elegans*. *Evolution* 35:489–509.

———. 1981b. Behavioral variation in natural populations. II. The inheritance of a feeding response in crosses between geographic races of the garter snake, *Thamnophis elegans*. *Evo-lution* 35:510–15.

Askew, R. R., and M. R. Shaw. 1986. Parasitoid communities: their size, structure and develop-ment. Pages 225–64 in J. Waage and D. Greathead, eds., *Insect parasitoids*. Academic, New York.

Atsatt, P. R. 1981. Ant-dependent food plant selection by the mistletoe butterfly *Ogyris amryllis* (Lycaenidae). *Oecologia* 48:60–63.

Awahmukalah, D. S. T., and M. A. Brooks. 1985. Search for a sensitive stage for aposymbiosis induction in *Culex pipiens* by antibiotic treatment of immature stages. *Journal of Invertebrate Pathology* 45:54–59.

Axelrod, R., and W. D. Hamilton. 1981. The evolution of cooperation. *Science* 211:1390–96.

Baker, H. G. 1975. Sugar concentrations in nectars from hummingbird flowers. *Biotropica* 7:37–41.

———. 1983. An outline of the history of anthecology, or pollination biology. Pages 7–28 in L. Real, ed., *Pollination biology*. Academic, New York.

Baker, H. G., and I. Baker. 1983. Floral nectar sugar constituents in relation to pollinator type. Pages 117–41 in C. E. Jones and R. J. Little, eds., *Handbook of experimental pollination biol-ogy*. Van Nostrand Reinhold, New York.

Baker, H. G., and P. D. Hurd, Jr. 1968. Intrafloral ecology. *Annual Review of Entomology* 13:385–414.

Bakker, J., R. T. Folkertsma, J. N. A. M. R. van der Voort, J. M. de Boer, and F. J. Gommers. 1993. Changing concepts and molecular approaches in the management of virulence genes in potato cyst nematodes. *Annual Review of Phytopathology* 31:171–92.

Barbosa, P., V. Krischik, and D. Lance. 1989. Life-history traits of forest-inhabiting flightless Lepidoptera. *American Midland Naturalist* 122:262–74.

Barbosa, P., P. Martinat, and M. Waldvogel. 1986. Development, fecundity and survival of the herbivore *Lymantria dispar* and the number of plant species in its diet. *Ecological Entomol-ogy* 11:1–6.

Bariana, H. S , and R. A. McIntosh. 1993. Cytogenetic studies in wheat. XV. Location of rust resistance genes in VPM1 and their genetic linkage with other disease resistance genes in chromosome 2A. *Genome* 36:476–82.

Barker, J. S. F. 1992. Genetic variation in cactophilic *Drosophila* for oviposition on natural yeast substrates. *Evolution* 46:1070–83.

Barr, A. R. 1980. Cytoplasmic incompatibility in natural populations of mosquito, *Culex pipiens* L. *Nature* 283:71–72.

Barrett, J. A. 1988. Frequency-dependent selection in plant-fungal interactions. *Philosophical Transactions of the Royal Society of London B* 319:473–83.

Barrow, D. A., and R. S. Pickard. 1984. Size-related selection of food plants by bumblebees. *Ecological Entomology* 9:369–73.

Barton, N. H. 1992. On the spread of new gene combinations in the third phase of Wright's shifting-balance. *Evolution* 46:551–57.

Barton, N. H., and G. M. Hewitt. 1985. Analysis of hybrid zones. *Annual Review of Ecology and Systematics* 16:113–48.

Basset, Y. 1992. Host specificity of arboreal and free-living insect herbivores in rain forests. *Biological Journal of the Linnean Society* 47:115–33.

Bates, H. W. 1862. XXXII. Contributions to an insect fauna of the Amazon valley. LEPIDOPTERA: HELICONIDAE. *Transactions of the Linnean Society of London* 23:495–566.

Baur, R., P. Feeny, and E. Städler. 1993. Oviposition stimulants for the black swallowtail butterfly: Identification of electrophysiologically active compounds in carrot volatiles. *Journal of Chemical Ecology* 19:919–37.

Bawa, K. S. 1990. Plant-pollinator interactions in tropical rain forests. *Annual Review of Ecology and Systematics* 21:399–422.

Bazzaz, F. A. 1990a. Plant-plant interactions in successional environments. Pages 239–63 in J. B. Grace and D. Tilman, eds., *Perspectives on plant competition.* Academic, New York.

———. 1990b. The response of natural ecosystems to the rising global CO_2 levels. *Annual Review of Ecology and Systematics* 21:167–96.

Bazzaz, F. A., and K. Garbutt. 1988. The response of annuals in competitive neighborhoods: Effects of elevated CO_2. *Ecology* 69:937–46.

Beattie, A. J. 1985. *The evolutionary ecology of ant-plant mutalisms.* Cambridge University Press, Cambridge.

Beattie, A. J., C. Turnbull, T. Hough, S. Jobson, and R. B. Knox. 1985. The vulnerability of pollen and fungal spores to ant secretions: Evidence and some evolutionary implications. *American Journal of Botany* 72:606–14.

Beattie, A. J., C. Turnbull, T. Hough, and R. B. Knox. 1986. Antibiotic production: A possible function for the metapleural glands of ants (Hymenoptera: Formicidae). *Annals of the Entomological Society of America* 79:448–50.

Beattie, A. J., C. Turnbull, B. Knox, and E. G. Williams. 1984. Ant inhibition of pollen function: A possible reason why ant pollination is rare. *American Journal of Botany* 71:421–26.

Beaver, P. C., and R. C. Jung, eds. 1985. *Animal agents and vectors of human disease.* 5th edition. Lea & Febiger, Philadelphia.

Beaver, R. A. 1989. Insect-fungus relationships in the bark and ambrosia beetles. Pages 121–43 in N. Wilding, N. M. Collins, P. M. Hammond, and J. F. Webber, eds., *Insect-fungus interactions.* Academic, New York.

Beccari, O. [1904] 1989. *Wanderings in the great forests of Borneo: Travels and researches of a naturalist in Sarawak.* Reprint. Oxford University Press, New York.

Becerra, J. X. I., and D. L. Venable. 1989. Extrafloral nectaries: A defense against ant-Homoptera mutualisms? *Oikos* 55:276–80.

Beddard, F. E. 1892. *Animal coloration.* Swan, Sonnenschein, & Co., London.

Beehler, B. M. 1983a. Frugivory and polygamy in birds of paradise. *Auk* 100:1–12.

———. 1983b. Notes on the behaviour and ecology of MacGregor's bird of paradise. *Emu* 83:28–30.

———. 1987. Ecology and behavior of the buff-tailed sicklebill (Paradisaeidae: *Epimachus albertisi*). *Auk* 104:48–55.

Beissinger, S. R. 1983. Hunting behavior, prey selection, and energetics of snail kites in Guyana: Consumer choice by a specialist. *Auk* 100:84–92.

———. 1990. Alternative foods of a diet specialist, the snail kite. *Auk* 107:327–33.

Bell, G., and J. Maynard Smith. 1987. Short-term selection for recombination among mutually antagonistic species. *Nature* 328:66–68.

Belt, T. [1874] 1928. *The naturalist in Nicaragua*. Reprint. J. M. Dent, London.

Benkman, C. W. 1987a. Crossbill foraging behavior, bill structure, and patterns of food profitability. *Wilson Bulletin* 99:351–68.

———. 1987b. Food profitability and the foraging ecology of crossbills. *Ecological Monographs* 57:251–67.

———. 1988a. On the advantages of crossed mandibles: An experimental approach. *Ibis* 130:288–93.

———. 1988b. Seed handling ability, bill structure, and the cost of specialization for crossbills. *Auk* 105:715–19.

———. 1989. On the evolution and ecology of island populations of crossbills. *Evolution* 43:1324–30.

———. 1993. Adaptation to single resources and the evolution of crossbill (*Loxia*) diversity. *Ecological Monographs* 63:305–25.

Benkman, C. W., and A. K. Lindholm. 1991. The advantages and evolution of a morphological novelty. *Nature* 349:519–20.

Bennetzen, J. L., and S. H. Hulbert. 1992. Extramarital sex amongst the beets—Organization, instability, and evolution of plant disease resistance genes. *Plant Molecular Biology* 20:575–80.

Bennetzen, J. L., M. Qin, S. Ingels, and A. H. Ellingboe. 1988. Allele-specific and *Mutator*-associated instability at the *Rp1* disease-resistance locus of maize. *Nature* 332:369–70.

Benson, W. W. 1985. Amazon ant plants. Pages 239–66 in G. T. Prance and T. E. Lovejoy, eds., *Amazonia*. Pergamon, Oxford.

Benzing, D. H. 1970. An investigation of two bromeliad myrmecophytes: *Tillandsia butzii* Mez, *T. caput-medusae* E. Morren and their ants. *Bulletin of the Torrey Botanical Club* 97: 109–15.

———. 1991. Myrmecotrophy: Origins, operation, and importance. Pages 353–73 in C. R. Huxley and D. F. Cutler, eds., *Ant-plant interactions*. Oxford University Press, Oxford.

Berenbaum, M. R. 1983. Coumarins and caterpillars: A case for coevolution. *Evolution* 37:163–79.

———. 1990. Evolution of specialization in insect-umbellifer associations. *Annual Review of Entomology* 35:319–43.

Berenbaum, M. R., A. R. Zangerl, and J. K. Nitao. 1986. Constraints on chemical coevolution: Wild parsnip and the parsnip webworm. *Evolution* 40:1215–28.

Bermudes, D., and L. Margulis. 1988. Symbiont acquisition as neoseme: Origin of species and higher taxa. *Symbiosis* 5:185–97.

Bernays, E. A., and K. L. Bright. 1991. Dietary mixing in grasshoppers: Switching induced by nutritional imbalances in foods. *Entomologia Experimentalis et Applicata* 61:247–53.

Bernays, E. A., K. Bright, J. J. Howard, D. Raubenheimer, and C. Champagne. 1992. Variety is the spice of life: Frequent switching between foods in the polyphagous grasshopper *Taeniopoda eques* Burmeister (Orthoptera: Acrididae). *Animal Behavior* 44:721–31.

Bernays, E. A., and J. C. Lee. 1988. Food aversion learning in the polyphagous grasshopper *Schistocerca americana. Physiological Entomology* 13:131–37.

Berrie, A. D. 1976. Detritus, microorganisms and animals in freshwater. Pages 323–38 in J. M. Anderson and A. Macfadyen, eds., *The role of terrestrial and aquatic organisms in decomposition processes*. Blackwell, Oxford.

Beveridge, I. 1982. Specificity and evolution of the anocephalate cestodes of marsupials. *Mémoires du Muséum National d'Histoire Naturelle Paris,* Nouvelle série, Série A, Zoology, 123:103–9.

Bieman, C. F. M. den. 1987. Host plant relations in the planthopper genus *Ribautodelphax* (Homoptera, Delphacidae). *Ecological Entomology* 12:163–72.

Biffen, R. H. 1905. Mendel's laws of inheritance and wheat breeding. *The Journal of Agricultural Science* 1:4–48.

Birch, L. C. 1960. The genetic factor in population ecology. *American Naturalist* 94:5–24.

Bjorndal, K. A. 1991. Diet mixing: Nonadditive interactions of diet items in an omnivorous freshwater turtle. *Ecology* 72:1234–41.

Blaney, W. M., and M. S. J. Simmonds. 1985. Food selection by locusts: The role of learning in rejection behaviour. *Entomologia Experimentalis et Applicata* 39:273–78.

Blank, R. J., and R. K. Trench. 1985. Speciation and symbiotic dinoflagellates. *Science* 229: 656–58.

Boller, E. F., K. Russ, V. Vallo, and G. L. Bush. 1976. Incompatible races of European cherry fruit fly, *Rhagoletis cerasi* (Diptera: Tephritidae), their origin and potential use in biological control. *Entomologia Experimentalis et Applicata* 20:237–47.

Boorstin, D. J. 1948. *The lost world of Thomas Jefferson.* Henry Holt & Co., New York.

Boucher, D. H. 1985. The idea of mutualism, past and future. Pages 1–28 in D. H. Boucher, ed., *The biology of mutualism: Ecology and evolution.* Oxford University Press, New York.

Boucot, A. J., ed. 1990. *Evolutionary paleobiology of behavior and coevolution.* Elsevier, New York.

Bouma, J. E., and R. E. Lenski. 1988. Evolution of a bacteria/plasmid association. *Nature* 335:351–52.

Bourne, G. R. 1985. The role of profitability in snail kite foraging. *Journal of Animal Ecology* 54:697–709.

Bowers, M. D., N. E. Stamp, and S. K. Collinge. 1992. Early stage of host range expansion by a specialist herbivore, *Euphydryas phaeton* (Nymphalidae). *Ecology* 73:526–36.

Boyle, L., S. L. O'Neill, H. M. Robertson, and T. L. Karr. 1993. Interspecific and intraspecific horizontal transfer of *Wolbachia* in *Drosophila*. *Science* 260:1796–99.

Bradshaw, A. D. 1952. Population of *Agrostis tenuis* resistant to lead and zinc poisoning. *Nature* 169:1098.

Breeuwer, J. A. J., and J. H. Werren. 1990. Microorganisms associated with chromosome destruction and reproductive isolation between two insect species. *Nature* 346:558–60.

Breton, L. M., and J. F. Addicott. 1992. Density-dependent mutualism in an aphid-ant interaction. *Ecology* 73:2175–80.

Briskie, J. V., S. G. Sealy, and K. A. Hobson. 1992. Behavioral defenses against avian brood parasitism in sympatric and allopatric host populations. *Evolution* 46:334–40.

Brodie, E. D. 1968. Investigations on the skin toxin of the adult rough-skinned newt, *Taricha granulosa. Copeia* 1968:307–13.

Brodie, E. D., III, and E. D. Brodie, Jr. 1990. Tetrodotoxin resistance in garter snakes: An evolutionary response of predators to dangerous prey. *Evolution* 44:651–59.

———. 1991. Evolutionary response of predators to dangerous prey: Reduction of toxicity of newts and resistance of garter snakes in island populations. *Evolution* 45:221–24.

Bronstein, J. L. 1987. Maintenance of species-specificity in a Neotropical fig-pollinator wasp mutualism. *Oikos* 48:39–46.

———. 1988. Mutualism, antagonism, and the fig-pollinator interaction. *Ecology* 69:1298–1302.

———. 1989. A mutualism at the edge of its range. *Experientia* 45:622–37.

Bronstein, J. L., P.-H. Gouyon, C. Gliddon, F. Kjellberg, and G. Michaloud. 1990. The ecological consequences of flowering asynchrony in monoecious figs: A simulation study. *Ecology* 71:2145–56.

Brooker, M. G., and L. C. Brooker. 1989. The comparative breeding behaviour of two sympatric cuckoos, Horsfield's bronze-cuckoo *Chrysococcyx basalis* and the shining bronze-cuckoo *C. lucidus,* in Western Australia: A new model for the evolution of egg morphology and host specificity in avian brood parasites. *Ibis* 131:528–47.

Brooks, D. R., and D. A. McLennan. 1991. *Phylogeny, ecology, and behavior.* University of Chicago Press, Chicago.

Brower, L. P. 1984. Chemical defense in butterflies. Pages 109–34 in R. I. Vane-Wright and P. R. Ackery, eds., *The biology of butterflies.* Academic, New York.

Brown, J. S., and T. L. Vincent. 1992. Organization of predator-prey communities as an evolutionary game. *Evolution* 46:1269–83.

Brown, R. J., M. N. Brown, M. de L. Brooke, and N. B. Davies. 1990. Reactions of parasitized and unparasitized populations of *Acrocephalus* warblers to model cuckoo eggs. *Ibis* 132:109–11.

Brown, W. L., Jr., and E. O. Wilson. 1956. Character displacement. *Systematic Zoology* 5:49–64.

Brues, C. T. 1920. The selection of food-plants by insects, with special reference to lepidopterous larvae. *American Naturalist* 54:313–32.

———. 1924. The specificity of food-plants in the evolution of phytophagous insects. *American Naturalist* 58:127–44.

Bryant, J. P. 1987. Feltleaf willow–snowshoe hare interactions: Plant carbon/nutrient balance and floodplain succession. *Ecology* 68:1319–27.

Bryant, J. P., F. S. Chapin III, P. B. Reichardt, and T. P. Clausen. 1987. Response of winter chemical defense in Alaska paper birch and green alder to manipulation of plant carbon/nutrient balance. *Oecologia* 72:510–14.

Bryant, J. P., I. Heitkonig, P. Kuropat, and N. Owen-Smith. 1991. Effects of severe defoliation on the long-term resistance to insect attack and on leaf chemistry in six woody species of the southern African savanna. *American Naturalist* 137:50–63.

Buchmann, S. L. 1987. The ecology of oil flowers and their bees. *Annual Review of Ecology and Systematics* 18:343–69.

Bull, J. J., and I. J. Molineux. 1992. Molecular genetics of adaptation in an experimental model of cooperation. *Evolution* 46:882–95.

Bull, J. J., I. J. Molineux, and W. R. Rice. 1991. Selection of benevolence in a host-parasite system. *Evolution* 45:875–82.

Bull, J. J., and W. R. Rice. 1991. Distinguishing mechanisms for the evolution of cooperation. *Journal of Theoretical Biology* 149:63–74.

Burdon, J. J. 1987. *Diseases and plant population biology.* Cambridge University Press, Cambridge.

Burdon, J. J., A. H. D. Brown, and A. M. Jarosz. 1990. The spatial scale of genetic interactions in host-pathogen coevolved systems. Pages 233–47 in J. J. Burdon and S. R. Leather, eds., *Pests, pathogens, and plant communities.* Blackwell, Oxford.

Burdon, J. J., and A. M. Jarosz. 1991. Host-pathogen interactions in natural populations of *Linum marginale* and *Melampsora lini.* I. Patterns of resistance and racial variation in a large host population. *Evolution* 45:205–17.

Burdon, J. J., A. M. Jarosz, and G. C. Kirby. 1989. Pattern and patchiness in plant-pathogen interactions—causes and consequences. *Annual Review of Ecology and Systematics* 20:119–36.

Burdon, J. J., J. D. Oates, and D. R. Marshall. 1983. Interactions between *Avena* and *Puccinia* species. I. The wild hosts: *Avena barbata* Pott ex Link, *A. fatua* L., and *A. ludoviciana* Durieu. *Journal of Applied Ecology* 20:571–84.

Burgdorfer, W. 1984. Vertical transmission of spotted fever group and scrub typhus rickettsiae. Pages 77–92 in K. F. Harris, ed., *Current topics in vector research,* volume II. Praeger, New York.

Bush, G. L. 1969. Sympatric host race formation and speciation in frugivorous flies of the genus *Rhagoletis. Evolution* 23:237–51.

Butlin, R. 1987. Speciation by reinforcement. *Trends in Ecology and Evolution* 2:8–13.

Calaby, J. H. 1960. Observations on the banded anteater *Myrmecobius f. fasciatus* Waterhouse (Marsupialia), with particular reference to its food habits. *Proceedings of the Zoological Society of London* 135:183–207.

Calver, M. C., D. A. Saunders, and B. D. Porter. 1987. The diet of nestling rainbow bee-eaters, *Merops ornatus,* on Rottnest Island, Western Australia, and observations on a non-destructive method of diet analysis. *Australian Wildlife Research* 14:541–50.

Cane, J. H., G. C. Eickwort, F. R. Wesley, and J. Speilholz. 1983. Foraging, grooming, and mate-seeking behaviors of *Macropis nuda* (Hymenoptera: Melittidae) and use of *Lysimachia ciliata* (Primulaceae) oils in larval provisions and cell linings. *American Midland Naturalist* 110:257–64.

Carlquist, S. 1970. *Hawaii: A natural history.* Natural History Press, New York.

Carlton, J. T., G. J. Vermeij, D. R. Lindberg, D. A. Carlton, and E. C. Dudley. 1991. The first historical extinction of a marine invertebrate in an ocean basin: The demise of the eelgrass limpet *Lottia alveus. Biological Bulletin* 180:72–80.

Carroll, C. R., and C. A. Hoffman. 1980. Chemical feeding deterrent mobilized in response to insect herbivory and counteradaptation by *Epilachna tredecimnotata. Science* 209:414–16.

Carroll, S. P., and C. Boyd. 1992. Host race radiation in the soapberry bug: Natural history with the history. *Evolution* 46:1052–69.

Case, T. J. 1979. Character displacement and coevolution in some *Cnemidophorus* lizards. *Fortschritte der Zoologie* 25:235–82.

Caughley, G., D. Grice, R. Barker, and B. Brown. 1988. The edge of the range. *Journal of Animal Ecology* 57:771–85.

Chai, P. 1986. Field observations and feeding experiments on the responses of rufous-tailed jacamars (*Galbula ruficauda*) to free-flying butterflies in a tropical rainforest. *Biological Journal of the Linnean Society* 29:161–89.

Chanway, C. P., F. B. Holl, and R. Turkington. 1989. Effect of *Rhizobium leguminosarum* biovar. *trifolii* genotype on specificity between *Trifolium repens* and *Lolium perenne. Journal of Ecology* 77:1150–60.

Charlesworth, B., J. A. Coyne, and N. H. Barton. 1987. The relative rates of evolution of sex chromosomes and autosomes. *American Naturalist* 130:113–46.

Charlesworth, D., and B. Charlesworth. 1976. Theoretical genetics of Batesian mimicry. II. Evolution of supergenes. *Journal of Theoretical Biology* 55:305–24.

Cherrett, J. M. 1989. Key concepts, the results of a survey of our members' opinions. Pages 1–16 in J. M. Cherrett, ed., *Ecological concepts: The contribution of ecology to an understanding of the natural world.* Blackwell, Oxford.

Cherrett, J. M., R. J. Powell, and D. J. Stradling. 1989. The mutualism between leaf-cutting ants and their fungus. Pages 93–120 in N. Wilding, N. M. Collins, P. M. Hammond, and J. F. Webber, eds., *Insect-fungus interactions.* Academic, New York.

Christ, B. J., C. O. Person, and D. D. Pope. 1987. The genetic determination of variation in pathogenicity. Pages 7–19 in M. S. Wolfe and C. E. Caten, eds., *Populations of plant pathogens: Their dynamics and genetics.* Blackwell, Oxford.

Cittadino, E. 1990. *Nature as the laboratory: Darwinian plant ecology in the German empire, 1880–1900.* Cambridge University Press, Cambridge.

Claridge, M. F., and J. Den Hollander. 1982. Virulence to rice cultivars and selection for virulence in populations of the brown planthopper, *Nilaparvata lugens. Entomologia Experimentalis et Applicata* 32:213–21.

Claridge, M. F., J. Den Hollander, and J. C. Morgan. 1988. Variation in hostplant relations and courtship signals of weed-asssociated populations of the brown planthopper, *Nilaparvata lugens* (Stål), from Australia and Asia: A test of the recognition species concept. *Biological Journal of the Linnean Society* 35:79–93.

Claridge, M. F., and G. A. Nixon. 1986. *Oncopis flavicollis* (L.) associated with tree birches (*Betula*): A complex of biological species or a host plant utilization polymorphism? *Biological Journal of the Linnean Society* 27:381–97.

Clark, D. B., and D. A. Clark. 1991. Herbivores, herbivory, and plant phenology: Patterns and consequences in a tropical rain-forest cycad. Pages 209–225 in P. W. Price, T. M. Lewinsohn, G. W. Fernandes, and W. W. Benson, eds., *Plant-animal interactions: Evolutionary ecology in tropical and temperate regions.* Wiley, New York.

Clarke, C. A., and P. M. Sheppard. 1960. The evolution of mimicry in the butterfly *Papilio dardanus. Heredity* 14:163–73.

———. 1963. Interactions between major genes and polygenes in the determination of the mimetic patterns of *Papilio dardanus. Evolution* 17:404–13.

———. 1971. Further studies on the genetics of the mimetic butterfly *Papilio memnon* L. *Philosophical Transactions of the Royal Society of London B* 263:35–70.

Clarke, J. F. G. 1952. Host relationships of moths of the genera *Depressaria* and *Agonopterix,* with descriptions of new species. *Smithsonian Miscellaneous Collections* 117(7):1–20.

Clarkson, R. W., and W. L. Minckley. 1988. Morphology and foods of Arizona catostomid fishes: *Catostomus insignis, Pantosteus clarki,* and their putative hybrids. *Copeia* 1988:422–33.

Clay, K. 1988. Clavicipitaceous fungal endophytes of grasses: Coevolution and the change from parasitism to mutualism. Pages 79–105 in K. Pirozynski and D. L. Hawksworth, eds., *Coevolution of fungi with plants and animals.* Academic, New York.

———. 1990. Comparative demography of three graminoids infected by systemic, clavicipitaceous fungi. *Ecology* 71:558–70.

Clayton, D. H., R. D. Gregory, and R. D. Price. 1992. Comparative ecology of Neotropical bird lice (Insecta: Phthiraptera). *Journal of Animal Ecology* 61:781–95.

Clements, F. E. 1905. *Research methods in ecology.* University Publishing Co., Lincoln, Nebr.

Clements, F. E., and V. E. Shelford. 1939. *Bio-ecology.* Wiley Interscience, New York.

Collins, J. P. 1981. Distribution, habitats and life history variation in the tiger salamander, *Ambystoma tigrinum,* in east central and southeast Arizona. *Copeia* 1981:666–75.

———. 1986. Evolutionary ecology and the use of natural selection in ecological theory. *Journal of the History of Biology* 19:257–88.

Collins, N. M., and M. G. Morris. 1985. *Threatened swallowtail butterflies of the world: The IUCN red data book.* International Union for Conservation of Nature and Natural Resources, Gland, Switzerland.

Colwell, R. K. 1985. The evolution of ecology. *American Zoologist.* 25:771–77.

Comins, H. N., M. P. Hassell, and R. M. May. 1992. The spatial dynamics of host-parasitoid systems. *Journal of Animal Ecology* 61:735–48.

Connell, J. H. 1980. Diversity and the coevolution of competitors, or the ghost of competition past. *Oikos* 35:131–38.

Conway Morris, S., and D. W. T. Crompton. 1982. Origins and evolution of acanthocephalan worms. *Mémoires du Muséum National d'Histoire Naturelle Paris,* Nouvelle série, Série A, Zoology, 123:61–71.

Cope, E. D. 1896. *The primary factors of organic evolution.* Open Court Publishing Co., Chicago.

Counce, S. J., and D. F. Poulson. 1966. The expression of maternally-transmitted sex ratio condition (SR) in two strains of *Drosophila melanogaster. Genetica* 37:364–90.

Courtney, S. P., G. K. Chen, and A. Gardner. 1989. A general model for individual host selection. *Oikos* 55:55–65.

Cox, T. S., and J. H. Hatchett. 1986. Genetic model for wheat/hessian fly (Diptera: Cecidomyiidae) interaction: Strategies for deployment of resistance genes in wheat cultivars. *Environmental Entomology* 15:24–31.

Coyne, J. A., H. A. Orr, and D. J. Futuyma. 1988. Do we need a new species concept? *Systematic Zoology* 37:190–200.

Crawley, M. J., and J. R. Krebs. 1992. Foraging theory. Pages 90–114 in M. J. Crawley, ed., *Natural enemies: The population biology of predators, parasites and diseases.* Blackwell, Oxford.

Crome, F. H. J. 1975. The ecology of fruit pigeons in tropical northern Queensland. *Australian Wildlife Research* 2:155–85.

Crome, F. H. J., and A. K. Irvine. 1986. "Two bob each way": The pollination and breeding system of the Australian rain forest tree *Syzygium cormiflorum* (Myrtaceae). *Biotropica* 18:115–25.

Crow, J. F., W. R. Engels, and D. Denniston. 1990. Phase three of Wright's shifting-balance theory. *Evolution* 44:233–47.

Crowcroft, P. 1991. *Elton's ecologists: A history of the Bureau of Animal Population.* University of Chicago Press, Chicago.

Curtis, L. A. 1990. Parasitism and the movements of intertidal gastropod individuals. *Biological Bulletin* 179:105–12.

Cushman, J. H., and J. F. Addicott. 1991. Conditional interactions in ant-plant-herbivore mutualisms. Pages 92–103 in C. R. Huxley and D. F. Cutler, eds., *Ant-plant interactions.* Oxford University Press, Oxford.

Cushman, J. H., and A. J. Beattie. 1991. Mutualisms: Assessing the benefits to hosts and visitors. *Trends in Ecology and Evolution* 6:193–95.

Cushman, J. H., and T. G. Whitham. 1989. Conditional mutualism in a membracid-ant association: Temporal, age-specific, and density dependent effects. *Ecology* 70:1040–47.

———. 1991. Competition mediating the outcome of a mutualism: Protective services of ants as a limiting resource for membracids. *American Naturalist* 138:851–65.

D'Alessandro-Bacigalupo, A. 1985. Flagellate protozoa (Mastigophora). Pages 13–35 in P. C. Beaver and R. C. Jung, eds., *Animal agents and vectors of human disease.* 5th ed. Lea & Febiger, Philadelphia.

Dangl, J. L. 1992. The major histocompatibility complex à la carte: Are there analogies to plant disease resistance genes on the menu? *Plant Journal* 2:3–11.

Darwin, C. 1859. *On the origin of species by means of natural selection, or the preservation of favoured races in the struggle for life.* Facsimile of the 1st ed. Harvard University Press, Cambridge.

———. [1862] 1979. *On the various contrivances by which British and foregin orchids are fertilised by insects, and on the good effects of intercrossing.* Facsimile ed. Earl M. Coleman, Standfordville, N.Y.

———. [1863] 1977. [A review of H. W. Bates's paper on mimicry in butterflies.] Pages 87–92 in P. H. Barrett, ed., *The collected papers of Charles Darwin,* volume 2. University of Chicago Press, Chicago.

————. [1871] 1981. *The descent of man, and selection in relation to sex.* Reprint. Princeton University Press, Princeton.

————. 1877. *The various contrivances by which orchids are fertilised by insects.* 2d ed. Murray, London.

Darwin, F., ed. 1896. *The life and letters of Charles Darwin.* Volumes I and II. Appleton, New York.

Da Silva, H. R., M. C. de Britto-Pereira, and U. Carmaschi. 1989. Frugivory and seed dispersal by *Hyla truncata,* a Neotropical treefrog. *Copeia* 1989:781–83.

Davidar, P. 1983. Similarity between flowers and fruits in some flowerpecker pollinated mistletoes. *Biotropica* 15:32–37.

Davidson, D. W. 1988. Ecological studies of Neotropical ant gardens. *Ecology* 69:1138–52.

Davidson, D. W., and B. L. Fisher. 1991. Symbiosis of ants with *Cecropia* as a function of light regime. Pages 289–309 in C. R. Huxley and D. F. Cutler, eds., *Ant-plant interactions.* Oxford University Press, Oxford.

Davidson, D. W., R. B. Foster, R. R. Snelling, and P. W. Lozada. 1991. Variable composition of some tropical ant-plant symbioses. Pages 145–62 in P. W. Price, T. M. Lewinsohn, G. W. Fernandes, and W. W. Benson, eds., *Plant-animal interactions: Evolutionary ecology in tropical and temperate regions.* Wiley, New York.

Davies, K. G., M. P. Robinson, and V. Laird. 1992. Proteins involved in the attachment of a hyperparasite, *Pasteuria penetrans,* to its plant-parasitic nematode host, *Meloidogyne incognita. Journal of Invertebrate Pathology* 59:18–23.

Davies, N. B., and M. de L. Brooke. 1989a. An experimental study of co-evolution between the cuckoo, *Cuculus canorus,* and its hosts. I. Host egg discrimination. *Journal of Animal Ecology* 58:207–24.

————. 1989b. An experimental study of co-evolution between the cuckoo, *Cuculus canorus,* and its hosts. II. Host egg markings, chick discrimination and general discussion. *Journal of Animal Ecology* 58:225–36.

Davis, D. R. 1967. A revision of the moths of the subfamily Prodoxinae (Lepidoptera: Incurvariidae). *U.S. Natural History Museum, Bulletin* 255:1–170.

Davis, D. R., O. Pellmyr, and J. N. Thompson. 1992. Biology and systematics of *Greya* Busck and *Tetragma,* new genus (Lepidoptera: Prodoxidae). *Smithsonian Contributions to Zoology* 524:1–88.

Debener, T., M. Lehnackers, M. Arnold, and J. L. Dangl. 1991. Identification and molecular mapping of a single *Arabidopsis thaliana* locus determining resistance to a phytopathogenic *Pseudomonas syringae* isolate. *Plant Journal* 1:289–302.

De Meester, L. 1993. Genotype, fish-mediated chemicals, and phototactic behavior in *Daphnia magna. Ecology* 74:1467–74.

Dénarié, J., F. Debellé, and C. Rosenberg. 1992. Signaling and host range variation in nodulation. *Annual Review of Microbiology* 46:497–531.

Den Hollander, J., and P. K. Pathak. 1981. The genetics of the 'biotypes' of the rice brown planthopper, *Nilaparvata lugens. Entomologia Experimentalis et Applicata* 29:76–86.

Denno, R. F., S. Larsson, and K. L. Olmstead. 1990. Role of enemy-free space and plant quality in host-plant selection by willow beetles. *Ecology* 71:124–37.

DeSalle, R., and A. R. Templeton. 1987. Comments on "The significance of asymmetrical sexual isolation." *Evolutionary Biology* 21:21–27.

Dethier, V. G. 1941. Chemical factors determining the choice of food plants by *Papilio* larvae. *American Naturalist* 75:61–73.

————. 1980. Food-aversion learning in two polyphagous caterpillars, *Diacrisia virginica* and *Estigmene congrua. Physiological Entomology* 5:321–25.

————. 1988. The feeding behavior of a polyphagous caterpillar (*Diacrisia virginica*) in its natural habitat. *Canadian Journal of Zoology* 66:1280–88.

————. 1989. Patterns of locomotion of polyphagous arctiid caterpillars in relation to foraging. *Ecological Entomology* 14:375–86.

Dethier, V. G., and M. T. Yost. 1979. Oligophagy and absence of food-aversion learning in tobacco hornworms, *Manduca sexta*. *Physiological Entomology* 4:125–30.

de Wit, P. J. G. M. 1992. Molecular characterization of gene-for-gene systems in plant-fungus interactions and the application of avirulence genes in control of plant pathogens. *Annual Review of Phytopathology* 30:391–418.

Diamond, J. 1986a. Biology of birds of paradise and bowerbirds. *Annual Review of Ecology and Systematics* 17:17–37.

————. 1986b. Evolution of ecological segregation in the New Guinea montane avifauna. Pages 98–125 in J. Diamond and T. J. Case, eds., *Community ecology*. Harper & Row, New York.

Diamond, J., S. L. Pimm, M. E. Gilpin, and M. LeCroy. 1989. Rapid evolution of character displacement in myzomelid honeyeaters. *American Naturalist* 134:675–708.

Dickinson, M. J., D. A. Jones, and J. D. G. Jones. 1993. Close linkage between the *Cf-2/Cf-5* and *Mi* resistance loci in tomato. *Molecular Plant-Microbe Interactions* 6:341–47.

Diehl, S. R., and G. L. Bush. 1984. An evolutionary and applied perspective of insect biotypes. *Annual Review of Entomology* 29:471–504.

Dixey, F. A. 1894. On the phylogeny of the Pierinae, as illustrated by their wing-markings and geographical distribution. *Transactions of the Entomological Society of London* 1894:249–334.

————. 1896. On the relation of mimetic patterns to the original form. *Transactions of the Entomological Society of London* 1896:65–79.

————. 1897. Mimetic attraction. *Transactions of the Entomological Society of London* 1897:317–32.

————. 1906. Record of proceedings for October 3rd, 1906. *Proceedings of the Entomological Society of London* 1906:lxix–lxxi.

————. 1909. On Müllerian mimicry and diaposematism. *Transactions of the Entomological Society of London* 1909:559–83.

Dixon, A. F. G., P. Kendlmann, J. Leps, and J. Holman. 1987. Why are there so few species of aphids, especially in the tropics? *American Naturalist* 129:580–92.

Dobson, A. P. 1988. The population biology of parasite-induced changes in host behavior. *Quarterly Review of Biology* 63:139–65.

Dobson, A. P., P. J. Hudson, and A. M. Lyles. 1992. Macroparasites: Worms and others. Pages 329–48 in M. J. Crawley, ed., *Natural enemies: The population biology of predators, parasites and diseases*. Blackwell, Oxford.

Dobzhansky, T. 1941. *Genetics and the origin of species*. Columbia University Press, New York.

Dodson, S. 1988. The ecological role of chemical stimuli for the zooplankton: Predator-avoidance behavior in *Daphnia*. *Limnology and Oceanography* 33:1431–39.

————. 1989. Predator-induced reaction norms. *BioScience* 39:447–52.

Dussourd, D. E., and R. F. Denno. 1991. Deactivation of plant defense: Correspondence between insect behavior and secretory canal architecture. *Ecology* 72:1383–96.

Dussourd, D. E., and T. Eisner. 1987. Vein-cutting behavior: Insect counterploy to the latex defense of plants. *Science* 237:898–901.

Dwyer, G., S. A. Levin, and L. Buttel. 1990. A simulation model of the population dynamics and evolution of myxomatosis. *Ecological Monographs* 60:423–47.

Eastop, V. F. 1986. Aphid-plant associations. Pages 35–54 in A. R. Stone and D. L. Hawksworth, eds., *Coevolution and systematics*. Clarendon, Oxford.

Ebba, T., and C. Person. 1975. Genetic control of virulence in *Ustilago hordei*. IV. Duplicate genes for virulence and genetic and environmental modification of a gene-for-gene relationship. *Canadian Journal of Genetics and Cytology* 17:631–36.

Ebbert, M. A. 1991. The interaction phenotype in the *Drosophila willistoni*-spiroplasma symbiosis. *Evolution* 45:971–88.

———. 1992. Endosymbiotic sex ratio distorters in insects and mites. Pages 150–91 in D. L. Wrench and M. A. Ebbert, eds., *Evolution and diversity of sex ratio in haplodiploid insects and mites.* Chapman & Hall, New York.

Ehlinger, T. J. 1990. Habitat choice and phenotype-limited feeding efficiency in bluegill: Individual differences and trophic polymorphism. *Ecology* 71:886–96.

Ehrlich, P. R. 1992. Population biology of checkerspot butterflies and the preservation of global biodiversity. *Oikos* 63:6–12.

Ehrlich, P. R., D. D. Murphy, M. C. Singer, C. B. Sherwood, R. R. White, and I. L. Brown. 1980. Extinction, reduction, stability and increase: The responses of checkerspot butterfly (*Euphydryas*) populations to the California drought. *Oecologia* 46:101–5.

Ehrlich, P. R., and P. H. Raven. 1964. Butterflies and plants: A study in coevolution. *Evolution* 18:586–608.

Ehrlich, P. R., and E. O. Wilson. 1991. Biodiversity studies: Science and policy. *Science* 253:758–62.

Ehrman, L., J. R. Factor, N. Somerson, and P. Manzo. 1989. The *Drosophila paulistorum* endosymbiont in an alternative species. *American Naturalist* 134:890–96.

Ehrman, L., and M. Wasserman. 1987. The significance of asymmetrical sexual isolation. *Evolutionary Biology* 21:1–20.

Eldredge, N., and S. J. Gould. 1972. Punctuated equilibria: An alternative to phyletic gradualism. Pages 82–115 in T. J. M. Schopf, ed., *Models in paleobiology.* Freeman, San Francisco.

Ellison, R. L., and J. N. Thompson. 1987. Variation in seed and seedling size: The effects of seed herbivores on *Lomatium grayi* (Umbelliferae). *Oikos* 49:269–80.

Elton, C. 1927. *Animal ecology.* Sidgwick & Jackson, London.

Emlen, J. M. 1966. The role of time and energy in food preference. *American Naturalist* 100:611–17.

Emmel, J. F., and O. Shields. 1978. Larval foodplant records for *Papilio zelicaon* in the western United States, and further evidence for the conspecificity of *P. zelicaon* and *P. gothica. Journal of Research on Lepidoptera* 17:56–67.

Erwin, T. L. 1982. Tropical forests: Their richness in Coleoptera and other arthropod species. *Coleopterists Bulletin* 36:74–75.

Erwin, T. L., and J. C. Scott. 1980. Seasonal and size patterns, trophic structure, and richness of Coleoptera in the tropical arboreal ecosystem: The fauna of the tree *Luehea seemanii* Triana and Planch in the Canal Zone of Panama. *Coleopterists Bulletin* 34:305–22.

Evans, D. R., J. Hill, T. A. Williams, and I. Rhodes. 1985. Effects of coexistence on the performance of white clover-perennial ryegrass mixtures. *Oecologia* 66:536–39.

Evans, R. C., and R. Turkington. 1988. Maintenance of morphological variation in a biotically patchy environment. *New Phytologist* 109:369–76.

Ewald, P. 1983. Host-parasite relations, vectors, and the evolution of disease severity. *Annual Review of Ecology and Systematics* 14:465–85.

———. 1987. Transmission modes and evolution of the parasitism-mutualism continuum. *Annals of the New York Academy of Sciences* 503:295–306.

Faegri, K., and L. van der Pijl. 1971. *The principles of pollination ecology.* Pergamon, Oxford.

Fagerström, T. 1988. Lotteries in communities of sessile organisms. *Trends in Ecology and Evolution* 3:303–6.

Fajer, E. D., M. D. Bowers, and F. A. Bazzaz. 1989. The effects of enriched carbon dioxide atmospheres on plant-insect herbivore interactions. *Science* 243:1198–1200.

Farrell, B. D., D. E. Dussourd, and C. Mitter. 1991. Escalation of plant defense: Do latex and resin canals spur plant diversification? *American Naturalist* 138:881–900.

Feder, J. L., C. A. Chilcote, and G. L. Bush. 1988. Genetic differentiation between sympatric host races of the apple maggot fly *Rhagoletis pomonella*. *Nature* 336:61–64.

———. 1990a. The geographic pattern of genetic differentiation between host associated populations of *Rhagoletis pomonella* (Diptera: Tephritidae) in the eastern United States and Canada. *Evolution* 44:570–94.

———. 1990b. Regional, local and microgeographic allele frequency variation between apple and hawthorn populations of *Rhagoletis pomonella* in western Michigan. *Evolution* 44:595–608.

Feeny, P. 1991. Chemical constraints on the evolution of swallowtail butterflies. Pages 315–40 in P. W. Price, T. M. Lewinsohn, G. W. Fernandes, and W. W. Benson, eds., *Plant-animal interactions: Evolutionary ecology in tropical and temperate regions.* Wiley, New York.

Feeny, P., K. Sachdev, L. Rosenberry, and M. Carter. 1988. Luteolin 7-O-(6″-O-Malonyl)-b-D-glucoside and Trans-chlorogenic acid: Oviposition stimulants for the black swallowtail butterfly. *Phytochemistry* 27:3439–48.

Feeny, P., E. Städler, I. Åhman, and M. Carter. 1989. Effects of plant odor on oviposition by the black swallowtail butterfly, *Papilio polyxenes* (Lepidoptera: Papilionidae). *Journal of Insect Behavior* 2:803–27.

Feinsinger, P. 1978. Ecological interactions between plants and hummingbirds in a successional tropical community. *Ecological Monographs* 48:269–87.

———. 1980. Asynchronous migration patterns and the coexistence of tropical hummingbirds. Pages 411–19 in A. Keast and E. S. Morton, eds., *Migrant birds in the Neotropics: Ecology, behavior, distribution, and conservation.* Smithsonian Institution Press, Washington, D.C.

———. 1983. Coevolution and pollination. Pages 282–310 in D. J. Futuyma and M. Slatkin, eds., *Coevolution.* Sinauer Associates, Sunderland, Mass.

———. 1987. Approaches to nectarivore-plant interactions in the New World. *Revista Chilena de Historia Natural* 60:285–319.

Feinsinger, P., and R. K. Colwell. 1978. Community organization among Neotropical nectar-feeding birds. *American Zoologist* 18:779–95.

Feinsinger, P., and H. M. Tiebout III. 1991. Competition among plants sharing hummingbird pollinators: Laboratory experiments on a mechanism. *Ecology* 72:1946–52.

Feinsinger, P., H. M. Tiebout III, and B. E. Young. 1991. Do tropical bird-pollinated plants exhibit density-dependent interactions? Field experiments. *Ecology* 72:1953–63.

Felsenstein, J. 1988. Phylogenies and quantitive characters. *Annual Review of Ecology and Systematics* 19:445–71.

Fenner, F., and K. Myers. 1978. Myxoma virus and myxomatosis in retrospect: The first quarter century of a new disease. Pages 539–70 in E. Kurstak and K. Maromorosch, eds., *Viruses and environment.* Academic, New York.

Fiala, B., U. Maschwitz, and T. Y. Pong. 1991. The association between *Macaranga* trees and ants in South-East Asia. Pages 263–70 in C. R. Huxley and D. F. Cutler, eds., *Ant-plant interactions.* Oxford University Press, Oxford.

Fisher, R. A. [1930] 1958. *The genetical theory of natural selection.* 2d rev. ed. Reprint. Dover, New York.

Fishlyn, D. A., and D. W. Phillips. 1980. Chemical camouflaging and behavioral defenses against a predatory seastar by three species of gastropods from the surfgrass *Phyllospadix* community. *Biological Bulletin* 158:34–48.

Fleming, T. H. 1988. *The short-tailed fruit bat: A study in plant-animal interactions.* University of Chicago Press, Chicago.

————. 1991. Fruiting plant-frugivore mutualism: The evolutionary theater and the ecological play. Pages 119–44 in P. W. Price, T. M. Lewinsohn, G. W. Fernandes, and W. W. Benson, eds., *Plant-animal interactions: evolutionary ecology in tropical and temperate regions*. Wiley, New York.

Flor, H. H. 1942. Inheritance of pathogenicity in *Melampsora lini*. *Phytopathology* 32:653–69.

————. 1955. Host-parasite interaction in flax rust—its genetics and other implications. *Phytopathology* 45:680–85.

————. 1956. The complementary genic systems in flax and flax rust. *Advances in Genetics* 8:29–54.

Foley, W. J., and I. D. Hume. 1987a. Passage of digesta markers in two species of arboreal folivorous marsupials—the greater glider (*Petauroides volans*) and the brushtail possum (*Trichosurus vulpecula*). *Physiological Zoology* 60:103–13.

————. 1987b. Nitrogen requirements and urea metabolism in two arboreal folivorous marsupials, the greater glider (*Petauroides volans*) and the brushtail possum (*Trichosurus vulpecula*), fed *Eucalyptus* foliage. *Physiological Zoology* 60:241–50.

Forbes, H. O. 1885. *A naturalist's wanderings in the Eastern Archipelago: A narrative of travel and exploration from 1878 to 1883*. Harper & Brothers, New York.

Ford, E. B. 1964. *Ecological genetics*. Methuen, London.

Ford, J. 1980. Morphological and ecological divergence and convergence in isolated populations of the red-tailed black-cockatoo. *Emu* 80:103–20.

Fox, L. R., and P. A. Morrow. 1981. Specialization: Species property or local phenomenon? *Science* 211:887–93.

Frank, S. A. 1991. Ecological and genetic models of host-pathogen coevolution. *Heredity* 67:73–83.

————. 1992. Models of plant-pathogen coevolution. *Trends in Genetics* 8:213–19.

————. 1993. Coevolutionary genetics of plants and pathogens. *Evolutionary Ecology* 7:45–75.

Freeland, W. J., and D. H. Janzen. 1974. Strategies in herbivory by mammals: The role of plant secondary compounds. *American Naturalist* 108:269–89.

Freeland, W. J., and J. W. Winter. 1975. Evolutionary consequences of eating: *Trichosurus vulpecula* (Marsupialia) and the genus *Eucalyptus*. *Journal of Chemical Ecology* 1:439–55.

Friedmann, H., L. F. Kiff, and S. I. Rothstein. 1977. A further contribution to knowledge of the host relations of the parasitic cowbirds. *Smithsonian Contributions to Zoology* 235:1–75.

Friend, J. A., and J. E. Kinnear. [1983] 1991. Numbat. Pages 84–85 in R. Strahan, ed., *The Australian Museum complete book of Australian mammals*. Reissue. Cornstalk Publishing, North Ryde, Australia.

Fritz, R. S., and P. W. Price. 1988. Genetic variation among plants and insect community structure: Willows and sawflies. *Ecology* 69:845–56.

Fritz, R. S., and E. L. Simms, eds. 1992. *Plant resistance to herbivores and pathogens: Ecology, evolution, and genetics*. University of Chicago Press, Chicago.

Fry, C. H. 1984. *The bee-eaters*. T. & A. D. Poyser, London.

Futuyma, D. J. 1970. Variation in genetic response to interspecific competition in laboratory populations of *Drosophila*. *American Naturalist* 104:239–52.

————. 1979. *Evolutionary biology*. Sinauer Associates, Sunderland, Mass.

————. 1986. Reflections on reflections: Ecology and evolutionary biology. *Journal of the History of Biology* 19:303–12.

————. 1991. Evolution of host specificity in herbivorous insects: Genetic, ecological, and phylogenetic aspects. Pages 431–54 in P. W. Price, T. M. Lewinsohn, G. W. Fernandes, and W. W. Benson, eds., *Plant-animal interactions: Evolutionary ecology in tropical and temperate regions*. Wiley, New York.

Futuyma, D. J., R. P. Cort, and I. van Noordwijk. 1984. Adaptation to host plants in the fall cankerworm (*Alsophila pometaria*) and its bearing on the evolution of host affiliation in phytophagous insects. *American Naturalist* 123:287–96.

Futuyma, D. J., and S. S. McCafferty. 1990. Phylogeny and the evolution of host plant associations in the leaf beetle genus *Ophraella* (Coleoptera, Chrysomelidae). *Evolution* 44:1885–1913.

Futuyma, D. J., and G. Moreno. 1988. The evolution of ecological specialization. *Annual Review of Ecology and Systematics* 19:207–33.

Futuyma, D. J., and T. E. Philippi. 1987. Genetic variation and covariation in responses to host plants by *Alsophila pometaria* (Lepidoptera: Geometridae). *Evolution* 41:269–79.

Futuyma, D. J., and M. Slatkin, eds. 1983. *Coevolution*. Sinauer Associates, Sunderland, Mass.

Gaffney, P. M. 1975. Roots of the niche concept. *American Naturalist* 109:490.

Gagné, R. J. 1989. *The plant-feeding gall midges of North America*. Cornell University Press, Ithaca, N.Y.

Gaston, K. J., and D. Reavey. 1989. Patterns in the life histories and feeding strategies of British macrolepidoptera. *Biological Journal of the Linnean Society* 37:367–81.

Gaston, K. J., D. Reavey, and G. R. Valladares. 1992. Intimacy and fidelity: Internal and external feeding by the British microlepidoptera. *Ecological Entomology* 17:86–88.

Gauld, I. D. 1986a. Latitudinal gradients in ichneumonid species richness in Australia. *Ecological Entomology* 11:155–61.

———. 1986b. Taxonomy, its limitations and its role in understanding parasitoid biology. Pages 1–21 in J. Waage and D. Greathead, eds., *Insect parasitoids*. Academic, New York.

Gautier-Hion, A., and G. Michaloud. 1989. Are figs always keystone resources for tropical frugivorous vertebrates? A test in Gabon. *Ecology* 70:1826–33.

Gebhardt, C., and F. Salamini. 1992. Restriction fragment length polymorphism analysis of plant genomes and its application to plant breeding. *International Review of Cytology* 135:201–37.

Gerritsen, J., and J. R. Strickler. 1977. Encounter probabilities and community structure in zooplankton: A mathematical model. *Journal of the Fisheries Research Board of Canada*. 34:73–82.

Ghiselin, M. T. 1984. Pages xi–xix in C. Darwin, *The various contrivances by which orchids are fertilized by insects,* 2d ed., 1877. Reprint. University of Chicago Press, Chicago.

Gilbert, L. E. 1979. Development of theory in the analysis of insect-plant interactions. Pages 117–54 in D. J. Horn, R. D. Mitchell, and G. R. Stairs, eds., *Analysis of ecological systems*. Ohio State University Press, Columbus.

———. 1980. Food web organization and the conservation of neotropical diversity. Pages 11–33 in M. E. Soulé and B. A. Wilcox, eds., *Conservation biology: An evolutionary-ecological perspective*. Sinauer Associates, Sunderland, Mass.

———. 1983. Coevolution and mimicry. Pages 263–81 in D. J. Futuyma and M. Slatkin, eds., *Coevolution*. Sinauer Associates, Sunderland, Mass.

———. 1984. The biology of butterfly communities. Pages 41–54 in R. Vane-Wright and P. Ackery, eds., *The biology of butterflies*. Academic, New York.

———. 1991. Biodiversity of a Central American *Heliconius* community: Pattern, process, and problems. Pages 403–27 in P. W. Price, T. M. Lewinsohn, G. W. Fernandes, and W. W. Benson, eds., *Plant-animal interactions: Evolutionary ecology in tropical and temperate regions*. Wiley, New York.

Gilbert, L. E., and P. H. Raven, eds. 1975. *Coevolution of animals and plants*. University of Texas Press, Austin.

Gill, D. E. 1972. Intrinsic rates of increase, saturation densities and competitive ability. I. An experiment with *Paramecium*. *American Naturalist* 106:461–71.

―――. 1974. Intrinsic rate of increase, saturation density, and competitive ability. II. The evolution of competitive ability. *American Naturalist* 108:103–16.

―――. 1978. The metapopulation ecology of the red-spotted newt, *Notophthalmus viridescens* (Rafinesque). *Ecological Monographs* 48:145–66.

Gill, D. E., and B. A. Mock. 1985. Ecological and evolutionary dynamics of parasites: The case of *Trypanosoma diemyctyli* in the red-spotted newt *Notophthalmus viridescens*. Pages 157–83 in D. Rollinson and R. M. Anderson, eds., *Ecology and genetics of host-parasite interactions.* Academic, New York.

Gilpin, M. E., and I. Hanski. 1991. *Metapopulation dynamics: Empirical and theoretical investigations.* Academic, New York.

Gochfeld, M., and J. Burger. 1984. Age differences in foraging behaviour of the American robin (*Turdus migratorius*). *Behaviour* 88:227–39.

Goldthwaite, R. O., R. G. Coss, and D. H. Owings. 1990. Evolutionary dissipation of an antisnake system: Differential behavior by California and Arctic ground squirrels in above- and below-ground contexts. *Behaviour* 112:246–69.

Gómez, J. M., and R. Zamora. 1992. Pollination by ants: Consequences of the quantitative effects on a mutualistic system. *Oecologia* 91:410–18.

Gould, F. 1983. Genetics of plant-herbivore systems: Interactions between applied and basic studies. Pages 599–653 in R. F. Denno and M. S. McClure, eds., *Variable plants and herbivores in natural and managed systems.* Academic, New York.

Gould, S. J. 1977. *Ontogeny and phylogeny.* Harvard University Press, Cambridge.

―――. 1980a. G. G. Simpson, paleontology, and the modern synthesis. Pages 153–72 in E. Mayr and W. B. Provine, eds., *The evolutionary synthesis: Perspectives on the unification of biology.* Harvard University Press, Cambridge.

―――. 1980b. *The panda's thumb: More reflections in natural history.* Norton, New York.

―――. 1989. *Wonderful life: The Burgess Shale and the nature of history.* Norton, New York.

Gould, S. J., and R. C. Lewontin. 1979. The spandrals of San Marco and the Panglossian paradigm: A critique of the adaptationist programme. *Proceedings of the Royal Society of London B* 205:581–98.

Govers, F., J.-P. Nap, A. Van Kammen, and T. Bisseling. 1987. Nodulins in the developing root nodule. *Plant Physiology and Biochemistry* 25:309–22.

Grant, B. R., and P. R. Grant. 1989. *Evolutionary dynamics of a natural population: The large cactus finch of the Galápagos.* University of Chicago Press, Chicago.

Grant, P. R. 1972. Convergent and divergent character displacement. *Biological Journal of the Linnean Society* 4:39–68.

―――. 1986. *Ecology and evolution of Darwin's finches.* Princeton University Press, Princeton.

Grant, V. 1983. The systematic and geographical distribution of hawkmoth flowers in the temperate North American flora. *Botanical Gazette* 144:439–49.

Grant, V., and K. A. Grant. 1965. *Flower pollination in the phlox family.* Columbia University Press, New York.

Greber, R. S. 1984. Leafhopper and aphid-borne viruses affecting subtropical cereal and grass crops. Pages 141–83 in K. F. Harris, ed., *Current topics in vector research,* volume II. Praeger, New York.

Green, R. E., M. R. W. Rands, and S. J. Moreby. 1987. Species differences in diet and the development of seed digestion in partridge chicks *Perdix perdix* and *Alectoris rufa. Ibis* 129:511–14.

Greenwood, P. H. 1965. Environmental effects on the pharyngeal gill of a cichlid fish, *Astatoreochromis alluaudi,* and their taxonomic implications. *Proceedings of the Linnean Society of London* 176:1–10.

Grinnell, J. 1917. The niche relationships of the California thrasher. *Auk* 34:427–33.

Grudzien, T. A., and B. J. Turner. 1984. Direct evidence that the *Ilyodon* morphs are a single biological species. *Evolution* 38:402–7.

Guégan, J.-F., A. Lambert, C. Lévêque, C. Combes, and L. Euzet. 1992. Can host body size explain the parasite species richness in tropical freshwater fishes? *Oecologia* 90:197–204.

Haber, W. A., and G. W. Frankie. 1989. A tropical hawkmoth community: Costa Rican dry forest Sphingidae. *Biotropica* 21:155–72.

Hagen, R. H., R. C. Lederhouse, J. L. Bossart, and J. M. Scriber. 1991. *Papilio canadensis* and *P. glaucus* (Papilionidae) are distinct species. *Journal of the Lepidopterists' Society* 45:245–58.

Hagen, R. H., and J. M. Scriber. 1994. Sex chromosomes and speciation in tiger swallowtails. In J. M. Scriber, Y. Tsubaki, and R. C. Lederhouse, eds., *The ecology and evolutionary biology of the Papilionidae.* In press.

Hairston, N. G., Sr., K. C. Nishikawa, and S. L. Stenhouse. 1987. The evolution of competing species of terrestrial salamanders: Niche partitioning or interference? *Evolutionary Ecology* 1:247–62.

Hairston, N. G., Sr., R. H. Wiley, C. K. Smith, and K. A. Kneidel. 1992. The dynamics of two hybrid zones in Appalachian salamanders of the genus *Plethodon. Evolution* 46:930–38.

Hajek, A. E., R. A. Humber, S. R. A. Walsh, and J. C. Silver. 1991. Sympatric occurrence of two *Entomophaga aulicae* (Zygomycetes: Entomophthorales) complex species attacking forest Lepidoptera. *Journal of Invertebrate Pathology* 58:373–80.

Haldane, J. B. S. 1949. Disease and evolution. *La Ricerca Scientifica* (suppl.) 19:68–76.

Halford, D. A., D. T. Bell, and W. A. Loneragan. 1984. Diet of the western grey kangaroo (*Macropus fulginosus* Desm.) in a mixed pasture-woodland habitat of Western Australia. *Journal of the Royal Society of Western Australia* 66:119–28.

Hall, J. G. 1981. A field study of the Kaibab squirrel in Grand Canyon National Park. *Wildlife Monographs* 75:1–54.

Hamilton, K. G. A. 1990. Homoptera. In D. A. Grimaldi, ed., Insects from the Santana Formation, Lower Cretaceous, of Brazil. *Bulletin of the American Museum of Natural History* 195:82–122.

Hamilton, W. D. 1964. The genetical evolution of social behaviour. I and II. *Journal of Theoretical Biology* 7:1–52.

Hamilton, W. D., R. Axelrod, and R. Tanese. 1990. Sexual reproduction as an adaptation to resist parasites (a review). *Proceedings of the National Academy of Sciences, USA* 87:3566–73.

Hannemann, H. J. 1953. Natürliche Gruppierung der europäischen Arten der Gattung *Depressaria* s. l. (Lep. Oecoph.). *Mitteilungen aus dem Zoologischen Museum in Berlin* 29:269–373.

Hansen, M. C., and R. H. Mapes. 1990. A predator-prey relationship between sharks and cephalopods in the late Paleozoic. Pages 189–92 in A. J. Boucot, ed., *Evolutionary paleobiology of behavior and coevolution.* Elsevier, New York.

Hanski, I. 1989. Fungivory: fungi, insects, and ecology. Pages 25–68 in N. Wilding, N. M. Collins, P. M. Hammond, and J. F. Webber, eds., *Insect-fungus interactions.* Academic, New York.

———. 1992. Insectivorous mammals. Pages 163–87 in M. J. Crawley, ed., *Natural enemies: The population biology of predators, parasites and diseases.* Blackwell, Oxford.

Hardy, A. C. 1954. Escape from specialization. Pages 122–42 in J. Huxley, A. C. Hardy, and E. B. Ford, eds., *Evolution as a process.* George Allen & Unwin, London.

Hare, J. D. 1990. Ecology and management of the Colorado potato beetle. *Annual Review of Entomology* 35:81–100.

Hare, J. D., and G. G. Kennedy. 1986. Genetic variation in plant-insect associations: Survival of *Leptinotarsa decemlineata* populations on *Solanum carolinense. Evolution* 40:1031–43.

Harley, J. L. 1989. The significance of mycorrhiza. *Mycological Research* 92:129–39.

Harper, J. L. 1967. A Darwinian approach to plant ecology. *Journal of Ecology* 55:247–70.

Harrison, R. G. 1990. Hybrid zones: Windows on evolutionary process. *Oxford Surveys in Evolutionary Biology* 7:69–128.

Harrison, S., D. D. Murphy, and P. R. Ehrlich. 1988. Distribution of the bay checkerspot butterfly, *Euphydryas editha bayensis:* Evidence for a metapopulation model. *American Naturalist* 132:360–82.

Harvell, C. D. 1986. The ecology and evolution of inducible defenses in marine bryozoans: Cues, costs, and consequences. *American Naturalist* 128:810–23.

———. 1990a. The ecology and evolution of inducible defenses. *Quarterly Review of Biology* 65:323–40.

———. 1990b. The evolution of inducible defence. *Parasitology* 100 (Suppl.):53–62.

———. 1992. Inducible defenses and allocation shift in a marine bryozoan. *Ecology* 73:1567–76.

Hassell, M. P., H. N. Comins, and R. M. May. 1991. Spatial structure and chaos in insect population dynamics. *Nature* 353:255–58.

Havel, J. E. 1985. Cyclomorphosis of *Daphnia pulex* spined morphs. *Limnology and Oceanography* 30:853–61.

Hawkins, B. A. 1990. Global patterns of parasitoid assemblage size. *Journal of Animal Ecology* 59:57–72.

Hawksworth, D. L. 1988. The variety of fungal-algal symbioses, their evolutionary significance, and the nature of lichens. *Botanical Journal of the Linnean Society* 96:3–20.

———. 1991. The fungal dimension of biodiversity: Magnitude, significance, and conservation. *Mycological Research* 95:641–55.

Hay, M. E., J. E. Duffy, and W. Fenical. 1990. Host-plant specialization decreases predation on a marine amphipod: An herbivore in plant's clothing. *Ecology* 71:733–43.

Hay, M. E., J. E. Duffy, V. J. Paul, P. E. Renaud, and W. Fenical. 1990. Specialist herbivores reduce their susceptibility to predation by feeding on the chemically defended seaweed *Avrainvillea longicaulis. Limnology and Oceanography* 35:1734–43.

Hay, M. E., J. R. Pawlik, J. E. Duffy, and W. Fenical. 1989. Seaweed-herbivore-predator interactions: Host-plant specialization reduces predation on small herbivores. *Oecologia* 81:418–27.

Heie, O. E., and W. L. Friedrich. 1990. The hickory aphid from the Iceland Miocene. Pages 104–7 in A. J. Boucot, ed., *Evolutionary paleobiology of behavior and coevolution.* Elsevier, New York.

Heinrich, B. 1979. 'Majoring' and 'minoring' by foraging bumblebees, *Bombus vagans:* An experimental analysis. *Ecology* 60:245–55.

Heinrich, B., and P. H. Raven. 1972. Energetics and pollination ecology. *Science* 176:597–602.

Henderson, R. W. 1984. The diets of Hispaniolan colubrid snakes. I. Introduction and prey genera. *Oecologia* 62:234–39.

Henderson, R. W., and A. Schwartz. 1986. The diet of the Hispaniolan colubrid snake, *Darlingtonia haetiana. Copeia* 1986:529–31.

Hennig, W. 1966. *Phylogenetic systematics.* University of Illinois Press, Urbana.

Herre, E. A. 1989. Coevolution of reproductive characteristics in 12 species of New World figs and their pollinator wasps. *Experientia* 45:637–47.

———. 1993. Population structure and the evolution of virulence in nematode parasites of fig wasps. *Science* 259:1442–45.

Herrera, C. M. 1981. Are tropical fruits more rewarding to dispersers than temperate ones? *American Naturalist* 118:896–907.

———. 1984. A study of avian frugivores, bird-dispersed plants, and their interaction in Mediterranean scrublands. *Ecological Monographs.* 54:1–23.

————. 1985. Determinants of plant-animal coevolution: The case of mutualistic dispersal of seeds by vertebrates. *Oikos* 44:132–41.

————. 1987a. Components of pollinator 'quality': Comparative analysis of a diverse insect assemblage. *Oikos* 50:79–90.

————. 1987b. Vertebrate-dispersed plants of the Iberian Peninsula: A study of fruit characteristics. *Ecological Monographs* 57:305–31.

————. 1988. Variation in mutualisms: The spatio-temporal mosaic of a pollinator assemblage. *Biological Journal of the Linnean Society* 35:95–125.

Hindar, K., N. Ryman, and G. Ståhl. 1986. Genetic differentiation among local populations and morphotypes of Arctic charr, *Salvelinus alpinus. Biological Journal of the Linnean Society* 27:269–85.

Hindell, M. A., K. A. Handasyde (née Lithgow), and A. K. Lee. 1985. Tree species selection by free-ranging koala populations in Victoria. *Australian Wildlife Research* 12:137–44.

Hodges, R. W. 1974. Gelechioidea; Oecophoridae (in part). Fasc. 6.2 in R. B. Dominick et al., eds., *The moths of America north of Mexico.* E. W. Classey, London.

Hodkinson, I. D., and D. Casson. 1991. A lesser predilection for bugs: Hemiptera (Insecta) diversity in tropical rain forests. *Biological Journal of the Linnean Society* 43:101–9.

Hofbauer, J., P. Schuster, and K. Sigmund. 1979. A note on evolutionarily stable strategies and game dynamics. *Journal of Theoretical Biology* 81:609–12.

Hoffmann, A. A. 1988. Partial cytoplasmic incompatibility between two Australian populations of *Drosophila melanogaster. Entomologia Experimentalis et Applicata* 48:61–67.

Hoffmann, A. A., M. Turelli, and G. M. Simmons. 1986. Undirectional incompatibility between populations of *Drosophila simulans. Evolution* 40:692–701.

Hölldobler, B., and E. O. Wilson. 1990. *The ants.* Belknap Press, Harvard University Press, Cambridge.

Honda, K. 1986. Flavanone glycosides as oviposition stimulants in a papilionid butterfly, *Papilio protenor. Journal of Chemical Ecology* 12:1999–2010.

Honma, M. A., and F. M. Ausubel. 1987. *Rhizobium meliloti* has three functional copies of the *nodD* symbiotic regulatory gene. *Proceedings of the National Academy of Sciences, USA* 84:8558–62.

Horvitz, C. C., and D. W. Schemske. 1984. Effects of ants and an ant-tended herbivore on seed production of a Neotropical herb. *Ecology* 65:1369–78.

————. 1988. A test of the pollinator limitation hypothesis for a Neotropical herb. *Ecology* 69:200–206.

Horvitz, C. C., C. Turnbull, and D. J. Harvey. 1987. Biology of the immature stages of *Eurybia elvina* (Lepidoptera: Riodinidae), a myrmecophilous metalmark butterfly. *Annals of the Entomological Society of America* 80:513–19.

Howe, H. F. 1984. Constraints on the evolution of mutualisms. *American Naturalist* 123:764–77.

————. 1993. Specialized and generalized dispersal systems: Where does 'the paradigm' stand? *Vegetatio* 107/108:3–13.

Howe, H. F., and J. Smallwood. 1982. Ecology of seed dispersal. *Annual Review of Ecology and Systematics* 13:201–28

Hsiao, C., and T. H. Hsiao. 1985. Rickettsia as the cause of cytoplasmic incompatibility in the alfalfa weevil, *Hypera postica. Journal of Invertebrate Zoology* 45:244–46.

Hsiao, T. H. 1978. Host plant adaptations among geographic populations of the Colorado potato beetle. *Entomologia Experimentalis et Applicata* 24:437–47.

————. 1988. Host specificity, seasonality and bionomics of *Leptinotarsa* beetles. Pages 581–99 in P. Jolivet, E. Petitipierre, and T. H. Hsiao, eds., *Biology of Chrysomelidae.* Kluwer, Dordrecht.

Hsiao, T. H., and C. Hsiao. 1985. Hybridization and cytoplasmic incompatibility among alfalfa weevil strains. *Entomologia Experimentalis et Applicata* 37:155–59.

Hugueny, B. 1989. West African rivers as biogeographic islands: Species richness of fish communities. *Oecologia* 79:236–43.

Hulbert, S. H., and J. L. Bennetzen. 1991. Recombination at the *Rp1* locus of maize. *Molecular and General Genetics* 226:377–82.

Hulbert, S. H., P. C. Lyons, and J. L. Bennetzen. 1991. Reactions of maize lines carrying *Rp* resistance genes to isolates of the common rust pathogen, *Puccinia sorghi*. *Plant Disease* 75: 1130–33.

Hulbert, S. H., M. A. Sudupak, and K. S. Hong. 1993. Genetic relationships between alleles of the *Rp1* rust resistance locus of maize. *Molecular Plant-Microbe Interactions* 6:387–92.

Hume, I. D. 1982. *Digestive physiology and nutrition of marsupials*. Cambridge University Press, Cambridge.

Hunt, R. S., and G. A. Van Sickle. 1984. Variation in susceptibility to sweet fern rust among *Pinus contorta* and *P. banksiana*. *Canadian Journal of Forest Research* 14:672–75.

Hutchinson, G. E. 1959. Homage to Santa Rosalia *or* Why are there so many kinds of animals? *American Naturalist* 93:145–59.

———. 1965. *The ecological theater and the evolutionary play*. Yale University Press, New Haven.

Huxley, C. R. 1978. The ant-plants *Myrmecodia* and *Hydnophytum* (Rubiaceae) and the relationships between their morphology, ant occupants, physiology and ecology. *New Phytologist* 80:231–68.

———. 1980. Symbiosis between ants and epiphytes. *Biological Reviews* 55:321–40.

———. 1986. Evolution of benevolent ant-plant relationships. Pages 257–82 in B. E. Juniper and Sir Richard Southwood, eds., *Insects and the plant surface*. Edward Arnold, London.

Huxley, J. 1942. *Evolution: The modern synthesis*. George Allen & Unwin, London.

Innis, G. J. 1989. Feeding ecology of fruit pigeons in subtropical rainforests of south-eastern Queensland. *Australian Wildlife Research* 16:365–94.

Irving-Bell, R. J. 1983. Cytoplasmic incompatibility within and between *Culex molestus* and *Cx. quinquefasciatus* (Diptera: Culicidae). *Journal of Medical Entomology* 20:44–48.

Istock, C. A. 1967. The evolution of complex life cycle phenomena: An ecological perspective. *Evolution* 21:592–605.

Jackson, R. R., and A. van Olphen. 1991. Prey-capture techniques and prey preferences of *Corythalia canosa* and *Pystira orbiculata*, ant-eating jumping spiders (Araneae, Salticidae). *Journal of Zoology, London* 223:577–91.

Jacob, F. 1977. Evolution as tinkering. *Science* 196:1161–66.

Jaenike, J. 1978. Resource predictability and niche breadth in the *Drosophila quinaria* species group. *Evolution* 32:676–78.

———. 1987. Genetics of oviposition-site preference in *Drosophila tripunctata*. *Heredity* 59:363–69.

———. 1989a. Genetic population structure of *Drosophila tripunctata:* patterns of variation and covariation of traits affecting resource use. *Evolution* 43:1467–82.

———. 1989b. Genetics of butterfly-host associations. *Trends in Ecology and Evolution* 4:34–35.

———. 1990. Host specialization in phytophagous insects. *Annual Review of Ecology and Systematics* 21:243–73.

———. 1993. Rapid evolution of host specificity in a parasitic nematode. *Evolutionary Ecology* 7:103–8.

Jaenike, J., and R. K. Selander. 1980. On the question of host races in the fall webworm, *Hyphantria cunea*. *Entomologia Experimentalis et Applicata* 27:31–37.

Janssen, R., J. Bakker, and F. J. Gommers. 1991. Mendelian proof for a gene-for-gene relationship between virulence of *Globodera rostochiensis* and the H_1 resistance gene in *Solanum tuberosum* ssp. *andigena* CPC 1673. *Revue Nématologie* 14:213–19.

Janzen, D. H. 1966. Coevolution of mutualism between ants and acacias in Central America. *Evolution* 20:249–75.

———. 1967a. Fire, vegetation structure, and the ant × acacia interaction in Central America. *Ecology* 48:26–35.

———. 1967b. Interaction of the bull's-horn acacia (*Acacia cornigera* L.) with an ant inhabitant (*Pseudomyrmex ferruginea* F. Smith) in eastern Mexico. *University of Kansas Science Bulletin* 47:315–558.

———. 1973. Dissolution of mutualism between *Cecropia* and its *Azteca* ants. *Biotropica* 5:15–28.

———. 1974a. Epiphytic myrmecophytes in Sarawak: Mutualism through the feeding of plants by ants. *Biotropica* 6:237–59.

———. 1974b. Swollen-thorn acacias of Central America. *Smithsonian Contributions to Botany* 13:1–131.

———. 1974c. Tropical blackwater rivers, animals, and mast fruiting by the Dipterocarpaceae. *Biotropica* 6:69–103.

———. 1975. *Pseudomyrmex nigropilosa:* A parasite of a mutualism. *Science* 188:936–37.

———. 1976. Why bamboos wait so long to flower. *Annual Review of Ecology and Systematics* 7:347–91.

———. 1977. Why fruits rot, seeds mold, and meat spoils. *American Naturalist* 111:691–713.

———. 1979. How to be a fig. *Annual Review of Ecology and Systematics* 10:13–51.

———. 1980. When is it coevolution? *Evolution* 34:611–12.

———. 1984a. Natural history of *Hylesia lineata* (Saturniidae: Hemileucinae) in Santa Rosa National Park, Costa Rica. *Journal of the Kansas Entomological Society* 57:490–514.

———. 1984b. Two ways to be a tropical big moth: Santa Rosa saturniids and sphingids. *Oxford Surveys in Evolutionary Biology* 1:85–140.

———. 1985. Coevolution as a process: What parasites of animals and plants do not have in common. Pages 83–99 in K. C. Kim, ed., *Coevolution of parasitic arthropods and mammals*. Wiley, New York.

———. 1988. Ecological characterization of a Costa Rican dry forest caterpillar fauna. *Biotropica* 20:120–35.

———. 1992. How to save tropical biodiversity. *American Entomologist* 37: 159–71.

Jarosz, A. M., and J. J. Burdon. 1991. Host-pathogen interactions in natural populations of *Linum marginale* and *Melampsora lini*. II. Local and regional variation in patterns of resistance and racial structure. *Evolution* 45:1618–27.

———. 1992. Host-pathogen interactions in natural populations of *Linum marginale* and *Melampsora lini*. III. Influence of pathogen epidemics on host survivorship and flower production. *Oecologia* 89:53–61.

Jebb, M. 1991. Cavity structure and function in the tuberous Rubiaceae. Pages 374–89 in C. R. Huxley and D. F. Cutler, eds., *Ant-plant interactions*. Oxford University Press, Oxford.

Jeffords, M. R., J. G. Sternburg, and G. P. Waldbauer. 1979. Batesian mimicry: Field demonstration of the survival value of pipevine swallowtail and monarch color patterns. *Evolution* 33:275–86.

Jeffords, M. R., G. P. Waldbauer, and J. G. Sternburg. 1980. Determination of the time of day at which diurnal moths painted to resemble butterflies are attacked by birds. *Evolution* 34:1205–11.

Jeon, K. W. 1972. Development of cellular dependence on infective organisms: Micrurgical studies in amoebas. *Science* 176:1122–23.

―――. 1991. Amoeba and x-Bacteria: Symbiont acquisition and possible species change. Pages 118–31 in L. Margulis and R. Fester, eds., *Symbiosis as a source of evolutionary innovation: Speciation and morphogenesis*. MIT Press, Cambridge.

Jeon, K. W., and M. S. Jeon. 1976. Endosymbiosis in amoebae: Recently established endosymbionts have become required cytoplasmic components. *Journal of Cellular Physiology* 89:337–44.

Jermy, T. 1976. Insect-host-plant relationship—co-evolution or sequential evolution? Pages 109–13 in T. Jermy, ed., *The host-plant in relation to insect behaviour and reproduction*. Plenum, New York.

―――. 1984. Evolution of insect/host plant relationships. *American Naturalist* 124:609–30.

―――. 1987. The role of experience in the host selection of phytophagous insects. Pages 143–57 in R. F. Chapman, E. A. Bernays, and J. G. Stoffolano, Jr., eds., *Perspectives in chemoreception and behavior*. Springer-Verlag, New York.

―――. 1993. Evolution of insect-plant relationships—a devil's advocate approach. *Entomologia Experimentalis et Applicata* 66:3–12.

John, A. M. 1990. Endemic disease in host populations with fully specified demography. *Theoretical Population Biology* 37:455–71.

Johnson, K. G., G. B. Schaller, and J. Hu. 1988. Comparative behavior of red and giant pandas in the Wolong Reserve, China. *Journal of Mammalogy* 69:552–64.

Johnson, R. A. 1986. Intraspecific resource partitioning in the bumble bees *Bombus ternarius* and *B. pennsylvanicus*. *Ecology* 67:133–38.

Jones, C. G., and J. S. Coleman. 1988. Plant stress and insect behavior: Cottonwood, ozone and the feeding and oviposition preference of a beetle. *Oecologia* 76:51–56.

Jones, C. G., and J. H. Lawton. 1991. Plant chemistry and insect species richness of British umbellifers. *Journal of Animal Ecology* 60:767–77.

Jordano, P. 1993. Geographical ecology and variation of plant-seed disperser interactions: Southern Spanish junipers and frugivorous thrushes. *Vegetatio* 107:85–104.

Joseph, L. 1982a. The glossy black-cockatoo on Kangaroo Island. *Emu* 82:46–49.

―――. 1982b. The red-tailed black-cockatoo in south-eastern Australia. *Emu* 82:42–45.

―――. 1986. Seed-eating birds of southern Australia. Pages 85–93 in H. A. Ford and D. C. Paton, eds., *The dynamic partnership: Birds and plants in southern Australia*. Handbook of the Flora and Fauna of South Australia Series. South Australian Government, South Australia.

Karban, R., A. K. Brody, and W. C. Schnathorst. 1989. Crowding and a plant's ability to defend itself against herbivores and diseases. *American Naturalist* 134:749–60.

Karban, R., and J. H. Myers. 1989. Induced plant responses to herbivory. *Annual Review of Ecology and Systematics* 20:331–48.

Kavanagh, R. P., and M. J. Lambert. 1990. Food selection by the greater glider, *Petauroides volans:* Is foliar nitrogen a determinant of habitat quality? *Australian Wildlife Research* 17:285–99.

Keen, N. T., S. Tamaki, D. Kobayashi, D. Gerhold, M. Stayton, H. Shen, S. Gold, J. Lorang, H. Thordal-Christensen, D. Dahlbeck, and D. Staskawicz. 1990. Bacteria expressing avirulence gene D produce a specific elicitor of the soybean hypersensitive reaction. *Molecular Plant-Microbe Interactions* 3:112–21.

Keith, J. O. 1965. The Abert squirrel and its dependence on ponderosa pine. *Ecology* 46:150–63.

Kellen, W. R., and D. F. Hoffman. 1981. *Wolbachia* sp. (Rickettsiales: Rickettsiaceae), a symbiont of the almond moth, *Ephestia cautella:* Ultrastructure and influence on host fertility. *Journal of Invertebrate Pathology* 37:273–83.

Kennedy, C. E. J. 1986. Attachment may be a basis for specialization in oak aphids. *Ecological Entomology* 11:291–300.

Kennedy, C. E. J., and T. R. E. Southwood. 1984. The number of species of insects associated with British trees: A re-analysis. *Journal of Animal Ecology* 53:455–78.

Kerfoot, W. C. 1977. Competition in cladoceran communities: The cost of evolving defenses against copepod predation. *Ecology* 58:303–13.

Kettlewell, H. B. D. 1955. Selection experiments on industrial melanism in the Lepidoptera. *Heredity* 9:323–42.

Khush, G. S., and D. S. Brar. 1991. Genetics of resistance to insects in crop plants. *Advances in Agronomy* 45:223–74.

Kiester, A. R., R. Lande, and D. W. Schemske. 1984. Models of coevolution and speciation in plants and their pollinators. *American Naturalist* 124:220–43.

Kingsland, S. E. 1985. *Modeling nature: Episodes in the history of population ecology.* University of Chicago Press, Chicago.

———. 1986. Mathematical figments, biological facts: Population ecology in the thirties. *Journal of the History of Biology* 19:235–56.

Kingsolver, J. G., and D. W. Schemske. 1991. Path analysis of selection. *Trends in Ecology and Evolution* 6:276–80.

Kirk, W. D. J. 1984. Pollen-feeding in thrips (Insecta: Thysanoptera). *Journal of Zoology, London* 204:107–17.

———. 1985. Pollen-feeding and the host specificity and fecundity of flower thrips (Thysanoptera). *Ecological Entomology* 10:281–89.

Kitting, C. L. 1980. Herbivore-plant interactions of individual limpets maintaining a mixed diet of intertidal marine algae. *Ecological Monographs* 50:527–50.

Klebenow, D. A., and G. M. Gray. 1968. Food habits of juvenile sage grouse. *Journal of Range Management* 1968:8–83.

Kleinfeldt, S. E. 1978. Ant-gardens: The interaction of *Codonanthe crassifolia* (Gesneriaceae) and *Crematogaster longispina* (Formicidae). *Ecology* 59:449–56.

Knox, A. G. 1992. Species and pseudospecies: The structure of crossbill populations. *Biological Journal of the Linnean Society* 47:325–35.

Koopman, K. F. 1988. Systematics and distribution. Pages 7–17 in A. M. Greenhall and U. Schmidt, eds., *Natural history of vampire bats.* CRC Press, Boca Raton, Fla.

Koptur, S., and J. H. Lawton. 1988. Interactions among vetches bearing extrafloral nectaries, their biotic protective agents, and herbivores. *Ecology* 69:278–83.

Kornfield, I., D. C. Smith, P. S. Gagnon, and J. N. Taylor. 1982. The cichlid fish of Cuatro Ciénegas, Mexico: Direct evidence of conspecificity among distinct trophic morphs. *Evolution* 36:658–64.

Krebs, J. R., and M. I. Avery. 1984. Chick growth and prey quality in the European bee-eater (*Merops apiaster*). *Oecologia* 64:363–68.

Kritsky, G. 1991. Darwin's Madagascan hawkmoth prediction. *American Entomologist* 37:206–10.

Kruuk, H., and W. A. Sands. 1972. The aardwolf (*Proteles cristatus* Sparrman 1783) as predator of termites. *East African Wildlife Journal* 10:211–27.

Lack, D. 1944. Correlation between beak and food in the crossbill, *Loxia curvirostra*. *Ibis* 86:552–53.

———. [1947] 1983. *Darwin's finches.* Reprint. Cambridge University Press, Cambridge.

———. 1954. *The natural regulation of animal numbers.* Oxford University Press, Oxford.

Laidlaw, H. H., Jr., and R. E. Page, Jr. 1984. Polyandry in honeybees (*Apis mellifera* L.): Sperm utilization and intracolony genetic relationships. *Genetics* 108:985–97.

Lambert, F. R. 1989a. Fig-eating by birds in a Malaysian lowland rain forest. *Journal of Tropical Ecology* 5:401–12.

————. 1989b. Pigeons as seed predators and dispersers of figs in a Malaysian lowland forest. *Ibis* 131:521–27.

Lambert, F. R., and A. G. Marshall. 1991. Keystone characteristics of bird-dispersed *Ficus* in a Malaysian lowland rain forest. *Journal of Ecology* 79:793–809.

Lance, D., and P. Barbosa. 1982. Host tree influences on the dispersal of late instar gypsy moths, *Lymantria dispar. Oikos* 38:1–7.

Lande, R. 1986. The dynamics of peak shifts and the pattern of morphological evolution. *Paleobiology* 12:343–54.

Lanyon, S. M. 1992. Interspecific brood parasitism in blackbirds (Icterinae): A phylogenetic perspective. *Science* 255:77–79.

Law, R. 1985. Evolution in a mutualistic environment. Pages 145–70 in D. H. Boucher, ed., *The biology of mutualism: Ecology and evolution.* Oxford University Press, New York.

Law, R., and D. H. Lewis. 1983. Biotic environments and the maintenance of sex—some evidence from mutualistic symbioses. *Biological Journal of the Linnean Society* 20:249–76.

Lawrence, G. J., and J. J. Burdon. 1989. Flax rust from *Linum marginale:* Variation in a natural host-pathogen interaction. *Canadian Journal of Botany* 67:3192–98.

Lawton, J. H., and M. P. Hassell. 1981. Asymmetrical competition in insects. *Nature* 289:793–95.

Lawton, J. H., and S. McNeill. 1979. Between the devil and the deep blue sea: On the problem of being a herbivore. *Symposia of the British Ecological Society* 20:223–44.

Lawton, J. H., and P. W. Price. 1979. Species richness of parasites on hosts: Agromyzid flies on the British Umbelliferae. *Journal of Animal Ecology* 48:619–37.

Lawton, J. H., and D. Schröder. 1977. Effects of plant type, size of geographical range and taxonomic isolation on number of insect species associated with British plants. *Nature* 265:137–40.

Leather, S. R. 1991. Feeding specialisation and host distribution of British and Finnish *Prunus*-feeding macrolepidoptera. *Oikos* 60:40–48.

Le Brun, N., F. Renaud, P. Berrebi, and A. Lambert. 1992. Hybrid zones and host-parasite relationships: Effect on the evolution of parasitic specificity. *Evolution* 46:56–61.

Legrand, J. J., P. Juchault, D. Moraga, and E. Legrand-Hamelin. 1986. Microorganismes symbiotiques et speciation. *Bulletin Société Zoologique de France* 111:135–47.

Leighton, M., and D. R. Leighton. 1983. Vertebrate responses to fruiting seasonality within a Bornean rain forest. Pages 181–96 in S. L. Sutton, T. C. Whitmore, and A. C. Chadwick, eds., *Tropical rain forest: Ecology and management.* Blackwell, Oxford.

Lenski, R. E. 1988a. Experimental studies of pleiotropy and epistasis in *Escherichia coli.* I. Variation in competitive fitness among mutants resistant to virus T4. *Evolution* 42:425–32.

————. 1988b. Experimental studies of pleiotropy and epistasis in *Escherichia coli.* II. Compensation for maladaptive effects associated with resistance to virus T4. *Evolution* 42:433–40.

Lenski, R. E., and B. R. Levin. 1985. Constraints on the coevolution of bacteria and virulent phage: A model, some experiments and predictions for natural communities. *American Naturalist* 125:585–602.

Levey, D. J., and A. Grajal. 1991. Evolutionary implications of fruit processing limitations in cedar waxwings. *American Naturalist* 138:171–89.

Levin, S. A. 1983. Some approaches to the modelling of coevolutionary interactions. Pages 21–65 in M. H. Nitecki, ed., *Coevolution.* University of Chicago Press, Chicago.

Levins, R. 1968. *Evolution in changing environments.* Princeton University Press, Princeton.

Lewis, S. M., and B. Kensley. 1982. Notes on the ecology and behaviour of *Pseudamphitoides incurvaria* (Just) (Crustacea, Amphipoda, Ampithoidae). *Journal of Natural History* 16:267–74.

Li, C. C. 1976. *First course in population genetics*. Boxwood Press, Pacific Grove, Calif.

Lincoln, D. E., D. Couvet, and N. Sionit. 1986. Response of an insect herbivore to host plants grown in carbon dioxide enriched atmospheres. *Oecologia* 69:556–60.

Lindroth, R. L., J. M. Scriber, and M. T. S. Hsia. 1988. Chemical ecology of the tiger swallowtail. *Ecology* 69:814–22.

Linsley, E. G., J. W. MacSwain, and P. H. Raven. 1963. Comparative behavior of bees and Onagraceae. I. *Oenothera* bees of the Colorado Desert. II. *Oenothera* Bees of the Great Basin. *University of California Publications in Entomology* 33:1–58.

Linsley, E. G., C. M. Rick, and S. G. Stephens. 1966. Observations on the floral relationships of the Galápagos carpenter bee (Hymenoptera: Apidae). *Pan-Pacific Entomologist* 42:1–18.

Little, M. D. 1985. Cestodes (tapeworms). Pages 110–126 in P. C. Beaver and R. C. Jung, eds., *Animal agents and vectors of human disease*. 5th ed. Lea & Febiger, Philadelphia.

Lively, C. M. 1986. Competition, comparative life histories, and maintenance of shell dimorphism in a barnacle. *Ecology* 67:858–64.

———. 1987. Evidence from a New Zealand snail for the maintenance of sex by parasitism. *Nature* 328:519–21.

———. 1989. Adaptation by a parasitic trematode to local populations of its snail host. *Evolution* 43:1663–71.

Lockwood, J. A., and L. D. DeBrey. 1990. A solution for the sudden and unexplained extinction of the Rocky Mountain grasshopper (Orthoptera: Acrididae). *Environmental Entomology* 19:1194–1205.

Lofdahl, K. L. 1987. A genetic analysis of habitat selection in the cactophilic species, *Drosophila mojavensis*. Pages 153–62 in M. D. Huettel, ed., *Evolutionary genetics of invertebrate behavior: Progress and prospects*. Plenum, New York.

Longino, J. T. 1991. *Azteca* ants in *Cecropia* trees: Taxonomy, colony structure, and behaviour. Pages 271–88 in C. R. Huxley and D. F. Cutler, eds., *Ant-plant interactions*. Oxford University Press, Oxford.

Lorch, I. J., and K. W. Jeon. 1980. Resuscitation of amebae deprived of essential symbiotes: Micrurgical studies. *Journal of Protozoology* 27:423–26.

Loria, D. L., and F. R. Moore. 1990. Energy demands of migration on red-eyed vireos, *Vireo olicaceus*. *Behavioral Ecology* 1:24–35.

Losos, J. B. 1992. A critical comparison of the taxon-cycle and character-displacement models for size evolution of *Anolis* lizards in the Lesser Antilles. *Copeia* 1992:279–88.

Lowry, J. B. 1989. Green-leaf fractionation by fruit bats: Is this feeding behaviour a unique nutritional strategy for herbivores? *Australian Wildlife Research* 16:203–6.

Lubin, Y. D., and G. Montgomery. 1981. Defenses of *Nasutitermes* termites (Isoptera, Termitidae) against *Tamandua* anteaters (Edentata, Myrmecophagidae). *Biotropica* 13:66–76.

Luig, N. H. 1983. A survey of virulence genes in wheat stem rust, *Puccinia graminis* f. sp. *tritici*. Pages 1–199 in *Advances in plant breeding*. Supplement 11 to *Journal of Plant Breeding*. 1–199. Verlag Paul Parey, Berlin.

Lüscher, A., J. Connolly, and P. Jacquard. 1992. Neighbour specificity between *Lolium perenne* and *Trifolium repens* from a natural pasture. *Oecologia* 91:404–9.

MacArthur, R. H. 1965. Ecological consequences of natural selection. Pages 388–97 in T. H. Waterman and H. J. Morowitz, ed., *Theoretical and mathematical biology*. Blaisdell, New York.

MacArthur, R. H., and R. Levins. 1967. The limiting similarity, convergence, and divergence of coexisting species. *American Naturalist* 101:377–85.

MacArthur, R. H., and E. R. Pianka. 1966. On optimal use of a patchy environment. *American Naturalist* 100:603–9.

MacArthur, R. H., and E. O. Wilson. 1967. *The theory of island biogeography.* Princeton University Press, Princeton.

McCarthy, A. M. 1990. Speciation of echinostomes: Evidence for the existence of two sympatric sibling species in the complex *Echinoparyphium recurvatum* (von Linstow 1873) (Digenea: Echinostomatidae). *Parasitology* 101:35–42.

McCauley, D. E. 1991. The effect of host plant patch size variation on the population structure of a specialist insect, *Tetraopes tetraophthalmus. Evolution* 45:1675–84.

McCrea, K. D., and W. G. Abrahamson. 1987. Variation in herbivore infestation: Historical vs. genetic factors. *Ecology* 68:822–27.

McIntosh, R. P. 1980. The background and some current problems of theoretical ecology. *Synthese* 43:195–255.

———. 1985. *The background of ecology: Concept and theory.* Cambridge University Press, Cambridge.

McIver, J., M. A. Djordjevic, J. J. Weinman, G. L. Bender, and B. G. Rolfe. 1989. Extension of host range of *Rhizobium leguminosarum* bv. *trifolii* caused by point mutations in *nodD* that result in alterations in regulatory function and recognition of inducer molecules. *Molecular Plant-Microbe Interactions* 2:97–106.

McKey, D. 1991. Phylogenetic analysis of the evolution of a mutualism: *Leonardoxa* (Caesalpiniaceae) and its associated ants. Pages 310–34 in C. R. Huxley and D. F. Cutler, eds., *Ant-plant interactions.* Oxford University Press, Oxford.

McKey, D. B., J. S. Gartlan, P. G. Waterman, and G. M. Choo. 1981. Food selection by black colobus monkeys (*Colobus satanas*) in relation to plant chemistry. *Biological Journal of the Linnean Society* 16:115–46.

McNab, B. K. 1984. Physiological convergence amongst ant-eating and termite-eating mammals. *Journal of Zoology, London* 203:485–510.

McNaughton, S. J. 1986. On plants and herbivores. *American Naturalist* 128:765–70.

McPheron, B. A., D. Courtney Smith, and S. H. Berlocher. 1988. Genetic differences between host races of *Rhagoletis pomonella. Nature* 336:64–66.

Maddox, G. D., and N. Cappuccino. 1986. Genetic determination of plant susceptibility to an herbivorous insect depends on environmental context. *Evolution* 40:863–66.

Maddox, G. D., and R. B. Root. 1987. Resistance to 16 diverse species of herbivorous insects within a population of goldenrod, *Solidago altissima:* Genetic variation and heritability. *Oecologia* 72:8–14.

Maggenti, A. R. 1983. Nematode higher classification as influenced by species and family concepts. Pages 25–40 in A. R. Stone, H. M. Platt, and L. F. Khalil, eds., *Concepts in nematode systematics.* Academic, New York.

Malek, E. A. 1985. Blood flukes or schistosomes. Pages 97–109 in P. C. Beaver and R. C. Jung, eds., *Animal agents and vectors of human disease.* 5th ed. Lea & Febiger, Philadelphia.

Mallet, J. 1989. The genetics of warning colour in Peruvian hybrid zones of *Heliconius erato* and *H. melpomene. Proceedings of the Royal Society of London B* 236:163–85.

Mallet, J., and N. H. Barton. 1989. Strong natural selection in a warning-color hybrid zone. *Evolution* 43:421–31.

Mallet, J., N. Barton, G. M. Lamas, J. Santisteban C., M. Muedas M., and H. Eeley. 1990. Estimates of selection and gene flow from measures of cline width and linkage disequilibrium in *Heliconius* hybrid zones. *Genetics* 124:921–36.

Malmborg, P. K., and M. F. Willson. 1988. Foraging ecology of avian frugivores and some consequences for seed dispersal in an Illinois woodlot. *Condor* 90:173–86.

Malmquist, H. J. 1992. Phenotype-specific feeding behaviour of two Arctic charr *Salvelinus alpinus* morphs. *Oecologia* 92:354–61.

Malmquist, H. J., S. S. Snorrason, S. Skúlason, B. Jonsson, O. T. Sandlund, and P. M. Jónasson. 1992. Diet differentiation in polymorphic Arctic charr in Thingvallavatn, Iceland. *Journal of Animal Ecology* 61:21–35.

Mangel, M., and C. W. Clark. 1988. *Dynamic modeling in behavioral ecology.* Princeton University Press, Princeton.

Margulis, L., and D. Bermudes. 1985. Symbiosis as a mechanism of evolution: Status of the cell symbiosis theory. *Symbiosis* 1:101–24.

Margulis, L., and R. Fester, ed. 1991. *Symbiosis as a source of evolutionary innovation: Speciation and morphogenesis.* MIT Press, Cambridge.

Marquis, R. J. 1991. Herbivore fauna of *Piper* (Piperaceae) in a Costa Rican wet forest: Diversity, specificity, and impact. Pages 179–208 in P. W. Price, T. M. Lewinsohn, G. W. Fernandes, and W. W. Benson, eds., *Plant-animal interactions: Evolutionary ecology in tropical and temperate regions.* Wiley, New York.

Marshall, G. A. K. 1908. On diaposematism, with reference to some limitations of the Müllerian hypothesis of mimicry. *Transactions of the Entomological Society of London* 1908:93–142.

Martin, M. M. 1992. The evolution of insect-fungal associations: From contact to stable symbiosis. *American Zoologist* 32:593–605.

Martínez del Rio, C. 1990a. Dietary, phylogenetic, and ecological correlates of intestinal sucrase and maltase activity in birds. *Physiological Zoology* 63:987–1011.

———. 1990b. Sugar preferences in hummingbirds: The influence of subtle chemical differences on food choice. *Condor* 92:1022–30.

Martínez del Rio, C., H. G. Baker, and I. Baker. 1992. Ecological and evolutionary implications of digestive processes: Bird preferences and the sugar constituents of floral nectar and fruit pulp. *Experientia* 48:544–51.

Martínez del Rio, C., B. R. Stevens, D. Daneke, and P. T. Andreadis. 1988. Physiological correlates of preference and aversion for sugars in three species of birds. *Physiological Zoology* 61:222–29.

Marzluff, J. M., and R. P. Balda. 1992. *The pinyon jay: Behavioral ecology of a colonial and cooperative corvid.* T. & A. D. Poyser, London.

Mason, P. 1986. Brood parasitism in a host generalist, the shining cowbird. I. The quality of different species of hosts. *Auk* 103:52–60.

Mathias, M. E. 1965. Distribution patterns of certain Umbelliferae. *Annals of the Missouri Botanical Garden* 52:387–98.

Mathias, M. E., and L. Constance. 1965. A revision of the genus *Bowlesia* Ruiz & Pav. (Umbelliferae-Hydrocolyloideae) and its relatives. *University of California Publications in Botany* 38:1–73.

May, R. M. 1985. Host-parasite associations: Their population biology and population genetics. Pages 243–62 in D. Rollinson and R. M. Anderson, eds., *Ecology and genetics of host-parasite interactions.* Academic, New York.

———. 1990. How many species? *Philosophical Transactions of the Royal Society of London B* 330:293–304.

———. 1991. A fondness for fungi. *Nature* 352:475–76.

May, R. M., and R. M. Anderson. 1990. Parasite-host coevolution. *Parasitology* 100:S89–S101.

Maynard Smith, J. 1964. Group selection and kin selection. *Nature* 201:1145–47.

Mayr, E. 1942. *Systematics and the origin of species.* Columbia University Press, New York.

———. 1959. The emergence of evolutionary novelties. Pages 349–80 in S. Tax, ed., *Evolution after Darwin,* volume 1. University of Chicago Press, Chicago.

———. 1963. *Animal species and evolution.* Harvard University Press, Cambridge.

Mayr, E., and A. L. Rand. 1937. Results of the Archbold Expeditions, no. 14: The birds of the

1933–1934 Papuan expedition. *Bulletin of the American Museum of Natural History* 73:1–248.

Menken, S. B. J., W. M. Herrebout, and J. T. Wiebes. 1992. Small ermine moths (*Yponomeuta*): Their host relations and evolution. *Annual Review of Entomology* 37:41–66.

Merrick, M. J. 1992. Regulation of nitrogen fixation genes in free-living and symbiotic bacteria. Pages 835–76 in G. Stacey, R. H. Burris, and H. J. Evans, eds., *Biological nitrogen fixation*. Chapman & Hall, New York.

Meyer, A. 1987. Phenotypic plasticity and heterochrony in *Cichlasoma managuense* (Pisces, Cichlidae) and their implications for speciation in cichlid fishes. *Evolution* 41:1357–69.

———. 1989. Cost of morphological specialization: Feeding performance of the two morphs in the trophically polymorphic cichlid fish, *Cichlasoma citrinellum*. *Oecologia* 80:431–36.

———. 1990a. Ecological and evolutionary consequences of the trophic polymorphism in *Cichlasoma citrinellum* (Pisces: Cichlidae). *Biological Journal of the Linnean Society* 39:279–99.

———. 1990b. Morphometrics and allometry in the trophically polymorphic cichlid fish *Cichlasoma citrinellum:* Alternative adaptations and autogenetic changes in shape. *Journal of Zoology, London* 221:237–60.

Michener, C. D. 1981. Classification of the bee family Melittidae, with a review of species of Meganomiinae. *Contributions of the American Entomological Institute* 18(3):1–135.

Milbrath, L. R., M. J. Tauber, and C. A. Tauber. 1993. Prey specificity in *Chrysopa:* An interspecific comparison of larval feeding and defensive behavior. *Ecology* 74:1384–93.

Miles, N. J. 1983. Variation and host specificity in the yucca moth, *Tegeticula yuccasella* (Incurvariidae): A morphometric approach. *Journal of the Lepidopterists' Society* 37:207–16.

Miller, J. S. 1987a. Host plant associations in the Papilionidae: Parallel cladogenesis or colonization? *Cladistics* 3:105–20.

———. 1987b. Phylogenetic studies in the Papilioninae (Lepidoptera: Papilionidae). *Bulletin of the American Museum of Natural History* 186:367–512.

Mitchison, N. A. 1990. The evolution of acquired immunity to parasites. *Parasitology* 100 (suppl.): 27–34.

Mitman, G. 1988. From the population to society: The cooperative metaphors of W. C. Allee and A. E. Emerson. *Journal of the History of Biology* 21:173–94.

Mitter, C., B. Farrell, and B. Wiegmann. 1988. The phylogenetic study of adaptive zones: Has phytophagy promoted insect diversification? *American Naturalist* 132:107–28.

Mock, B. A., and D. E. Gill. 1984. The infrapopulation dynamics of trypanosomes in red-spotted newts. *Parasitology* 88:267–82.

Mode, C. J. 1958. A mathematical model for the co-evolution of obligate parasites and their hosts. *Evolution* 12:158–65.

Modi, R. I., and J. Adams. 1991. Coevolution in bacterial-plasmid populations. *Evolution* 45:656–67.

Moksnes, A., E. Røskaft, and A. T. Braa. 1991. Rejection behavior by common cuckoo hosts towards artificial brood parasite eggs. *Auk* 108:348–54.

Montague, J. R., and J. Jaenike. 1985. Nematode parasitism in natural populations of mycophagous drosophilids. *Ecology* 66:624–26.

Montgomery, S. L. 1983. Biogeography of the moth genus *Eupithecia* in Oceania and the evolution of ambush predation in Hawaiian caterpillars (Lepidoptera: Geometridae). *Entomologia generalis* 8:27–34.

Moore, F. R., and P. A. Simm. 1985. Migratory disposition and choice of diet by the yellow-rumped warbler (*Dendroica coronata*). *Auk* 102:820–26.

Mopper, S., J. B. Mitton, T. G. Whitham, N. S. Cobb, and K. M. Christensen. 1991. Genetic differentiation and heterozygosity in pinyon pine associated with resistance to herbivory and environmental stress. *Evolution* 45:989–99.

Mopper, S., T. G. Whitham, and P. W. Price. 1990. Plant phenotype and interspecific competition between insects determine sawfly performance and density. *Ecology* 71:2135–44.

Moran, N. A. 1988. The evolution of host-plant alternation in aphids: evidence that specialization is a dead end. *American Naturalist* 132:681–706.

———. 1989. A 48-million-year-old aphid-host plant association and complex life cycle: Biogeographic evidence. *Science* 245:173–75.

———. 1991. Phenotype fixation and genotypic diversity in the complex life cycle of the aphid *Pemphigus betae*. *Evolution* 45:957–70.

———. 1992. The evolution of aphid life cycles. *Annual Review of Entomology* 37:321–48.

Moran, N. A., and T. G. Whitham. 1988. Evolutionary reduction of complex life cycles: Loss of host alternation in *Pemphigus* (Homoptera: Aphididae). *Evolution* 42:717–28.

Mousseau, T. A., and H. Dingle. 1991. Maternal effects in insect life histories. *Annual Review of Entomology* 36:511–34.

Müller, F. 1879. *Ituna* and *Thyridia;* a remarkable case of mimicry in butterflies. *Proceedings of the Entomological Society of London* 1879:xx–xxix.

Müller, H. 1873–1877. On the fertilisation of flowers by insects and on the reciprocal adaptations of both. *Nature* 8:187–89, 205–6, 433–35; 9:44–46, 164–66; 10:129–30; 11:32–33, 110–12, 169–71; 12:50–51, 190–91; 13:210–12, 289–92; 14:173–75; 15:317–19, 473–75; 16:507–9.

———. 1883. *The fertilisation of flowers.* Translated by D'Arcy W. Thompson. Macmillan & Co., London.

Mulvey, M., J. M. Aho, C. Lydeard, P. L. Leberg, and M. H. Smith. 1991. Comparative population genetic structure of a parasite (*Fascioloides magna*) and its definitive host. *Evolution* 45:1628–40.

Nault, L. R. 1980. Maize bushy stunt and corn stunt: A comparison of disease symptoms, pathogen host ranges and vectors. *Phytopathology* 70:659–62.

Newman, E. I., and P. Reddell. 1987. The distribution of mycorrhizas among families of vascular plants. *New Phytologist* 106:745–51.

Ng, D. 1988. A novel level of interactions in plant-insect systems. *Nature* 334:611–13.

Nicholson, A. J. 1927. A new theory of mimicry in insects. *Australian Zoologist* 5:10–104.

Nielsen, E. S., and D. R. Davis. 1985. The first Southern Hemisphere prodoxid and the phylogeny of the Incurvariodea (Lepidoptera). *Systematic Entomology* 10:307–22.

Nijhout, H. F., G. A. Wray, and L. E. Gilbert. 1990. An analysis of the phenotypic effects of certain colour pattern genes in *Heliconius* (Lepidoptera: Nymphalidae). *Biological Journal of the Linnean Society* 40:357–72.

Nilsson, L. A. 1988. The evolution of flowers with deep corolla tubes. *Nature* 334:147–49.

Nilsson, L. A., L. Jonsson, L. Ralison, and E. Randrianjohany. 1985. Monophily and pollination mechanisms in *Angraecum arachnites* Schltr. (Orchidaceae) in a guild of long-tongued hawkmoths (Sphingidae) in Madagascar. *Biological Journal of the Linnean Society* 26:1–19.

———. 1987. Angraecoid orchids and hawkmoths in central Madagascar: Specialized pollination systems and generalist foragers. *Biotropica* 19:310–18.

Nishida, R., T. Ohsugi, S. Kokubo, and H. Fukami. 1987. Oviposition stimulants of a *Citrus*-feeding swallowtail butterfly, *Papilio xuthus* L. *Experientia* 43:342–44.

Nitao, J. K., M. P. Ayres, R. C. Lederhouse, and J. M. Scriber. 1991. Larval adaptation to lauraceous hosts: Geographic divergence in the spicebush swallowtail butterfly. *Ecology* 72:1428–35.

Nitecki, M. H., ed. 1983. *Coevolution.* University of Chicago Press, Chicago.

Noda, H. 1984. Cytoplasmic incompatibility in a rice planthopper. *Journal of Heredity* 75:345–48.

Nordeng, H. 1983. Solution to the "char problem" based on Arctic char (*Salvelinus alpinus*) in Norway. *Canadian Journal of Fisheries and Aquatic Sciences* 40:1372–87.

Nyström, K. G. K., O. Pehrsson, and D. Broman. 1991. Food of juvenile common eiders (*Somateria molissima*) in areas of high and low salinity. *Auk* 108:250–56.

Oates, J. D., J. J. Burdon, and J. B. Brouwer. 1983. Interactions between *Avena* and *Puccinia* species. II. The pathogens: *Puccinia coronata* Cda and *P. graminis* Pers. f. sp. *avenae* Eriks. & Henn. *Journal of Applied Ecology* 20:585–96.

O'Brien, W. J., and G. L. Vinyard. 1978. Polymorphism and predation: The effect of invertebrate predation on the distribution of two varieties of *Daphnia carinata* in South India ponds. *Limnology and Oceanography* 23:452–60.

O'Dowd, D. J., C. R. Brew, D. C. Christophel, and R. A. Norton. 1991. Mite-plant associations from the Eocene of southern Australia. *Science* 252:99–101.

O'Dowd, D. J., and M. F. Willson. 1989. Leaf domatia and mites on Australasian plants: Ecological and evolutionary implications. *Biological Journal of the Linnean Society* 37:191–236.

Ohsugi, T., R. Nishida, and H. Fukami. 1985. Oviposition stimulant of *Papilio xuthus*, a *Citrus*-feeding swallowtail butterfly. *Agricultural and Biological Chemistry* 49:1897–1900.

Oldroyd, B. P., T. E. Rinderer, and S. M. Buco. 1992. Intra-colonial foraging specialism by honey bees (*Apis mellifera*) (Hymenoptera: Apidae). *Behavioral Ecology and Sociobiology* 30:291–95.

Oliphant, L. W., and S. McTaggart. 1977. Prey utilized by urban merlins. *Canadian Field-Naturalist* 91:190–92.

O'Neill, S. L. 1989. Cytoplasmic symbionts in *Tribolium confusum*. *Journal of Invertebrate Pathology* 53:132–34.

O'Neill, S. L., and T. L. Karr. 1990. Bidirectional incompatibility between conspecific populations of *Drosophila simulans*. *Nature* 348:178–80.

Opler, P. A. 1973. Fossil lepidopterous leaf mines demonstrate the age of some insect-plant relationships. *Science* 179:1321–23.

Orians, G. H. 1962. Natural selection and ecological theory. *American Naturalist* 96:257–63.

Orihel, T. C. 1985. Filariae. Pages 171–91 in P. C. Beaver and R. C. Jung, eds., *Animal agents and vectors of human disease*. 5th ed. Lea & Febiger, Philadelphia.

Osborn, H. F. 1917. *The origin and evolution of life: On the theory of action, reaction and interaction of energy*. Charles Scribner's Sons, New York.

Osborn, T. C., D. C. Alexander, S. S. M. Sun, C. Cardona, and F. A. Bliss. 1988. Insecticidal activity and lectin homology of arcelin seed protein. *Science* 240:207–10.

Overmeer, W. P. J., and A. Q. Van Zon. 1976. Partial reproductive incompatibility between populations of spider mites (Acarina: Tetranychidae). *Entomologia Experimentalis et Applicata* 20:225–36.

Pahl, L. I. 1987. Feeding behaviour and diet of the common ringtail possum, *Pseudocheirus peregrinus*, in *Eucalyptus* woodlands and *Leptospermum* thickets in southern Victoria. *Australian Journal of Zoology* 35:487–506.

Pahl, L. I., and I. D. Hume. 1990. Preferences for *Eucalyptus* species of the New England tablelands and initial development of an artificial diet for koalas. Pages 123–28 in A. K. Lee, K. A. Handasyde, and G. D. Sanson, eds., *Biology of the koala*. Surrey Beatty & Sons, Sydney.

Paige, K. N. 1992. Overcompensation in response to mammalian herbivory: From mutualistic to antagonistic interactions. *Ecology* 73:2076–85.

Paige, K. N., and W. C. Capman. 1993. The effects of host-plant genotype, hybridization, and environment on gall-aphid attack and survival in cottonwood: The importance of genetic studies and the utility of RFLPs. *Evolution* 47:36–45.

Paige, K. N., and T. G. Whitham. 1987. Overcompensation in response to mammalian herbivory: The advantage of being eaten. *American Naturalist* 129:407–16.

Paine, R. T. 1966. Food web complexity and species diversity. *American Naturalist* 100:65–76.

———. 1980. Food webs: Linkage, interaction strength and community infrastructure. *Journal of Animal Ecology* 49:667–85.

Papaj, D. R., and R. J. Prokopy. 1989. Ecological and evolutionary aspects of learning in phytophagous insects. *Annual Review of Entomology* 34:315–50.

Park, T. 1948. Experimental studies of interspecies competition. I. Competition between populations of flour beetles, *Tribolium confusum* Duval and *Tribolium castaneum* Herbst. *Ecological Monographs* 18:265–307.

———. 1954. Experimental studies of interspecies competition. II. Temperature, humidity, and competition in two species of *Tribolium*. *Physiological Zoology* 27:177–238.

———. 1962. Beetles, competition, and populations. *Science* 138:1369–75.

Park, T., P. H. Leslie, and D. B. Mertz. 1964. Genetic strains and competition in populations of *Tribolium*. *Physiological Zoology* 37:97–162.

Parker, M. A. 1985. Local population differentiation for compatibility in an annual legume and its host-specific fungal pathogen. *Evolution* 39:713–23.

———. 1991. Nonadaptive evolution of disease resistance in an annual legume. *Evolution* 45:1209–17.

Parrott, D. M. 1981. Evidence for gene-for-gene relationships between resistance gene H_1 from *Solanum tuberosum* ssp. *andigena* and a gene in *Globodera rostochiensis,* and between H_2 from *S. multidissectum* and a gene in *G. pallida. Nematologica* 27:372–84.

Pashley, D. P. 1986. Host-associated genetic differentiation in fall armyworm (Lepidoptera: Noctuidae): A sibling species complex? *Annals of the Entomological Society of America* 79:898–904.

Pashley, D. P., and J. A. Martin. 1987. Reproductive incompatibility between host strains of the fall armyworm (Lepidoptera: Noctuidae). *Annals of the Entomological Society of America* 80:731–33.

Pasteels, J. M., M. Rowell-Rahier, and M. J. Raupp. 1988. Plant-derived defense in chrysomelid beetles. Pages 235–72 in P. Barbosa and D. K. Letourneau, eds., *Novel aspects of insect-plant interactions.* Wiley, New York.

Patch, E. M. 1938. Food-plant catalogue of the aphids of the world: Including the Phylloxeridae. *Maine Agricultural Experiment Station Bulletin* 393:35–431.

Paterson, H. E. H. 1982. Perspective on speciation by reinforcement. *South African Journal of Science* 78:53–57.

Paul, V. J., and K. L. Van Alstyne. 1988. Use of ingested algal diterpenoids by *Elysia halimedae* Macnae (Opisthobranchia: Ascoglossa) as antipredator defenses. *Journal of Experimental Marine Biology and Ecology* 119:15–29.

Peakall, R., C. J. Angus, and A. J. Beattie. 1990. The significance of ant and plant traits for pollination in *Leporella fimbriata. Oecologia* 84:457–60.

Peakall, R., and A. J. Beattie. 1989. Pollination of the orchid *Microtis parviflora* R. Br. by flightless worker ants. *Functional Ecology* 3:515–22.

———. 1991. The genetic consequences of worker ant pollination in a self-compatible, clonal orchid. *Evolution* 45:1837–48.

Peakall, R., A. J. Beattie, and S. H. James. 1987. Pseudocopulation of an orchid by male ants: A test of two hypotheses accounting for the rarity of ant pollination. *Oecologia* 73:522–24.

Peakall, R., S. N. Handel, and A. J. Beattie. 1991. The evidence for, and importance of, ant pollination. Pages 421–29 in C. R. Huxley and D. F. Cutler, eds., *Ant-plant interactions.* Oxford University Press, Oxford.

Peakall, R., and S. H. James. 1989. Outcrossing in an ant pollinated clonal orchid. *Heredity* 62:161–67.

Pedhazur, E. J. 1982. *Multiple regression in behavioral research.* Holt, Rinehart & Winston, New York.

Pellmyr, O. 1989. The cost of mutualism: Interactions between *Trollius europaeus* and its pollinating parasites. *Oecologia* 78:53–59.

———. 1992. The phylogeny of a mutualism: Evolution and coadaptation between *Trollius* and its seed-parasitic pollinators. *Biological Journal of the Linnean Society* 47:337–65.

Pellmyr, O., and J. N. Thompson. 1992. Multiple occurrences of mutualism in the yucca moth lineage. *Proceedings of the National Academy of Sciences, USA* 89:2927–29.

Pennings, S. C., M. T. Nadeau, and V. J. Paul. 1993. Selectivity and growth of the generalist herbivore *Dolabella auricularia* feeding upon complementary resources. *Ecology* 74:879–90.

Pérez-Rivera, R. A. 1991. Change in diet and foraging behavior of the Antillean euphonia in Puerto Rico after Hurricane Hugo. *Journal of Field Ornithology* 62:474–78.

Perring, T. M., A. D. Cooper, R. J. Rodriquez, C. A. Farrar, and T. S. Bellows, Jr. 1993. Identification of a whitefly species by genomic and behavioral studies. *Science* 259:74–77.

Perry, C. C., and M. A. Fraser. 1991. Silica deposition and ultrastructure in the cell wall of *Equisetum arvense:* The importance of cell wall structures and flow control in biosilicification. *Philosophical Transactions of the Royal Society of London B* 334:149–57.

Peterson, J. G. 1970. The food habits and summer distribution of juvenile sage grouse in central Montana. *Journal of Wildlife Management* 34:147–55.

Pfennig, D. W. 1992. Polyphenism in spadefoot toad tadpoles as a locally adjusted evolutionarily stable strategy. *Evolution* 46:1408–20.

Pickett, S. T. A., and P. S. White, eds. 1985. *The ecology of natural disturbance and patch dynamics.* Academic, New York.

Pierce, N. E. 1987. The evolution and biogeography of associations between lycaenid butterflies and ants. *Oxford Surveys in Evolutionary Biology* 4:89–116.

Pierce, N. E., and M. A. Elgar. 1985. The influence of ants on host plant selection by *Jalmenus evagoras,* a myrmecophilus lycaenid butterfly. *Behavioral Ecology and Sociobiology* 16:209–22.

Pimentel, D. 1961. Animal population regulation by the genetic feed-back mechanism. *American Naturalist* 95:65–79.

———. 1988. Herbivore population feeding pressure on plant hosts: Feedback evolution and host conservation. *Oikos* 53:289–302.

Pimentel, D., and A. C. Bellotti. 1976. Parasite-host population systems and genetic stability. *American Naturalist* 110:877–88.

Pimentel, D., E. H. Feinberg, P. W. Wood, and J. T. Hayes. 1965. Selection, spatial distribution and the coexistence of competing fly species. *American Naturalist* 99:97–109.

Pimentel, D., S. A. Levin, and D. Olson. 1978. Coevolution and the stability of exploiter-victim systems. *American Naturalist* 112:119–25.

Pimentel, D., U. Stachow, D. A. Takacs, H. W. Brubaker, A. R. Dumas, J. J. Meaney, J. A. S. O'Neil, D. E. Onsi, and D. B. Corzilius. 1992. Conserving biological diversity in agricultural/ forestry systems. *Bioscience* 42:354–62.

Piper, J. K. 1986. Seasonality of fruit characters and seed removal by birds. *Oikos* 46:303–10.

Poulton, E. B. 1890. *The colours of animals.* Appleton, New York.

———. 1896. *Charles Darwin and the theory of natural selection.* Macmillan, New York.

———. 1909. *Charles Darwin and the* Origin of Species: *Addresses, etc., in America and England in the year of the two anniversaries.* Longmans, Green & Co., London.

Pound, R. 1893. Symbiosis and mutualism. *American Naturalist* 27:509–20.

Powell, J. A. 1980. Evolution of larval food preference in Microlepidoptera. *Annual Review of Entomology* 25:133–59.

———. 1992. Interrelationships of yuccas and yucca moths. *Trends in Ecology and Evolution* 7:10–15.

Powell, J. R. 1982. Genetic and nongenetic mechanisms of speciation. Pages 67–74 in C. Barigozzi, ed., *Mechanisms of speciation.* Alan R. Liss, New York.

Pratt, T. K., and E. W. Stiles. 1985. The influence of fruit size and structure on composition of frugivore assemblages in New Guinea. *Biotropica* 17:314–21.

Price, P. W. 1977. General concepts in the evolutionary biology of parasites. *Evolution* 31:403–20.

———. 1980. *Evolutionary biology of parasites.* Princeton University Press, Princeton.

———. 1987. Evolution in parasite communities. *International Journal for Parasitology* 17:209–14.

———. 1990. Host populations as resources defining parasite community organization. Pages 21–40 in G. W. Esch, A. O. Bush, and J. M. Aho, eds., *Parasite communities: Patterns and processes.* Chapman & Hall, New York.

———. 1991a. Patterns in communities along latitudinal gradients. Pages 51–69 in P. W. Price, T. M. Lewinsohn, G. W. Fernandes, and W. W. Benson, eds., *Plant-animal interactions: Evolutionary ecology in tropical and temperate regions.* Wiley, New York.

———. 1991b. The web of life: Development over 3.8 billion years of trophic relationships. Pages 262–72 in L. Margulis and R. Fester, eds., *Symbiosis as a source of evolutionary innovation: Speciation and morphogenesis.* MIT Press, Cambridge.

———. 1992. Evolutionary perspectives on host plants and their parasites. *Advances in Plant Pathology* 8:1–30.

Price, P. W., C. E. Bouton, P. Gross, B. A. McPheron, J. N. Thompson, and A. E. Weis. 1980. Interactions among three trophic levels: Influence of plants on interactions between insect herbivores and natural enemies. *Annual Review of Ecology and Systematics* 11:41–65.

Price, P. W., and K. M. Clancy. 1983. Patterns in number of helminth parasite species in freshwater fishes. *Journal of Parasitology* 69:449–54.

Price, P. W., M. Westoby, and B. Rice. 1988. Parasite-mediated competition: Some predictions and tests. *American Naturalist* 131:544–55.

Price, P. W., M. Westoby, B. Rice, P. R. Atsatt, R. S. Fritz, J. N. Thompson, and K. Mobley. 1986. Parasite mediation in ecological interactions. *Annual Review of Ecology and Systematics* 17:487–505.

Price, T., M. Turelli, and M. Slatkin. 1993. Peak shifts produced by correlated response to selection. *Evolution* 47:280–90.

Prokopy, R. J., S. R. Diehl, and S. S. Cooley. 1988. Behavioral evidence for host races in *Rhagoletis pomonella* flies. *Oecologia* 76:138–47.

Provine, W. B. 1971. *The origins of theoretical population genetics.* University of Chicago Press, Chicago.

Pryor, T. 1987. The origin and structure of fungal disease resistance genes in plants. *Trends in Genetics* 3:157–61.

Punnett, R. C. 1915. *Mimicry in butterflies.* Cambridge University Press, Cambridge.

Puterka, G. J., and D. C. Peters. 1989. Inheritance of greenbug, *Schizaphis graminum* (Rondani), virulence to *Gb2* and *Gb3* resistance genes in wheat. *Genome* 32:109–14.

Pyke, D. A. 1987. Demographic responses of *Bromus tectorum* and seedlings of *Agropyron spicatum* to grazing by small mammals: The influence of grazing frequency and plant age. *Journal of Ecology* 75:825–35.

Quinnell, R. J., J. M. Behnke, and A. E. Keymer. 1991. Host specificity of and cross-immunity between two strains of *Heligmosomoides polygyrus. Parasitology* 102:419–27.

Raff, R. A., and T. C. Kaufman. 1983. *Embryos, genes, and evolution: The developmental-genetic basis of evolutionary change.* Macmillan, New York.

Ramirez, W. 1970. Host specificity of fig wasps (Agaonidae). *Evolution* 24:680–91.

Rand, A. L. 1940. Breeding habits of the birds of paradise *Macgregoria* and *Diphyllodes*. Results of the Archbold Expeditions no. 26. *American Museum Novitiates* 1073:1–14.

Rashbrook, V. K., S. G. Compton, and J. H. Lawton. 1992. Ant-herbivore interactions: Reasons for the absence of benefits to a fern with foliar nectaries. *Ecology* 73:2167–74.

Rausher, M. D. 1988. Is coevolution dead? *Ecology* 69:898–901.

Rausher, M. D., and E. L. Simms. 1989. The evolution of resistance to herbivory in *Ipomoea purpurea*. I. Attempts to detect selection. *Evolution* 43:563–72.

Real, L., and T. Caraco. 1986. Risk and foraging in stochastic environments. *Annual Review of Ecology and Systematics* 17:371–90.

Redford, K. H. 1985. Feeding and food preference in captive and wild giant anteaters (*Myrmecophaga tridactyla*). *Journal of Zoology, London* (Series A) 205:559–72.

Reid, D. G., H. Jinchu, D. Sai, W. Wei, and H. Yan. 1989. Giant panda *Ailuropoda melanoleuca* behaviour and carrying capacity following a bamboo die-off. *Biological Conservation* 49:85–104.

Reid, D. G., H. Jinchu, and H. Yan. 1991. Ecology of the red panda *Ailurus fulgens* in the Wolong Reserve, China. *Journal of Zoology, London* 225:347–64.

Reid, N. 1986. Pollination and seed dispersal of mistletoes (Loranthaceae) by birds in southern Australia. Pages 64–84 in H. A. Ford and D. C. Paton, eds., *The dynamic partnership: Birds and plants in southern Australia*. Handbook of the Flora and Fauna of South Australia Series. South Australian Government, South Australia.

———. 1990. Mutualistic interdependence between mistletoes (*Amyema quandang*) and spiny-cheeked honeyeaters and mistletoebirds in an arid woodland. *Australian Journal of Ecology* 15:175–90.

Remington, T. E., and C. E. Braun. 1985. Sage grouse food selection in winter, North Park, Colorado. *Journal of Wildlife Management* 49:1055–61.

Renaud, P. E., M. E. Hay, and T. M. Schmitt. 1990. Interactions of plant stress and herbivory: Intraspecific variation in the susceptibility of a palatable versus an unpalatable seaweed to sea urchin grazing. *Oecologia* 82:217–26.

Rensch, B. 1959. *Evolution above the species level*. Methuen, London.

———. 1980. Historical development of the present synthetic neo-Darwinism in Germany. Pages 284–303 in E. Mayr and W. B. Provine, eds., *The evolutionary synthesis: Perspectives on the unification of biology*. Harvard University Press, Cambridge.

Rice, K. J., and J. W. Menke. 1985. Competitive reversals and environment-dependent resource partitioning in *Erodium*. *Oecologia* 67:430–34.

Richards, W. R. 1976. A host index for species of Aphidoidea described during 1935 to 1969. *Canadian Entomologist* 108:499–550.

Richardson, K. C., R. D. Wooller, and B. G. Collins. 1986. Adaptations to a diet of nectar and pollen in the marsupial *Tarsipes rostratus* (Marsupialia: Tarsipedidae). *Journal of Zoology, London* (Series A) 208:285–97.

Richardson, P. M., W. P. Holmes, and G. B. Saul II. 1987. The effect of tetracycline on nonreciprocal cross incompatibility in *Mormoniella [= Nasonia] vitripennis*. *Journal of Invertebrate Pathology* 50:176–83.

Richardson, P. R. K. 1987. Aardwolf: The most specialized myrmecophagous mammal? *South African Journal of Science* 83:643–46.

Rickson, F. R. 1977. Progressive loss of ant-related traits of *Cecropia peltata* on selected Caribbean Islands. *American Journal of Botany* 64:585–92.

———. 1979. Absorption of animal tissue breakdown products into a plant stem—the feeding of a plant by ants. *American Journal of Botany* 66:87–90.

Rico-Gray, V., and L. R. Thien. 1989. Ant-mealybug interaction decreases fitness of *Schomburg-kia tibicinis* (Orchidaceae) in Mexico. *Journal of Tropical Ecology* 5:109–12.

Riddiford, N. 1986. Why do cuckoos *Cuculus canorus* use so many species of hosts? *Bird Study* 33:1–5.

Riessen, H. P. 1992. Cost-benefit model for the induction of an antipredator defense. *American Naturalist* 140:349–62.

Riessen, H. P., and W. G. Sprules. 1990. Demographic costs of antipredator defenses in *Daphnia pulex. Ecology* 71:1536–46.

Riley, C. V. 1873. On the oviposition of the yucca moth. *American Naturalist* 7:619–23.

———. 1892. The yucca moth and yucca pollination. *Third Annual Report of the Missouri Botanical Garden,* pp. 99–158.

Riska, B. 1989. Composite traits, selection response, and evolution. *Evolution* 43:1172–91.

Robbins, C. T. 1983. *Wildlife feeding and nutrition.* Academic, New York.

Roberson, J. A. 1986. Sage grouse–sagebrush relationships: A review. *USDA Forest Service Intermountain Research Station General Technical Report* INT-200:157–67.

Robertson, C. 1895. The philosophy of flower seasons, and the phaenological relations of the entomophilous flora and the anthophilous insect fauna. *American Naturalist* 29:97–117.

Roche, P., F. Debellé, F. Maillet, P. Lerouge, C. Faucher, G. Truchet, J. Dénarié, and J.-C. Promé. 1991. Molecular basis of symbiotic host specificity in *Rhizobium meliloti: nodH* and *nodPQ* genes encode the sulfation of lipo-oligosaccharide signals. *Cell* 67:1131–43.

Roff, D. A. 1990. The evolution of flightlessness in insects. *Ecological Monographs* 60:389–421.

Rohwer, S., and C. D. Spaw. 1988. Evolutionary lag versus bill-size constraints: A comparative study of the acceptance of cowbird eggs by old hosts. *Evolutionary Ecology* 2:27–36.

Rohwer, S., C. D. Spaw, and E. Røskaft. 1989. Costs to northern orioles of puncture-ejecting parasitic cowbird eggs from their nests. *Auk* 106:734–38.

Roininen, H., J. Vuorinen, J. Tahvanainen, and R. Julkunen-Tiitto. 1993. Host preference and allozyme differentiation in shoot galling sawfly, *Euura atra. Evolution* 47:300–308.

Romanes, G. J. 1892. *Darwin, and after Darwin: An exposition of the Darwinian theory and a discussion of post-Darwinian questions. I. The Darwinian theory.* Open Court Publishing Co., Chicago.

Romer, A. S. 1949. Time series and trends in animal evolution. Pages 103–20 in G. L. Jepsen, E. Mayr, and G. G. Simpson, eds., *Genetics, paleontology, and evolution.* Princeton University Press, Princeton.

Room, P. M., M. H. Julien, and I. W. Forno. 1989. Vigorous plants suffer most from herbivores: Latitude, nitrogen and biological control of the weed *Salvinia molesta. Oikos* 54:92–100.

Root, R. B. 1967. The niche exploitation pattern of the blue-grey gnatcatcher. *Ecological Monographs* 37:317–50.

———. 1973. Organization of a plant-arthropod association in simple and diverse habitats: The fauna of collards (*Brassica oleracea*). *Ecological Monographs* 43:95–124.

Rosenzweig, M. L., J. S. Brown, and T. L. Vincent. 1987. Red Queens and ESS: The coevolution of evolutionary rates. *Evolutionary Ecology* 1:59–94.

Rossiter, M. C. 1987. Use of a secondary host by non-outbreak populations of the gypsy moth. *Ecology* 68:857–68.

———. 1991a. Environmentally-based maternal effects: A hidden force in insect population dynamics? *Oecologia* 87:288–94.

———. 1991b. Maternal effects generate variation in life history: Consequences of egg weight plasticity in the gypsy moth. *Functional Ecology* 5:386–93.

Rothstein, S. I. 1990. A model system for coevolution: Avian brood parasitism. *Annual Review of Ecology and Systematics* 21:481–508.

Roubik, D. W. 1989. *Ecology and natural history of tropical bees.* Cambridge University Press, Cambridge.

Roughgarden, J. 1972. Evolution of niche width. *American Naturalist* 106:683–718.

———. 1974. Species packing and the competition function with illustrations from coral reef fish. *Theoretical Population Biology* 5:163–86.

———. 1976. Resource partitioning among competing species—a coevolutionary approach. *Theoretical Population Biology* 9:388–424.

———. 1979. *Theory of population genetics and evolutionary ecology: An introduction.* Macmillan, New York.

———. 1983a. The theory of coevolution. Pages 33–64 in D. J. Futuyma and M. Slatkin, eds., *Coevolution.* Sinauer Associates, Sunderland, Mass.

———. 1983b. Coevolution between competitors. Pages 383–403 in D. J. Futuyma and M. Slatkin, eds., *Coevolution.* Sinauer Associates, Sunderland, Mass.

Roughgarden, J., D. Heckel, and E. R. Fuentes. 1983. Coevolutionary theory and the biogeography and community structure of *Anolis.* Pages 371–410 in R. B. Juey, E. R. Pianka, and T. W. Schoener, eds., *Lizard ecology: Studies of a model organism.* Harvard University Press, Cambridge.

Roughgarden, J., and S. Pacala. 1989. Taxon cycle among *Anolis* lizard populations: Review of evidence. Pages 403–32 in D. Otte and J. A. Endler, eds., *Speciation and its consequences.* Sinauer Associates, Sunderland, Mass.

Rourke, J., and D. Wiens. 1977. Convergent floral evolution in South African and Australian Proteaceae and its possible bearing on pollination by nonflying mammals. *Annals of the Missouri Botanical Garden* 64:1–17.

Rousset, F., D. Vautrin, and M. Solignac. 1992. Molecular identification of *Wolbachia,* the agent of cytoplasmic incompatibility in *Drosophila simulans,* and variability in relation with host mitochondrial types. *Proceedings of the Royal Society of London B* 247:163–68.

Rowan, R., and D. A. Powers. 1991. A molecular genetic classification of zooanthellae and the evolution of animal-algal symbioses. *Science* 251:1348–51.

Rowell-Rahier, M. 1984a. The food plant preferences of *Phratora vitellinae* (Coleoptera: Chrysomelidae). A. Field observations. *Oecologia* 64:369–74.

———. 1984b. The food plant preferences of *Phratora vitellinae* (Coleoptera: Chrysomelinae). B. A laboratory comparison of geographically isolated populations and experiments on conditioning. *Oecologia* 64:375–80.

Roy, B. A. 1993. Patterns of rust infection as a function of host genetic diversity and host density in natural populations of the apomictic crucifer, *Arabis holboellii. Evolution* 47:111–24.

Rummel, J. D., and J. Roughgarden. 1985. A theory of faunal buildup for competition communities. *Evolution* 39:1009–33.

Russell, E. M. 1986. Observations on the behaviour of the honey possum, *Tarsipes rostratus* (Marsupialia: Tarsipedidae) in captivity. *Australian Journal of Zoology, Supplemental Series* 121:1–63.

Saffo, M. B. 1992. Invertebrates in endosymbiotic associations. *American Zoologist* 32:557–65.

Sage, R. D., D. Heyneman, K.-C. Lim, and A. C. Wilson. 1986. Wormy mice in a hybrid zone. *Nature* 324:60–63.

Sandlund, O. T., K. Gunnarsson, P. M. Jónasson, B. Jonsson, T. Lindem, K. P. Magnússon, H. J. Malmquist, H. Sigurjónsdóttir, S. Skúlason, and S. S. Snorrason. 1992. The Arctic charr *Salvelinus alpinus* in Thingvallavatn. *Oikos* 64:305–51

Savile, D. B. O. 1971. Coevolution of the rust fungi and their hosts. *Quarterly Review of Biology* 46:211–18.

———. 1972. Arctic adaptations in plants. *Canadian Department of Agriculture Monograph* no. 6:61–81.

———. 1976. Evolution of the rust fungi (Uredinales) as reflected by their ecological problems. *Evolutionary Biology* 9:137–207.

Saxena, R. C., and A. A. Barrion. 1987. Biotypes of insect pests of agricultural crops. *Insect Science and Its Application* 8:453–58.

Sbordoni, V., L. Bullini, G. Scarpelli, S. Forestiero, and M. Rampini. 1979. Mimicry in the burnet moth *Zygaena ephialtes:* Population studies and evidence of a Batesian-Müllerian situation. *Ecological Entomology* 4:83–93.

Schaffer, W. M., and M. L. Rosenzweig. 1978. Homage to the Red Queen. I. Coevolution of predators and their victims. *Theoretical Population Biology* 14:135–57.

Schaller, G. B., J. Hu, W. Pan, and J. Zhu. 1985. *The giant pandas of Wolong.* University of Chicago Press, Chicago.

Schemske, D. W. 1983. Limits to specialization and coevolution in plant-animal mutualisms. Pages 67–109 in M. H. Nitecki, ed., *Coevolution.* University of Chicago Press, Chicago.

Schemske, D. W., and C. C. Horvitz. 1984. Variation among floral visitors in pollination ability: A precondition for mutualism specialization. *Science* 225:519–21.

———. 1988. Plant-animal interactions and fruit production in a Neotropical herb: A path analysis. *Ecology* 69:1128–37.

Scheres, B., F. van Engelen, E. van der Knaap, C. van de Wiel, A. van Kammen, and T. Bisseling. 1990. Sequential induction of nodulin gene expression in the developing pea nodule. *Plant Cell* 2:687–700.

Schlaman, H. R. M., R. J. H. Okker, and B. J. J. Lugtenberg. 1992. Regulation of nodulation gene expression by NodD in rhizobia. *Journal of Bacteriology* 174:5177–82.

Schmalhausen, I. I. [1949] 1986. *Factors of evolution: The theory of stabilizing selection.* Reprint. University of Chicago Press, Chicago.

Schoener, T. W. 1971. Theory of feeding strategies. *Annual Review of Ecology and Systematics* 2:369–404.

———. 1974. Resource partitioning in ecological communities. *Science* 185:27–39.

———. 1987. A brief history of optimal foraging theory. Pages 5–65 in A. C. Kamil and H. R. Pulliam, eds., *Foraging behavior.* Plenum Press, New York.

Scriber, J. M. 1984. Larval foodplant utilization by the world Papilionidae (Lep.): Latitudinal gradients reappraised. *Tokurana (Acta Rhopalocerologica)* 6/7:1–50.

———. 1986a. Allelochemicals and alimentary ecology: Heterosis in a hybrid zone? Pages 43–71 in L. Brattsten and S. Admad, eds., *Molecular aspects of insect-plant associations.* Plenum, New York.

———. 1986b. Origins of the regional feeding abilities in the tiger swallowtail butterfly: Ecological monophagy and the *Papilio glaucus australis* subspecies in Florida. *Oecologia* 71:94–103.

———. 1988. Tale of the tiger: Beringial biogeography, binomial classification, and breakfast choices in the *Papilio glaucus* complex of butterflies. Pages 241–301 in K. C. Spencer, ed., *Chemical mediation of coevolution.* Academic, New York.

Scriber, J. M., B. L. Giebink, and D. Snider. 1991. Reciprocal latitudinal clines in oviposition behavior of *Papilio glaucus* and *P. canadensis* across the Great Lakes hybrid zone: Possible sex-linkage of oviposition preferences. *Oecologia* 87:360–68.

Scriber, J. M., R. L. Lindroth, and J. Nitao. 1989. Differential toxicity of a phenolic glycoside from quaking aspen to *Papilio glaucus* butterfly subspecies, hybrids and backcrosses. *Oecologia* 81:186–91.

Seger, J. 1988. Dynamics of some simple host-parasite models with more than two genotypes in each species. *Philosophical Transactions of the Royal Society of London B* 319:541–55.

———. 1992. Evolution of exploiter-victim relationships. Pages 3–25 in M. J. Crawley, ed.,

Natural enemies: The population biology of predators, parasites and diseases. Blackwell, Oxford.

Semlitsch, R. D. 1985. Reproductive strategy of a facultatively paedomorphic salamander *Ambystoma talpoideum*. *Oecologia* 65:305–13.

Semlitsch, R. D., and J. W. Gibbons. 1985. Phenotypic variation in metamorphosis and paedomorphosis in the salamander *Ambystoma talpoideum*. *Ecology* 66:1123–30.

Shapiro, A. M. 1976. Seasonal polyphenism. *Evolutionary Biology* 9:259–333.

Shapiro, A. M., and K. K. Masuda. 1980. The opportunistic origin of a new citrus pest. *California Agriculture* 34:4–5.

Shaw, J. H., J. Machado-Neto, and T. S. Carter. 1987. Behavior of free-living giant anteaters (*Myrmecophaga tridactyla*). *Biotropica* 19:255–59.

Sheehan, W., and B. A. Hawkins. 1991. Attack strategy as an indicator of host range in metopiine and pimpline Ichneumonidae (Hymenoptera). *Ecological Entomology* 16:129–31.

Shelford, R. W. 1902. Observations on some mimetic insects and spiders from Borneo and Singapore. *Proceedings of the Zoological Society of London* 1902:230–84.

———. 1912. Mimicry amongst the Blattidae; with a revision of the genus *Prosoplecta* Sauss., and the description of a new genus. *Proceedings of the Zoological Society of London* 1912:358–76.

———. [1916] 1985. *A naturalist in Borneo*. Reprint. Oxford University Press, Oxford.

Sheppard, P. M., J. R. G. Turner, K. S. Brown, W. W. Benson, and M. C. Singer. 1985. Genetics and the evolution of muellerian mimicry in *Heliconius* butterflies. *Philosophical Transactions of the Royal Society of London B* 308:433–610.

Shields, O., and J. L. Reveal. 1988. Sequential evolution of *Euphilotes* (Lycaenidae: Scolitantidini) on their plant host *Eriogonum* (Polygonaceae: Eriogonoideae). *Biological Journal of the Linnean Society* 33:51–93.

Shine, R. 1980. Reproduction, feeding and growth in the Australian burrowing snake *Vermicella annulata*. *Journal of Herpetology* 14:71–77.

———. 1981. Ecology of Australian elapid snakes of the genera *Furina* and *Glyphodon*. *Journal of Herpetology* 15:219–24.

Sidhu, G. S. 1981. The genetics of plant-nematode parasitic systems. *Botanical Review* 47:387–419.

———. 1984. Genetics of plant and animal parasitic systems. *Bioscience* 34:368–73.

Simberloff, D. 1983. Competition theory, hypothesis-testing, and other community ecological buzzwords. *American Naturalist* 122:626–35.

Simberloff, D., and T. Dayan. 1991. The guild concept and the structure of ecological communities. *Annual Review of Ecology and Systematics* 22:115–43.

Simms, E. L., and R. S. Fritz. 1990. The ecology and evolution of host-plant resistance to insects. *Trends in Ecology and Evolution* 5:356–60.

Simms, E. L., and M. D. Rausher. 1987. Costs and benefits of plant resistance to herbivory. *American Naturalist* 130:570–81.

———. 1989. The evolution of resistance to herbivory in *Ipomoea purpurea*. II. Natural selection by insects and costs of resistance. *Evolution* 43:573–85.

Simpson, B. B., J. L. Neff, and D. S. Seigler. 1983. Floral biology and floral rewards of *Lysimachia* (Primulaceae). *American Midland Naturalist* 110:249–56.

Simpson, G. G. 1944. *Tempo and mode in evolution*. Columbia University Press, New York.

———. 1951. *Horses: The story of the horse family in the modern world and through sixty million years of history*. Oxford University Press, New York.

———. 1953. *The major features of evolution*. Columbia University Press, New York.

Sims, S. L. 1980. Diapause dynamics and host plant suitability of *Papilio zelicaon* (Lepidoptera: Papilionidae). *American Midland Naturalist* 103:375–84.

Singer, M. C. 1982. Quantification of host preference by manipulation of oviposition behavior in the butterfly, *Euphydryas editha*. *Oecologia* 52:224–29.

———. 1983. Determinants of multiple host use by a phytophagous insect population. *Evolution* 37:389–403.

Singer, M. C., D. Ng, and R. A. Moore. 1991. Genetic variation in oviposition preference between butterfly populations. *Journal of Insect Behavior* 4:531–35.

Singer, M. C., D. Ng, and C. D. Thomas. 1988. Heritability of oviposition preference and its relationship to offspring performance within a single insect population. *Evolution* 42:977–85.

Singer, M. C., D. Ng, D. Vasco, and C. D. Thomas. 1992. Rapidly evolving associations among oviposition preferences fail to constrain evolution of insect diet. *American Naturalist* 139:9–20.

Sinha, R. C. 1984. Transmission mechanism of *Mycoplasma*-like organisms by leafhopper vectors. Pages 93–109 in K. F. Harris, ed., *Current topics in vector research,* volume II. Praeger, New York.

Skeate, S. T. 1987. Interactions between birds and fruits in a northern Florida hammock community. *Ecology* 68:297–309.

Skúlason, S., D. L. G. Noakes, and S. S. Snorrason. 1989. Ontogeny of trophic morphology in four sympatric morphs of Arctic charr *Salvelinus alpinus* in Thingvallavatn, Iceland. *Biological Journal of the Linnaean Society* 38:281–301.

Skutch, A. F. 1980. *A naturalist on a tropical farm.* University of California Press, Berkeley.

Slatkin, M. 1987. Gene flow and the geographic structure of natural populations. *Science* 236:787–92.

Smiley, J. T. 1978. Plant chemistry and evolution of host specificity: New evidence from *Heliconius* and *Passiflora*. *Science* 201:745–47.

———. 1985. Are chemical barriers necessary for evolution of butterfly-plant associations? *Oecologia* 65:580–83.

Smith, C. C. 1970. The coevolution of pine squirrels (*Tamiasciurus*) and conifers. *Ecological Monographs* 40:349–71.

Smith, C. H., ed. 1991. *Alfred Russel Wallace: An anthology of his shorter writings.* Oxford University Press, Oxford.

Smith, T. 1887. Parasitic bacteria and their relation to saprophytes. *American Naturalist* 21:1–9.

Snow, B. K. 1970. A field study of the bearded bellbird in Trinidad. *Ibis* 112:299–329.

Snow, D. W. 1962. The natural history of the oilbird, *Steatornis caripensis,* in Trinidad, West Indies. 2. Population, breeding, ecology and food. *Zoologica* 47:199–221.

Snow, D. W., and B. K. Snow. 1980. Relationships between hummingbirds and flowers in the Andes of Colombia. *Bulletin of the British Museum of Natural History (Zoology)* 38:105–39.

Snyder, M. A. 1992. Selective herbivory by Abert's squirrel mediated by chemical variability in ponderosa pine. *Ecology* 73:1730–41.

Snyder, N. F. R., and H. A. Snyder. 1969. A comparative study of mollusc predation by limpkins, everglade kites, and boat-tailed grackles. *Living Bird* 8:177–223.

Sodhi, N. S. 1992. Central place foraging and prey preparation by a specialist predator, the merlin. *Journal of Field Ornithology* 63:71–76.

Soler, M., and A. P. Müller. 1990. Duration of sympatry and coevolution between the great spotted cuckoo and its magpie host. *Nature* 343:748–50.

Somerson, M. L., L. Ehrman, J. P. Kocka, and F. J. Gottlieb. 1984. Streptococcal L-forms isolated from *Drosophila paulistorum* semispecies cause sterility in male progeny. *Proceedings of the National Academy of Sciences, USA* 81:282–85.

Soper, R. S., M. Shimazu, R. A. Humber, M. E. Ramos, and A. E. Hajek. 1988. Isolation and characterization of *Entomophaga maimaiga* sp. nov., a fungal pathogen of gypsy moth, *Lymantria dispar,* from Japan. *Journal of Invertebrate Pathology* 51:229–41.

Southgate, V. R., and D. Rollinson. 1987. Natural history of transmission and schistosome interactions. Pages 347–78 in D. Rollinson and A. J. G. Simpson, eds., *The biology of schistosomes: From genes to latrines*. Academic, New York.

Southwood, T. R. E. 1961. The number of species of insect associated with various trees. *Journal of Animal Ecology* 30:1–8.

———. 1973. The insect-plant relationship—an evolutionary perspective. *Symposia of the Royal Society of London* 6:143–55.

Spaw, C. D., and S. Rohwer. 1987. A comparative study of eggshell thickness in cowbirds and other passerines. *Condor* 89:307–18.

Spencer, H. J., and T. H. Fleming. 1989. Roosting and foraging behaviour of the Queensland tube-nosed bat, *Nyctimene robinsoni* (Pteropodidae): Preliminary radio-tracking observations. *Australian Wildlife Research* 16:413–20.

Spencer, K. A. 1990. *Host specialization in the world Agromyzidae (Diptera)*. Kluwer Academic Publishers, Boston.

Sperling, F. A. H. 1987. Evolution of the *Papilio machaon* species group in western Canada (Lepidoptera: Papilionidae). *Quaestiones Entomologicae* 23:198–315.

———. 1991. Mitochondrial DNA phylogeny, speciation, and hostplant coevolution of *Papilio* butterflies. Ph.D. dissertation, Cornell University.

Sprent, J. F. A. 1982. Host-parasite relationships of ascaridoid nematodes and their vertebrate hosts in time and space. *Mémoires du Muséum National d'Histoire Naturelle Paris,* Nouvelle Série, Série A, Zoology, 123:255–63.

———. 1983. Observations on the systematics of ascaridoid nematodes. Pages 303–19 in A. R. Stone, H. M. Platt, and L. F. Khalil, eds., *Concepts in nematode systematics*. Academic, New York.

Stebbins, G. L. 1950. *Variation and evolution in plants*. Columbia University Press, New York.

———. 1970. Adaptive radiation of reproductive characteristics in angiosperms. I. Pollination mechanisms. *Annual Review of Ecology and Systematics* 1:307–26.

———. 1981. Why are there so many species of flowering plants? *Bioscience* 31:573–76.

Stefani, R. 1956. Il problema della partenogenesi in "*Haploembia solieri*" Ramb. (Embioptera, Oligotomidae). *Atti dell'Accademia Nazionale de Lincei, Classe di Scienze Fisiche, Matematiche e Naturali,* Sez. IIIa, 5:127–201. [In White 1978; original not seen.]

———. 1960. *L. Artemia salina* partenogenetica a Cagliari. *Rivisti di Biologia*. [In White 1978; original not seen.]

Steiner, K. E., and V. B. Whitehead. 1988. The association between oil-producing flowers and oil-collecting bees in the Drakensberg of southern Africa. *Monographs in Systematic Botany from the Missouri Botanical Garden* 25:259–77.

———. 1990. Pollinator adaptation to oil-secreting flowers—*Rediviva* and *Diascia*. *Evolution* 44:1701–7.

———. 1991. Oil flowers and oil bees: Further evidence for pollinator adaptation. *Evolution* 45:1493–1501.

Steiner, W. W. M. 1981. Parasitization and speciation in mosquitoes: A hypothesis. Pages 91–119 in R. Pal, J. B. Kitzmiller, and T. Kanda, eds., *Cytogenetics and genetics of viruses*. Elsevier, New York.

Stephens, D. W., and J. R. Krebs. 1986. *Foraging theory*. Princeton University Press, Princeton.

Stevens, L., and M. J. Wade. 1990. Cytoplasmically inherited reproductive incompatibility in *Tribolium* flower beetles: The rate of spread and effect on population size. *Genetics* 124:367–72.

Stevens, L., and D. T. Wicklow. 1992. Multispecies interactions affect cytoplasmic incompatibility in *Tribolium* flour beetles. *American Naturalist* 140:642–53.

Stiles, F. G. 1975. Ecology, flowering phenology, and hummingbird pollination of some Costa Rican *Heliconia* species. *Ecology* 56:285–301.

———. 1976. Taste preferences, color preferences and flower choice in hummingbirds. *Condor* 78:10–26.

———. 1985. Seasonal patterns and coevolution in the hummingbird-flower community of a Costa Rican subtropical forest. Pages 757–85 in P. A. Buckley, M. S. Foster, E. S. Morton, R. S. Ridgley, and F. G. Buckley, eds., *Neotropical ornithology*. Ornithological Monographs, no. 36. American Ornithologists' Union, Washington, D.C.

Stiles, F. G., and C. E. Freeman. 1993. Patterns in floral nectar characteristics of some bird-visited plant species from Costa Rica. *Biotropica* 25:191–205.

Stone, A. R., and D. L. Hawksworth, eds. 1986. *Coevolution and systematics.* Clarendon Press, Oxford.

Stork, N. E. 1988. Insect diversity: Facts, fiction and speculation. *Biological Journal of the Linnean Society* 35:321–37.

———. 1993. How many species are there? *Biodiversity and Conservation* 2:215–32.

Stouthamer, R., J. A. J. Breeuwer, R. F. Luck, and J. H. Werren. 1993. Molecular identification of microorganisms associated with parthenogenesis. *Nature* 361:66–68.

Stouthamer, R., and R. F. Luck. 1993. Influence of microbe-associated parthenogenesis on the fecundity of *Trichogramma deion* and *T. pretiosum. Entomologia Experimentalis et Applicata* 67:183–92.

Stouthamer, R., R. F. Luck, and W. D. Hamilton. 1990. Antiobiotics cause parthenogenetic *Trichogramma* (Hymenoptera/Trichogrammatidae) to revert to sex. *Proceedings of the National Academy of Sciences, USA* 87:2424–27.

Stouthamer, R., J. D. Pinto, G. R. Platner, and R. F. Luck. 1990. Taxonomic status of thelytokous forms of *Trichogramma* (Hymenoptera: Trichogrammatidae). *Annals of the Entomological Society of America* 83:475–81.

Stradling, D. J. 1991. An introduction to the fungus-growing ants, Attini. Pages 15–18 in C. R. Huxley and D. F. Cutler, eds., *Ant-plant interactions.* Oxford University Press, Oxford.

Strong, D. R., Jr. 1974. Rapid asymptotic species accumulation in phytophagous insect communities: The pests of Cacao. *Science* 185:1064–66.

———. 1979. Biogeographic dynamics of insect-host plant communities. *Annual Review of Entomology* 24:89–119.

Strong, D. R., S. Larsson, and U. Gullberg. 1993. Heritability of host plant resistance to herbivory changes with gallmidge density during an outbreak on willow. *Evolution* 47:291–300.

Strong, D. R., J. H. Lawton, and Sir Richard Southwood. 1984. *Insects on plants: Community patterns and mechanisms.* Blackwell, Oxford.

Strong, D. R., Jr., and D. A. Levin. 1979. Species richness of plant parasites and growth form of their hosts. *American Naturalist* 114:1–22.

Strong, D. R., Jr., L. A. Szyska, and D. S. Simberloff. 1979. Tests of community-wide character displacement against null hypotheses. *Evolution* 33:897–913.

Sussman, R. W., and P. H. Raven. 1978. Pollination by lemurs and marsupials: An archaic coevolutionary system. *Science* 200:731–36.

Tabashnik, B. E. 1983. Host range evolution: The shift from native legume hosts to alfalfa by the butterfly *Colias philodice eriphyle. Evolution* 37:150–62.

Tahvanainen, J., R. Julkunen-Tiitto, and J. Kettunen. 1985. Phenolic glycosides govern the food selection patterns of willow feeding leaf beetles. *Oecologia* 67:52–56.

Tahvanainen, J., and P. Niemelä. 1987. Biogeographical and evolutionary aspects of insect herbivory. *Annales Zoologici Fennici* 24:239–47.

Tallamy, D. W. 1985. Squash beetle feeding behavior: An adaptation against induced cucurbit defenses. *Ecology* 66:1574–79.

Taper, M. L., and T. J. Case. 1992. Models of character displacement and the theoretical robustness of taxon cycles. *Evolution* 46:317–33.

Tauber, C. A., and M. J. Tauber. 1987. Food specificity in predacious insects: A comparative ecophysiological and genetic study. *Evolutionary Ecology* 1:175–86.

Taylor, R. J. 1984. Foraging in the eastern grey kangaroo and the wallaroo. *Journal of Animal Ecology* 53:65–74.

Tedla, S., and C. H. Fernando. 1970. Some remarks on the ecology of *Echinorhynchus salmonis* Muller 1784. *Canadian Journal of Zoology* 48:317–21.

Templeton, A. R. 1981. Mechanisms of speciation—a population genetic approach. *Annual Review of Ecology and Systematics* 12:23–48.

———. 1982. Adaptation and the integration of evolutionary forces. Pages 15–31 in R. Milkman, ed., *Perspectives on evolution*. Sinauer Associates, Sunderland, Mass.

Terborgh, J. 1986. Keystone plant resources in the tropical forest. Pages 330–44 in M. E. Soulé, ed., *Conservation biology: The science of scarcity and diversity*. Sinauer Associates, Sunderland, Mass.

Thomas, C. D., and S. Harrison. 1992. Spatial dynamics of a patchily distributed butterfly species. *Journal of Animal Ecology* 61:437–46.

Thomas, C. D., D. Ng, M. C. Singer, J. L. B. Mallet, C. Parmesan, and H. L. Billington. 1987. Incorporation of a European weed into the diet of a North American herbivore. *Evolution* 41:892–901.

Thomas, J. A., G. W. Elmes, J. C. Wardlaw, and M. Woyciechowski. 1989. Host specificity among *Maculinea* butterflies in *Myrmica* ant nests. *Oecologia* 79:452–57.

Thomas, J. A., M. L. Munguira, J. Martin, and G. W. Elmes. 1991. Basal hatching by *Maculinea* butterfly eggs: A consequence of advanced myrmecophily? *Biological Journal of the Linnean Society* 44:175–84.

Thompson, J. D., R. Turkington, and F. B. Holl. 1990. The influence of *Rhizobium leguminosarum* biovar. *trifolii* on the growth and neighbour relationships of *Trifolium repens* and three grasses. *Canadian Journal of Botany* 68:296–303.

Thompson, J. N. 1978. Within-patch structure and dynamics in *Pastinaca sativa* and resource availability to a specialized herbivore. *Ecology* 59:443–48.

———. 1980. Treefalls and colonization patterns of temperate forest herbs. *American Midland Naturalist* 104:176–84.

———. 1981a. Elaiosomes and fleshy fruits: Phenology and selection pressures for ant-dispersed seeds. *American Naturalist* 117:104–8.

———. 1981b. Reversed animal-plant interactions: The evolution of insectivorous and ant-fed plants. *Biological Journal of the Linnean Society* 16:147–55.

———. 1982. *Interaction and coevolution*. Wiley, New York.

———. 1983a. Partitioning of variance in demography: Within-patch differences in herbivory, survival, and flowering of *Lomatium farinosum* (Umbelliferae). *Oikos* 40:315–17.

———. 1983b. Selection of plant parts by *Depressaria multifidae* (Lep., Oecophoridae) on its seasonally-restricted hostplant, *Lomatium grayi* (Umbelliferae). *Ecological Entomology* 8:203–11.

———. 1983c. Selection pressures on phytophagous insects feeding on small host plants. *Oikos* 40:438–44.

———. 1983d. The use of ephemeral plant parts on small host plants: How *Depressaria leptotaeniae* (Lepidoptera: Oecophoridae) feeds on *Lomatium dissectum* (Umbelliferae). *Journal of Animal Ecology* 52:281–91.

———. 1984. Insect diversity and trophic structure of communities. Pages 591–606 in C. B. Huffaker and R. L. Rabb, eds., *Ecological entomology*. Wiley, New York.

————. 1985a. Postdispersal seed predation in *Lomatium* spp. (Umbelliferae): Variation among individuals and species. *Ecology* 66:1608–16.

————. 1985b. Within-patch dynamics of life histories, populations, and interactions: Selection over time in small spaces. Pages 253–64 in S. T. A. Pickett and P. S. White, eds., *The ecology of natural disturbance and patch dynamics.* Academic, New York.

————. 1986a. Constraints on arms races in coevolution. *Trends in Ecology and Evolution* 1:105–7.

————. 1986b. Oviposition behaviour and searching efficiency in a natural population of a braconid parasitoid. *Journal of Animal Ecology* 55:351–60

————. 1986c. Patterns in coevolution. Pages 119–43 in A. R. Stone and D. H. Hawksworth, eds., *Coevolution and systematics.* Clarendon Press, Oxford.

————. 1987a. The ontogeny of flowering and sex expression in divergent populations of *Lomatium grayi. Oecologia* 72:605–11.

————. 1987b. Symbiont-induced speciation. *Biological Journal of the Linnean Society* 32:385–93.

————. 1987c. Variance in number of eggs per patch: Oviposition behaviour and population dispersion in a seed parasitic moth. *Ecological Entomology* 12:311–20.

————. 1988a. Coevolution and alternative hypotheses on insect/plant interactions. *Ecology* 69:893–95.

————. 1988b. Evolutionary ecology of the relationship between oviposition preference and performance of offspring in phytophagous insects. *Entomologia Experimentalis et Applicata* 47:3–14.

————. 1988c. Evolutionary genetics of oviposition preference in swallowtail butterflies. *Evolution* 42:1223–34.

————. 1988d. Variation in interspecific interactions. *Annual Review of Ecology and Systematics* 19:65–87.

————. 1988e. Variation in preference and specificity in monophagous and oligophagous swallowtail butterflies. *Evolution* 42:118–28.

————. 1989. Concepts of coevolution. *Trends in Ecology and Evolution* 4:179–83.

————. 1990. Coevolution and the evolutionary genetics of interactions among plants and insects and pathogens. Pages 249–71 in J. J. Burdon and S. R. Leather, eds., *Pests, pathogens, and plant communities.* Blackwell, Oxford.

————. 1993. Preference hierarchies and the origin of geographic specialization in host use in swallowtail butterflies. *Evolution.* In press.

————. 1994. The geographic mosaic of evolving interactions. Pages 419–32 in S.R. Leather, A.D. Watt, N.J. Mills, and K.F.A. Walters, eds., *Individuals, populations and patterns in ecology.* Intercept Press, Andover, U.K.

Thompson, J. N., and J. J. Burdon. 1992. Gene-for-gene coevolution between plants and parasites. *Nature* 360:121–25.

Thompson, J. N., and M. E. Moody. 1985. Assessing probability of interaction in size-structured populations: *Depressaria* attack on *Lomatium.* Ecology 66:1597–1607.

Thompson, J. N., and O. Pellmyr. 1989. Origins of variance in seed number and mass: Interaction of sex expression and herbivory in *Lomatium salmoniflorum. Oecologia* 79:395–402.

————. 1991. Evolution of oviposition behavior and host preference in Lepidoptera. *Annual Review of Entomology* 36:65–89.

————. 1992. Mutualism with pollinating seed parasites amid co-pollinators: Constraints on specialization. *Ecology* 73:1780–91.

Thompson, J. N., W. Wehling, and R. Podolsky. 1990. Evolutionary genetics of host use in swallowtail butterflies. *Nature* 344:148–50.

Thompson, J. N., and M. F. Willson. 1978. Disturbance and the dispersal of fleshy fruits. *Science* 200:1161–63.

———. 1979. Evolution of temperate fruit/bird interactions: Phenological strategies. *Evolution* 33:973–82.

Tilman, D. 1988. *Plant strategies and the dynamics and structure of plant communities.* Princeton University Press, Princeton.

Tiritilli, M. E., and J. N. Thompson. 1988. Variation in swallowtail/plant interactions: Host selection and the shapes of survivorship curves. *Oikos* 53:153–60.

Tomback, D. F. 1982. Dispersal of whitebark pine seeds by Clark's nutcracker: A mutualism hypothesis. *Journal of Animal Ecology* 51:451–67.

———. 1983. Nutcrackers and pines: Coevolution or coadaptation? Pages 179–223 in M. H. Nitecki, ed., *Coevolution.* University of Chicago Press, Chicago.

Trelease, W. 1902. The Yucceae. *Annual Report of the Missouri Botanical Garden* 13:27–133.

Trowbridge, C. D. 1991. Diet specialization limits herbivorous sea slug's capacity to switch among food species. *Ecology* 72:1880–88.

Turelli, M., and A. A. Hoffmann. 1991. Rapid spread of an inherited incompatibility factor in California *Drosophila. Nature* 353:440–42.

Turkington, R. 1989. The growth, distribution and neighbour relationships of *Trifolium repens* in a permanent pasture. V. The coevolution of competitors. *Journal of Ecology* 77:717–33.

Turkington, R., and J. L. Harper. 1979. The growth, distribution and neighbour relationships of *Trifolium repens* in a permanent pasture. IV. Fine-scale biotic differentiation. *Journal of Ecology* 67:245–54.

Turner, J. R. G. 1980. Oscillations of frequency in batesian mimics, hawks and doves, and other simple frequency dependent polymorphisms. *Heredity* 45:113–26.

———. 1984. Mimicry: The palatability spectrum and its consequences. Pages 141–61 in R. I. Vane-Wright and P. R. Ackery, eds., *The biology of butterflies.* Academic, New York.

———. 1987. The evolutionary dynamics of batesian and muellerian mimicry: Similarities and differences. *Ecological Entomology* 12:81–95.

Turner, V. 1984. *Banksia* pollen as a source of protein in the diet of two Australian marsupials, *Cercartetus nanus* and *Tarsipes rostratus. Oikos* 43:53–61.

Tyler, H. A. 1975. *The swallowtail butterflies of North America.* Naturegraph Publishers, Healdsburg, Calif.

Tyre, A. J., and J. F. Addicott. 1993. Facultative non-mutualistic behaviour by an 'obligate' mutualist: 'cheating' by yucca moths. *Oecologia* 94:173–75.

Ule, E. 1901. Ameisengärten in Amazonasgebiet. *Engler's Botanische Jahrbuch* 30:45–51.

Valburg, L. K. 1992. Flocking and frugivory: The effect of social groupings on resource use in the common bush-tanager. *Condor* 94:358–63.

Vander Wall, S. B. 1990. *Food hoarding in animals.* University of Chicago Press, Chicago.

Vander Wall, S. B., and R. P. Balda. 1977. Coadaptations of the Clark's nutcracker and the piñon pine for efficient seed harvest and dispersal. *Ecological Monographs* 47:89–111.

van Riper, C., III, S. G. van Riper, M. L. Goff, and M. Laird. 1986. The epizootiology and ecological significance of malaria in Hawaiian land birds. *Ecological Monographs* 56:327–44.

Van Valen, L. 1973. A new evolutionary law. *Evolutionary Theory* 1:1–30.

Vaughan, T. A. 1982. Stephens' woodrat, a dietary specialist. *Journal of Mammalogy* 63:53–62.

Vaughan, T. A., and N. J. Czaplewski. 1985. Reproduction in Stephens' woodrat: The wages of folivory. *Journal of Mammalogy* 66:429–43.

Vermeij, G. J. 1987. *Evolution and escalation: An ecological history of life.* Princeton University Press, Princeton.

———. 1992. Time of origin and biogeographical history of specialized relationships between northern marine plants and herbivorous molluscs. *Evolution* 46:657–64.

Via, S. 1984. The quantitative genetics of polyphagy in an insect herbivore. I. Genotype-environment interaction in larval performance on different host plant species. *Evolution* 38:881–95.

———. 1986. Genetic covariance between oviposition preference and larval performance in an insect herbivore. *Evolution* 40:778–85.

———. 1990. Ecological genetics and host adaptation in herbivorous insects: the experimental study of evolution in natural and agricultural systems. *Annual Review of Entomology* 35:421–46.

———. 1991a. The genetic structure of host plant adaptation in a spatial patchwork: Demographic variability among reciprocally transplanted pea aphid clones. *Evolution* 45:827–52.

———. 1991b. Specialized host plant performance of pea aphid clones is not altered by experience. *Ecology* 72:1420–27.

Vijn, I., L. das Neves, A. van Kammen, H. Franssen, and T. Bisseling. 1993. Nod factors and nodulation in plants. *Science* 260:1764–65.

Vogel, S., and C. D. Michener. 1985. Long bee legs and oil-producing floral spurs, and a new *Rediviva* (Hymenoptera, Melittidae; Scrophulariaceae). *Journal of the Kansas Entomological Society* 58:359–64.

Wade, M. J. 1990. Genotype-environment interaction for climate and competition in a natural population of flour beetles, *Tribolium castaneum*. *Evolution* 44:2004–11.

Wade, M. J., and L. Stevens. 1985. Microorganism mediated reproductive isolation in flour beetles (genus *Tribolium*). *Science* 227:527–28.

Wagner, D. L., and J. A. Powell. 1988. A new *Prodoxus* from *Yucca baccata:* first report of a leaf-mining Prodoxine (Lepidoptera: Prodoxidae). *Annals of the Entomological Society of America* 81:547–53.

Wainwright, P. C., C. W. Osenberg, and G. G. Mittelbach. 1991. Trophic polymorphism in the pumpkinseed sunfish (*Lepomis gibbosus* Linnaeus): Effects of environment on ontogeny. *Functional Ecology* 5:40–55.

Waldvogel, M., and F. Gould. 1990. Variation in oviposition preference of *Heliothis virescens* in relation to macroevolutionary patterns of heliothine host range. *Evolution* 44:1326–37.

Wallace, A. R. 1867. Creation by law. *Quarterly Journal of Science* 4:471–88. [Reprinted in part in Smith 1991.]

———. 1870. *Contribution to the theory of natural selection.* Macmillan, London. [Combined with *Tropical nature and other essays* in 1895 and reprinted as *Natural selection and tropical nature.*]

———. [1889] 1905. *Darwinism: An exposition of the theory of natural selection with some of its applications.* (3d ed.) Macmillan, London.

———. 1895. *Natural selection and tropical nature: Essays on descriptive and theoretical biology.* Macmillan, London.

Walls, M., and M. Ketola. 1989. Effects of predator-induced spines on individual fitness in *Daphnia pulex. Limnology and Oceanography* 34:390–96.

Walsberg, G. E. 1975. Digestive adaptations of *Phainopepla nitens* associated with the eating of mistletoeberries. *Condor* 77:169–74.

———. 1977. Ecology and energetics of contrasting social systems in *Phainopepla nitens* (Aves: Ptilogonatidae). *University of California Publications in Zoology* 108:1–63.

Walter, G. H. 1983a. Differences in host relationships between male and female heteronomous parasitoids (Aphelinidae: Chalcidoidea): A review of host location, oviposition and pre-imaginal physiology and morphology. *Journal of the Entomological Society of South Africa* 46:261–82.

———. 1983b. 'Divergent male ontogenies' in Aphelinidae (Hymenoptera: Chalcidoidea): A simplified classification and a suggested evolutionary sequence. *Biological Journal of the Linnean Society* 19:63–82.

Ward, D. 1991. The size selection of clams by African black oystercatchers and kelp gulls. *Ecology* 72:513–22.

Waser, P. M. 1976. *Cercocebus albigena:* Site attachment, avoidance, and intergroup spacing. *American Naturalist* 110:911–35.

Wasserman, S. S. 1986. Genetic variation in adaptation to foodplants among populations of the southern cowpea weevil, *Callosobruchus maculatus:* Evolution of oviposition preference. *Entomologia Experimentalis et Applicata* 42:201–12.

Wasserman, S. S., and C. Mitter. 1978. The relationship of body size to breadth of diet in some Lepidoptera. *Ecological Entomology* 3:155–60.

Water, T. P. M., van de. 1983. A hostrace of the small ermine moth *Yponomeuta padellus* L. (Lepidoptera, Yponomeutidae) in northern Europe. *Netherlands Journal of Zoology* 33:276–82.

Watzin, M. C. 1985. Interactions among temporary and permanent meiofauna: Observations on the feeding and behavior of selected taxa. *Biological Bulletin* 169:397–416.

Weis, A. E., W. G. Abrahamson, and M. C. Andersen. 1992. Variable selection on *Eurosta*'s gall size, I: The extent and nature of variation in phenotypic selection. *Evolution* 46:1674–97.

Weis, A. E., W. G. Abrahamson, and K. D. McCrea. 1985. Host gall size and oviposition success by the parasitoid *Eurytoma gigantea*. *Ecological Entomology* 10:341–48.

Weis, A. E., and W. L. Gorman. 1990. Measuring selection on reaction norms: An exploration of the *Eurosta-Solidago* system. *Evolution* 44:820–31.

Weller, S. J., H. W. Ohm, F. L. Petterson, J. E. Foster, and P. L. Taylor. 1991. Genetics of resistance of CI 15160 durum wheat to biotype D of Hessian fly. *Crop Science* 31:1163–68.

Werner, E. E., and J. F. Gilliam. 1984. The ontogenetic niche and species interactions in size-structured populations. *Annual Review of Ecology and Systematics* 15:393–425.

Werren, J. H., S. W. Skinner, and A. M. Huger. 1986. Male-killing bacteria in a parasitic wasp. *Science* 231:990–92.

West, K., A. Cohen, and M. Baron. 1991. Morphology and behavior of crabs and gastropods from Lake Tanganyika, Africa: Implications for lacustrine predator-prey coevolution. *Evolution* 45:589–607.

West-Eberhard, M. J. 1983. Sexual selection, social competition and speciation. *Quarterly Review of Biology* 58:155–83.

Westoby, M. 1978. What are the biological bases of varied diets? *American Naturalist* 112:627–31.

Wheatley, B. P. 1980. Malaria as a possible selective factor in the speciation of macaques. *Journal of Mammalogy* 61:307–11.

Wheelwright, N. T. 1983. Fruits and the ecology of resplendent quetzals. *Auk* 100:286–301.

———. 1985. Fruit size, gape width, and the diets of fruit-eating birds. *Ecology* 66:808–18.

———. 1986. A seven-year study of individual variation in fruit production in tropical bird-dispersed tree species in the family Lauraceae. Pages 19–35 in A. Estrada and T. H. Fleming, eds., *Frugivores and seed dispersal*. Junk, Dordrecht.

Wheelwright, N. T., and C. H. Janson. 1985. Colors of fruit displays of bird-dispersed plants in two tropical forests. *American Naturalist* 126:777–99.

Wheelwright, N. T., and G. H. Orians. 1982. Seed dispersal by animals: Contrasts with pollen dispersal, problems of terminology, and constraints on coevolution. *American Naturalist* 119:402–13.

White, D. W., and E. W. Stiles. 1990. Co-occurrences of foods in stomachs and feces of fruit-eating birds. *Condor* 92:291–303.

White, M. J. D. 1978. *Modes of speciation.* Freeman, New York.

White, T. C. R. 1984. The abundance of invertebrate herbivores in relation to the availability of nitrogen in stressed food plants. *Oecologia* 63:90–105.

Whitham, T. G. 1981. Individual trees as heterogeneous environments: Adaptation to herbivory or epigenetic noise? Pages 9–27 in R. F. Denno and H. Dingle, eds., *Insect life history patterns: Habitat and geographic variation.* Springer-Verlag, New York.

———. 1989. Plant hybrid zones as sinks for pests. *Science* 244:1490–93.

Whitham, T. G., J. Maschinski, K. C. Larson, and K. N. Paige. 1991. Plant responses to herbivory: The continuum from negative to positive and underlying physiological mechanisms. Pages 227–56 in P. W. Price, T. M. Lewinsohn, G. W. Fernandes, and W. W. Benson, eds., *Plant-animal interactions: Evolutionary ecology in tropical and temperate regions.* Wiley, New York.

Whitham, T. G., and S. Mopper. 1985. Chronic herbivory: Impacts on architecture and sex expression of pinyon pine. *Science* 228:1089–91.

Wiens, D., and J. P. Rourke. 1978. Rodent pollination in southern African *Protea* spp. *Nature* 276:71–73.

Wiens, J. A. 1989. Spatial scaling in ecology. *Functional Ecology* 3:385–97.

Wiklund, C. 1981. Generalist vs. specialist oviposition behaviour in *Papilio machaon* (Lepidoptera) and functional aspects on the hierarchy of oviposition preferences. *Oikos* 36:163–70.

———. 1982. Generalist versus specialist utilization of host plants among butterflies. Pages 181–91 in J. H. Visser and A. K. Minks, eds., *Proceedings of the 5th International Symposium on Insect-Plant Relationships.* Centre for Agricultural Publications and Documentation, Wageningen, Netherlands.

Wilbur, H. M. 1980. Complex life cycles. *Annual Review of Ecology and Systematics* 11:67–93.

Wilding, N., N. M. Collins, P. M. Hammond, and J. F. Webber, eds. 1989. *Insect-fungus interactions.* Academic, New York.

Williams, E. E. 1972. The origins of faunas. Evolution of lizard congeners in a complex island fauna: A trial analysis. *Evolutionary Biology* 6:47–89.

Williams, G. C. 1966. *Adaptation and natural selection: A critique of some current evolutionary thought.* Princeton University Press, Princeton.

Willson, M. F. 1983. *Plant reproductive ecology.* Wiley, New York.

———. 1986. Avian frugivory and seed dispersal in eastern North America. *Current Ornithology* 3:223–79.

Willson, M. F., A. K. Irvine, and N. G. Walsh. 1989. Vertebrate dispersal syndromes in some Australian and New Zealand plant communities, with geographic comparisons. *Biotropica* 21:133–47.

Willson, M. F., and M. N. Melampy. 1983. The effect of bicolored fruit displays on fruit removal by avian frugivores. *Oikos* 41:27–31.

Willson, M. F., and J. N. Thompson. 1982. Phenology and ecology of color in bird-dispersed fruits, or why some fruits are red when they are 'green'. *Canadian Journal of Botany* 60:701–13.

Willson, M. F., and C. J. Whelan. 1990. The evolution of fruit color in fleshy-fruited plants. *American Naturalist* 136:790–809.

Wilson, E. O. 1959. Adaptive shift and dispersal in a tropical ant fauna. *Evolution* 13:122–44.

———. 1961. The nature of the taxon cycle in the Melanesian ant fauna. *American Naturalist* 95:169–93.

———. 1975. *Sociobiology: The new synthesis.* Harvard University Press, Cambridge.

Wood, T. K., and S. I. Guttman. 1983. *Enchenopa binotata* complex: Sympatric speciation? *Science* 220:310–12.

Wood, T. K., and M. C. Keese. 1990. Host-plant-induced assortative mating in *Enchenopa* tree-hoppers. *Evolution* 44:619–28.

Wood, T. K., K. L. Olmstead, and S. I. Guttman. 1990. Insect phenology mediated by host-plant water relations. *Evolution* 44:629–36.

Wooller, R. D., E. M. Russell, M. B. Renfree, and P. A. Towers. 1983. A comparison of seasonal changes in the pollen loads of nectarivorous marsupials and birds. *Australian Wildlife Research* 10:311–17.

Woolley, T. A. 1988. *Acarology: Mites and human welfare.* Wiley, New York.

Wright, S. 1932. The roles of mutation, inbreeding, crossbreeding and selection in evolution. *Proceedings of the VI International Genetics Congress* 1:356–66.

———. 1934. The method of path coefficients. *Annals of Mathematical Statistics* 5:161–215.

———. 1982. The shifting balance theory and macroevolution. *Annual Review of Genetics* 16:1–19.

Wynne-Edwards, V. C. 1962. *Animal dispersion in relation to social behaviour.* Oliver & Boyd, Edinburgh.

Yaeger, R. G. 1985. Amebae. Pages 36–46 in P. C. Beaver and R. C. Jung, eds., *Animal agents and vectors of human disease.* 5th ed. Lea & Febiger, Philadelphia.

Yang, W.-C., P. Katinakis, P. Hendriks, A. Smolders, F. de Vries, J. Spee, A. van Kammen, T. Bisseling, and H. Franssen. 1993. Characterization of *GmENOD40*, a gene showing novel patterns of cell-specific expression during soybean nodule development. *Plant Journal* 3:573–85.

Young, A. M. 1978. The biology of the butterfly *Aeria eurimedea agna* (Nymphalidae: Ithomiinae: Oleriini) in Costa Rica. *Journal of the Kansas Entomological Society* 51:1–10.

Young, J. P. W., and A. W. B. Johnston. 1989. The evolution of specificity in the legume-*Rhizobium* symbiosis. *Trends in Ecology and Evolution* 4:341–49.

Zangerl, A. R., and F. A. Bazzaz. 1984. Effects of short-term selection along environmental gradients on variation in populations of *Amaranthus retroflexus* and *Abutilon theophrasti.* *Ecology* 65:207–17.

Zaret, T. M. 1972. Predator-prey interaction in a tropical lacustrine ecosystem. *Ecology* 53:248–57.

Zhang, Z.-Q., J. P. Sanderson, and J. P. Nyrop. 1992. Foraging time and spatial patterns of predation in experimental populations: A comparative study of three mite predator-prey systems (Acari: Phytoseiidae, Tetranychidae). *Oecologia* 90:185–96.

Zwölfer, H. 1988. Evolutionary and ecological relationships of the insect fauna of thistles. *Annual Review of Entomology* 33:103–22.

Zwölfer, H., and M. Romstöck-Völkl. 1991. Biotypes and the evolution of niches in phytophagous insects on Cardueae hosts. Pages 487–507 in P. W. Price, T. M. Lewinsohn, G. W. Fernandes, and W. W. Benson, eds., *Plant-animal interactions: Evolutionary ecology in tropical and temperate regions.* Wiley, New York.

INDEX

Cherrett, J. M., 55, 247
Chilcote, C. A., 17, 61, 131
China, 146
Christ, B. J., 206
Christensen, K. M., 160
Christophel, D. C., 13
chromosomes, 78, 88, 90, 155–56
Chrysococcyx
	basalis, 266
	lucidus, 266
chrysomelid, 62, 65
Chrysopa
	quadripunctata, 91
	slossonae, 91, 139
Cichlasoma
	citrinellum, 113–14
	managuense, 114
cichlids, 113–14
Cirsium, 131
Citrus, 85
Cittadino, E., 34
cladocerans, 153, 154
cladogenesis, parallel, 277, 278
Cladosporium fulvum, 156, 207
Clamator glandarius, 268
clams, 247
Clancy, K. M., 157
Claridge, M. F., 131, 208
Clark, C. W., 51
Clark, D. A., 253
Clark, D. B., 253
Clarke, C. A., 212, 215
Clarke, J. F. G., 62, 72
Clarkson, R. W., 84
classification of interaction, 19–23
Clausen, T. P., 160, 224
Claviceps, 170
Clavicipitaceae, 170–71
Clay, K., 171
Clayton, D. H., 125
cleaner fish, 172, 175, 183
cleistogamy, 163
Clements, F. E., 34
clover, 24, 79
	common red, 24
	incarnate, 24
	white, 269–71
Cnemidophorus, 274
Cnidium, 74
coadaptation, 20, 253

coadapted gene complexes, 90, 110, 243
Cobb, N. S., 160
coccidiosis, 105
coccids, 112, 257
cockatoos, 147–49
cockroaches, 31
coevolution
	accumulating examples, 219–20
	and antiquity of interactions, 12–17
	ants and fungus gardens, 246–48
	ants and plants, 54, 224, 257–60
	approaches to study, 36
	arms races, 26–27, 47, 52, 54, 165, 183, 203, 204, 209, 216, 221, 261, 289
	asymmetries, 153, 208, 226, 234–35, 253–87
	bacteria and bacteriophage, 162–63
	bacteria and plasmids, 245–46
	biogeographic congruence, 223–25, 280, 290
	brood parasites and hosts, 261–68
	butterflies and plants, 276–79
	coevolutionary alternation, 221, 254–68, 274
	coevolutionary turnover, 221, 253, 271–74
	community context, 225, 285–86, 291
	and community organization, 255
	competitors, 51, 53, 221, 226–28, 248–52, 268–71
	and conservation, ix, 3, 12, 292–95
	cytoplasmic symbionts and hosts, 241–46
	definition, viii, 8, 47, 54, 70, 203
	toward decreased antagonism, 47–49
	and degree of specialization, 290
	diffuse, vii, 2, 234–38, 248, 253, 255, 259, 275, 282–85, 287, 290
	diversifying, 221, 239–52, 277–78
	Ehrlich and Raven hypothesis, 222, 276–79, 285
	escape-and-radiation, 222, 276–79, 285
	extinction, 223, 228–29, 272, 273
	gastropods and crabs, 14–27
	gene flow, 224–26
	gene-for-gene, 52–53, 77, 160, 204–11, 216–18, 221, 238, 293
	genetic drift, 224–28, 240
	genetic feedback, 20, 53, 221